Lecture Notes in Geoinformation and Cartography

The Lecture Notes in Geoinformation and Cartography series provides a contemporary view of current research and development in Geoinformation and Cartography, including GIS and Geographic Information Science. Publications with associated electronic media examine areas of development and current technology. Editors from multiple continents, in association with national and international organizations and societies bring together the most comprehensive forum for Geoinformation and Cartography.

The scope of Lecture Notes in Geoinformation and Cartography spans the range of interdisciplinary topics in a variety of research and application fields. The type of material published traditionally includes:

- proceedings that are peer-reviewed and published in association with a conference;
- post-proceedings consisting of thoroughly revised final papers; and
- research monographs that may be based on individual research projects.

The Lecture Notes in Geoinformation and Cartography series also includes various other publications, including:

- tutorials or collections of lectures for advanced courses;
- contemporary surveys that offer an objective summary of a current topic of interest; and
- emerging areas of research directed at a broad community of practitioners.

More information about this series at http://www.springer.com/series/7418

Ali Mansourian · Petter Pilesjö
Lars Harrie · Ron van Lammeren
Editors

Geospatial Technologies for All

Selected Papers of the 21st AGILE Conference on Geographic Information Science

Springer

Editors
Ali Mansourian
Department of Physical Geography and Ecosystem Science/GIS Centre
Lund University
Lund
Sweden

Lars Harrie
Department of Physical Geography and Ecosystem Science/GIS Centre
Lund University
Lund
Sweden

Petter Pilesjö
Department of Physical Geography and Ecosystem Science/GIS Centre
Lund University
Lund
Sweden

Ron van Lammeren
Laboratory of Geo-information Science and Remote Sensing
Wageningen University and Research
Wageningen, Gelderland
The Netherlands

ISSN 1863-2246 ISSN 1863-2351 (electronic)
Lecture Notes in Geoinformation and Cartography
ISBN 978-3-319-78207-2 ISBN 978-3-319-78208-9 (eBook)
https://doi.org/10.1007/978-3-319-78208-9

Library of Congress Control Number: 2018935230

Printed on acid-free paper

This Springer imprint is published by the registered company Springer International Publishing AG part of Springer Nature
The registered company address is: Gewerbestrasse 11, 6330 Cham, Switzerland

Preface

The Association of Geographic Information Laboratories in Europe (AGILE) has held annual conferences focusing on Geographic Information Science for more than two decades. The themes of the conferences have varied in response to changing research trends. In this way, the annual conference has always remained relevant and current and is an important meeting point for European as well as non-European researchers in the field. The annual AGILE conference is now widely held to be the most important academic Geographic Information Science conference in Europe.

The 21st AGILE conference took place in Lund, Sweden, on 12–15 June, 2018. The theme of the conference was *Geospatial Technologies for All.* The conference accepted full papers, short papers and posters and included a workshop day prior the main conference. For more than a decade, the full papers from the annual conference have been published as a book by Springer International Publishing AG. This year there were 46 full papers submitted to the conference, of which 19 were accepted for publication in this volume, its title taken from the conference theme. We would like to thank all of the authors for their contributions, which are a critical prerequisite for maintaining the quality of AGILE conference. We would also like to thank all the reviewers in the Scientific Committee for carefully reviewing the papers, which enabled us to make this selection.

The papers in this volume are divided into four parts. Part I deals with *Natural Resource Management and Earth Observation* and includes five studies about how geospatial technologies can contribute to a sustainable development of our environment. During recent years' AGILE conferences, there have been many submissions in the field of *Volunteered Geographic Information (VGI) and Participatory GIS.* This trend has continued this year, and the six papers in part two of this volume represent that work. Part III is called *Spatiotemporal Data Modelling and Data Mining*, and it includes six papers that describe new methods for understanding spatiotemporal phenomena. A notable reflection here is the increasing importance of the temporal dimension, as seen in several submissions; GIScience is about more than just space. The final Part IV of this volume deals with the issue of *Quality and Uncertainty of Geographic Information*, a key issue for everyone that wants to make a decision based on spatial analysis. To sum up, we think that the 19 contributions published in this volume give a broad overview of the field at this time and do a superb job in exemplifying the title of this book: *Geospatial Technologies for All.*

We are grateful to everyone that has helped us with editing this volume, as well as creating the programme for the 21st AGILE conference. The AGILE Council was of key assistance in the scientific organization of the conference. We also would like to thank our colleagues at Lund University for all their help with the various large and small tasks that enable a conference like this to succeed. We would also like to thank Springer International Publishing AG for their helpful cooperation, and for continuing to provide the opportunity to publish the selected full papers in their academic series, Springer Lecture Notes in Geoinformation and Cartography.

Lund, Sweden — Ali Mansourian
Lund, Sweden — Petter Pilesjö
Lund, Sweden — Lars Harrie
Wageningen, The Netherlands — Ron van Lammeren
February 2018

Organizing Committee

Scientific Programme Committee

Ali Mansourian, Lund University, Sweden (Chair)
Petter Pilesjö, Lund University, Sweden
Lars Harrie, Lund University, Sweden
Ron van Lammeren, Wageningen University and Research

Local Organizing Committee

Petter Pilesjö (Chair)
Lars Harrie (Workshop Chair)
David Tenenbaum
Eva Andersson
Andreas Persson
Roger Groth
Karin Larsson
Micael Runnstrom
Mitch Selander

Scientific Committee

Ana Paula Afonso, University of Lisbon, Portugal
Fernando Bacao, New University of Lisbon, Portugal
Marek Baranowski, Institute of Geodesy and Cartography, Poland
Melih Basaraner, Yildiz Technical University, Turkey
Giedrė Beconytė, Vilnius University, Lithuania

Itzhak Benenson, Tel Aviv University, Israel
Lars Bernard, Technical University Dresden, Germany
Michela Bertolotto, University College Dublin, Ireland
Ralf Bill, Rostock University, Germany
Sandro Bimonte, IRSTEA, France
Thomas Blaschke, University of Salzburg, Austria
Arnold Bregt, Wageningen University and Research, The Netherlands
Thomas Brinkhoff, Jade University Oldenburg, Germany
Pedro Cabral, New University of Lisbon, Portugal
Sven Casteleyn, University Jaume I of Castellon, Spain
Christophe Claramunt, Naval Academy Research Council, France
Serena Coetzee, University of Pretoria, South Africa
Lex Comber, University of Leeds, UK
Joep Crompvoets, KU Leuven, Belgium
Isabel Cruz, University of Illinois at Chicago, USA
Sytze de Bruin, Wageningen University and Research, The Netherlands
Cidalia Fonte, University of Coimbra, Portugal
Anders Friis-Christensen, European Commission, Joint Research Centre, Italy
Jerome Gensel, University of Grenoble, France
Michael Gould, Esri and University Jaume I, Spain
Carlos Granell, University Jaume I of Castellón, Spain
Henning Sten Hansen, Aalborg University, Denmark
Lars Harrie, Lund University, Sweden
Francis Harvey, University of Leipzig, Germany
Roberto Henriques, New University of Lisbon, Portugal
Gerard Heuvelink, Wageningen University and Research, The Netherlands
Stephen Hirtle, University of Pittsburgh, USA
Hartwig Hochmair, University of Florida, USA
Joaquín Huerta, University Jaume I of Castellon, Spain
Bashkim Idrizi, Mother Teresa University, Republic of Macedonia
Mike Jackson, University of Nottingham, UK
Bin Jiang, University of Gävle, Sweden
Didier Josselin, University of Avignon, France
Derek Karssenberg, Utrecht University, The Netherlands
Tomi Kauppinen, Aalto University, Finland
Marinos Kavouras, National Technical University of Athens, Greece
Dimitris Kotzinos, University of Cergy-Pontoise, France
Petr Kuba Kubicek, Masaryk University, Czech Republic
Patrick Laube, Zurich University of Applied Science, Switzerland
Robert Laurini, University of Lyon, France
Francisco J. Lopez-Pellicer, University of Zaragoza, Spain
Malgorzata Luc, Jagiellonian University, Poland
Ali Mansourian, Lund University, Sweden
Bruno Martins, University of Lisbon, Portugal
Filipe Meneses, University of Minho, Portugal

Peter Mooney, Maynooth University, Ireland
João Moura Pires, New University of Lisbon, Portugal
Beniamino Murgante, University of Basilicata, Italy
Javier Nogueras-Iso, University of Zaragoza, Spain
Juha Oksanen, Finnish Geospatial Research Institute, Finland
Toshihiro Osaragi, Tokyo Institute of Technology, Japan
Frank Ostermann, University of Twente, The Netherlands
Volker Paelke, Hochschule Ostwestfalen-Lippe, Germany
Marco Painho, New University of Lisbon, Portugal
Petter Pilesjö, Lund University, Sweden
Poulicos Prastacos, FORTH, Greece
Hardy Pundt, Harz University of Applied Sciences, Germany
Ross Purves, University of Zurich, Switzerland
Viktor Putrenko, National Technical University of Ukraine, Ukraine
Martin Raubal, ETH Zürich, Switzerland
Wolfgang Reinhardt, Bundeswehr University Munich, Germany
Claus Rinner, Ryerson University, Canada
Jorge Rocha, University of Minho, Portugal
Armanda Rodrigues, New University of Lisbon, Portugal
Maribel Yasmina Santos, University of Minho, Portugal
Tapani Sarjakoski, Finnish Geospatial Research Institute, Finland
L. Tiina Sarjakoski, Finnish Geospatial Research Institute, Finland
Sven Schade, European Commission—DG JRC, Belgium
Christoph Schlieder, University of Bamberg, Germany
Monika Sester, Leibniz University of Hanover, Germany
Takeshi Shirabe, Royal Institute of Technology, Sweden
Jantien Stoter, Delft University of Technology, The Netherlands
Maguelonne Teisseire, IRSTEA, France
Fred Toppen, Utrecht University, The Netherlands
Nico Van de Weghe, Ghent University, Belgium
Ron van Lammeren, Wageningen University and Research, The Netherlands
Jos van Orshoven, KU Leuven, Belgium
Danny Vandenbroucke, KU Leuven, Belgium
Lluis Vicens, University of Girona, Spain
Luis M. Vilches-Blázquez, Pontifical Xavierian University, Spain
Kirsi Virrantaus, Aalto University, Finland
Vít Voženílek, Palacky University Olomouc, Czech Republic
Monica Wachowicz, University of New Brunswick, Canada
Gudrun Wallentin, University of Salzburg, Austria
Robert Weibel, University of Zurich, Switzerland
Stephan Winter, University of Melbourne, Australia
F. Javier Zarazaga-Soria, University of Zaragoza, Spain
Alexander Zipf, Heidelberg University, Germany
Jonas Ardö, Lund University, Sweden
Dirk Burghardt, TU Dresden, Germany

Cécile Duchêne, IGN, France
Lars Eklundh, Lund University, Sweden
Sara Irina Fabrikant, University of Zurich, Switzerland
Jan-Henrik Haunert, University of Bonn, Germany
Fredrik Lindberg, Gothenburg University, Sweden
Martijn Meijers, Delft University of Technology, Netherlands
Peter van Oosterom, Delft University of Technology, Netherlands
Andreas Persson, Lund University, Sweden
Micael Runnström, Lund University, Sweden
David Tenenbaum, Lund University, Sweden

Contents

Part I
Natural Resource Management and Earth Observation

Evaluating Spatial Data Acquisition and Interpolation Strategies for River Bathymetries

Robert Krüger, Pierre Karrasch and Lars Bernard

Abstract The study implements a workflow to evaluate the effects of different data sampling methods and interpolation methods, when measuring and modelling a river bathymetry based on point data. Interpolation and sampling strategies are evaluated against a reference data set. The evaluation of the results includes critically discussing characteristics of the input data, the used methods and the transferability of the results. The results show that the decision for or against a particular sampling method and for a specific setting of the parameters can certainly have a great influence on the quality of the interpolation results. Further, some general guidelines for the acquisition of bathymetries are derived from the study results.

Keywords Spatial interpolation · Riverbed modelling · Spatial sampling
Water frame work directive · Bathymetry

1 Introduction and Motivation

It is almost two decades after the European Water Frame Directive (WFD) entered into force and required the European Commission and the Member States to develop a monitoring and reporting system to provide evidence on the progress made towards sustainable water use and (long term) protection of the available water resources in Europe (European Union 2000). Geoinformation technologies have always been core to the improvement of the data acquisition processes as well as to the data processing and integration efforts. Today, the Water Information System for Europe (WISE[1]) provides the European overview on the current state of the WFD implementation and progress towards the general WFD goal that all surface waters and groundwaters will be in good ecological status by 2027.

[1]http://water.europa.eu/.

R. Krüger (✉) · P. Karrasch · L. Bernard
Chair of Geoinformatics, Technische Universität Dresden, Dresden, Germany
e-mail: robert.krueger@tu-dresden.de

A. Mansourian et al. (eds.), *Geospatial Technologies for All*,
Lecture Notes in Geoinformation and Cartography,
https://doi.org/10.1007/978-3-319-78208-9_1

The assessment of the ecological status of the inland water builds on a compilation of indicator sets about the biological, chemical and hydro morphological states of the water resources. Considering recent developments in geodata acquisition and processing techniques, this paper proposes and evaluates new methods for the acquisition of the hydro morphological characteristics of rivers, in order to design more time and cost-efficient monitoring methods.

Assessing hydro-morphological characteristics of rivers requires information on river depths and their variances in longitudinal and cross profiles of a river. For this purpose, various methods for obtaining river bathymetry are available. Bathymetry can be obtained by traditional survey or GPS methods if the river is (Milne and Sear 1997; Casas et al. 2006; Merwade et al. 2008). Alternatively, bathymetry can be gathered by combining survey data collected during low water levels with sonar data collected during high water levels (Casas et al. 2006). For shallow waters, it is possible to obtain bathymetry data by airborne bathymetric LiDAR systems (Hilldale and Raff 2008), while for deeper rivers multi-beam sonar is an option. Both systems can generate high resolution bathymetries, but generate high costs due to the expensive equipment (Conner and Tonina 2014). In the recent past, various examples show that parameters to represent water body structures can be obtained with inexpensive/low-cost unmanned boat systems (Fig. 1).

As such Systems typically only have single-beam echo sounders, which can gather single depth information in the immediate environment of the boats or are expanded by a step motor, which allows the echo sounder to turn the beam in one direction (usually transversely to the direction of travel). However, all these methods have in common that they do not allow a uniform and gapless recording of the channel profile. Such a complete measurement can be used for determining the depth variances. As the acquired datasets contain several gaps, spatial interpolation methods are required to derive bathymetry coverage.

Current GIS and statistic packages offer number of well-established statistical (e.g. Kriging) and non-statistical spatial (e.g. Inverse Distance Weighting, TIN) interpolation methods to derive surfaces (Pebesma 2004). Additionally, considering the different possible parameterizations for each of these methods, a plethora of different interpolation results can be easily achieved. A qualitative assessment of the results can be done on the basis of the known fundamental advantages and

Fig. 1 Unmanned boat systems carrying laser scanners, cameras (RGB, infrared, 360°) and echo sounders (single beam)

disadvantages of the methods. A quantitative assessment, on the other hand, can only be done using real measurements and comparing them with the interpolation results. Additional to error analysis, Monte Carlo approaches or cross-validations are often used as an instrument to assess the performance and stability of an interpolation model (Pebesma 2004). With regard to the example of analyzing channel profiles, recommendations should help in choosing the most appropriate method, parameterization and the strategy for data sampling.

In a study testing several interpolation methods for calculating bathymetry from cross-sections, Glenn et al. (2016) concluded that the result is mainly influenced by the cross-section spacing (acquisition strategy) and the coordinate system used. Earlier, Heritage et al. (2009) stated, that the layout of survey strategy is more important than the choice of the interpolation methods. In a recent study, Santillan et al. (2016) tested a small set of acquisition strategies for the interpolation of river bathymetries and summarized that the survey configuration had a bigger effect on the interpolation accuracy than the choice of interpolation method. Nevertheless, the authors argued, that their results are limited to the survey configurations used, the small study area and the low resolution of the data.

2 Implementing a Workflow to Compare Bathymetry Acquisition and Interpolation Methods

The study presented here implements an evaluation workflow as depicted in Fig. 2. Based on an available reference channel profile, different acquisition strategies are simulated. The extracted depth information from the reference data set forms the

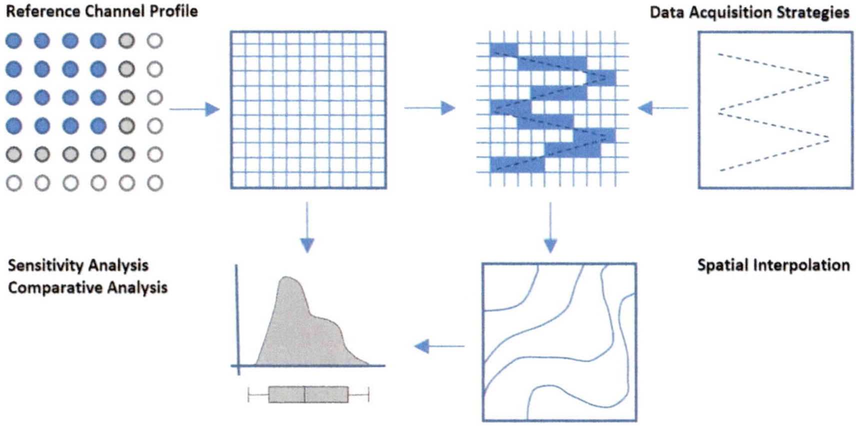

Fig. 2 Overview of the evaluation study workflow

input to different spatial interpolations. The interpolation results are evaluated against the reference data set using different statistical measures.

2.1 Creation of the Reference Dataset

To do the qualitative and quantitate analyses of the different interpolation methods, the points used for the interpolation are taken from a well-known surface. There are two possibilities to obtain such a surface: First, using a measured (real) channel and second by the creation of a synthetic channel. Both options have their pros and cons. Taking measured channel as a reference provides the opportunity to use a real-world scenario for benchmarking the different interpolation methods. On the other hand, this somewhat limits the analysis about specific characteristics of a river channel. The use of a synthetic channel offers the possibility to create specific channel characteristics and to test the interpolation methods on a wider range. Nevertheless, the conclusions drawn from such interpolations strongly rely on how well the synthetic channel resembles the characteristics of a real one. Thus, for this study, a dataset of real channel is used. By choosing different subsections of the channel, the influence of different river parameters (sinuosity, depth variance) is addressed.

The dataset used for this study contains bathymetric data of the Ohio River (USA) and covers the first 200 km downstream from Pittsburgh (US Army Corps of Engineers 2016). It results from several campaigns by ship-based sonar between August 2006 and September 2010. The dataset contains point elevation data of the channel in Cartesian *x, y, h*-coordinates, and is stored in several text-files. For the *x, y*-coordinates the UTM zone 17N NAD83 is used as spatial reference (US Survey feet). For the *h*-coordinate NAVD88 (US Survey feet) is used. Depending on the subset, the density (grid with 5, 10, 20 ft spacing) as well as the layout of the points (grid and cross-sections respectively) differs. Thus, considering the width of the Ohio River channel in the study area (800–1300 ft) the data set provides a quite high resolution.

The workflow to prepare the reference data has been implemented in ESRI ArcGIS 10.3. In a first step, the areas with the highest point density (5 ft point spacing) are selected from the dataset and imported into ArcGIS. After a visual inspection, four subsets with different characteristics have been selected (Fig. 3). While subsets 1, 2 and 4 are relatively straight (sinuosity about 1), subset 3 has a higher sinuosity (sinuosity = 1.46). These subsets also show different ranges of height variances, with 2.5 ft for subset 2–8.2 ft for subset 4.

As shown in prior works, a flow-oriented and curvilinear coordinate system is expected to yield better results (Merwade et al. 2006; Glenn et al. 2016). Thus the approach of Merwade et al. (2005) is adopted, which allows the transformation of points from the cartesian *x, y*-coordinate system into the *s, n*-coordinate system. The transformation process requires an arbitrary line within the river channel as a starting point. In the original approach, the thalweg derived from existing

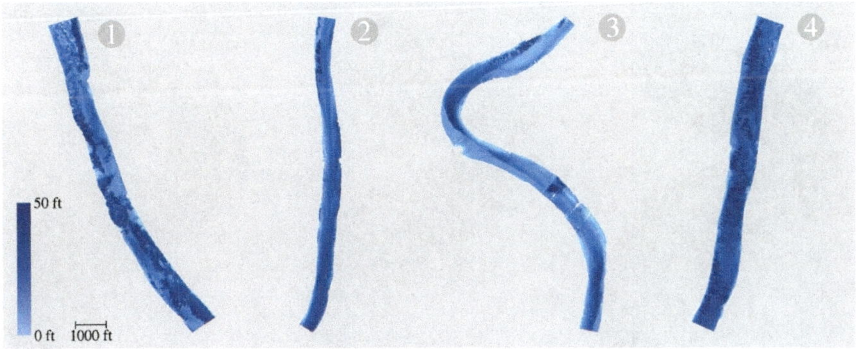

Fig. 3 River subsets used in this study (water depth)

bathymetry data, is used. As it is assumed, that the rivers' bathymetry is the not yet known target of the study, the centerline is used. Using satellite imagery (ArcGIS Basemap Imagery) the centerline was extracted.

2.2 *Data Acquisition Strategies*

For all interpolations executed in this study, the input values (i.e. measurements) are extracted from the reference surface. Different acquisition strategies, i.e. different layouts for extracting points have been applied. In general, these strategies can be categorized as orthogonal *cross-sections* or *cruising trajectories*.

Cross-sections are measurements taken in lines orthogonal to the flow direction of a river. A point spacing of 10 ft is chosen in n-direction (along the cross-section). Starting from 10 ft the spacing between the cross-sections is increased by doubling the value for each test case. The maximum interval is 2560 ft, so that depending on the subset, there are between five (subset 2) and nine (subset 3) cross-sections in the input data. A further increase would result in too few cross-sections for the interpolation.

Cruising trajectories for boat measurement campaigns are typically realized as either parallel to the river centerline or as zig-zag trajectory. The smallest wave-length of the zig-zag trajectory is set to 160 ft, as decreasing the wavelength more would effectively lead to a duplication of the cross-section geometries. Further wavelengths are generated by doubling the value for each test case.

Along the cruising trajectories and cross-sections, points from the reference surface are extracted with a spacing of 10 ft.

In order to evaluate the effect of an applied acquisition strategy, 56 acquisition strategies for each of the subsets (Fig. 3) are created. Each strategy defines the path for the data acquisition and consists of different cross sections or different trajectories (Fig. 4).

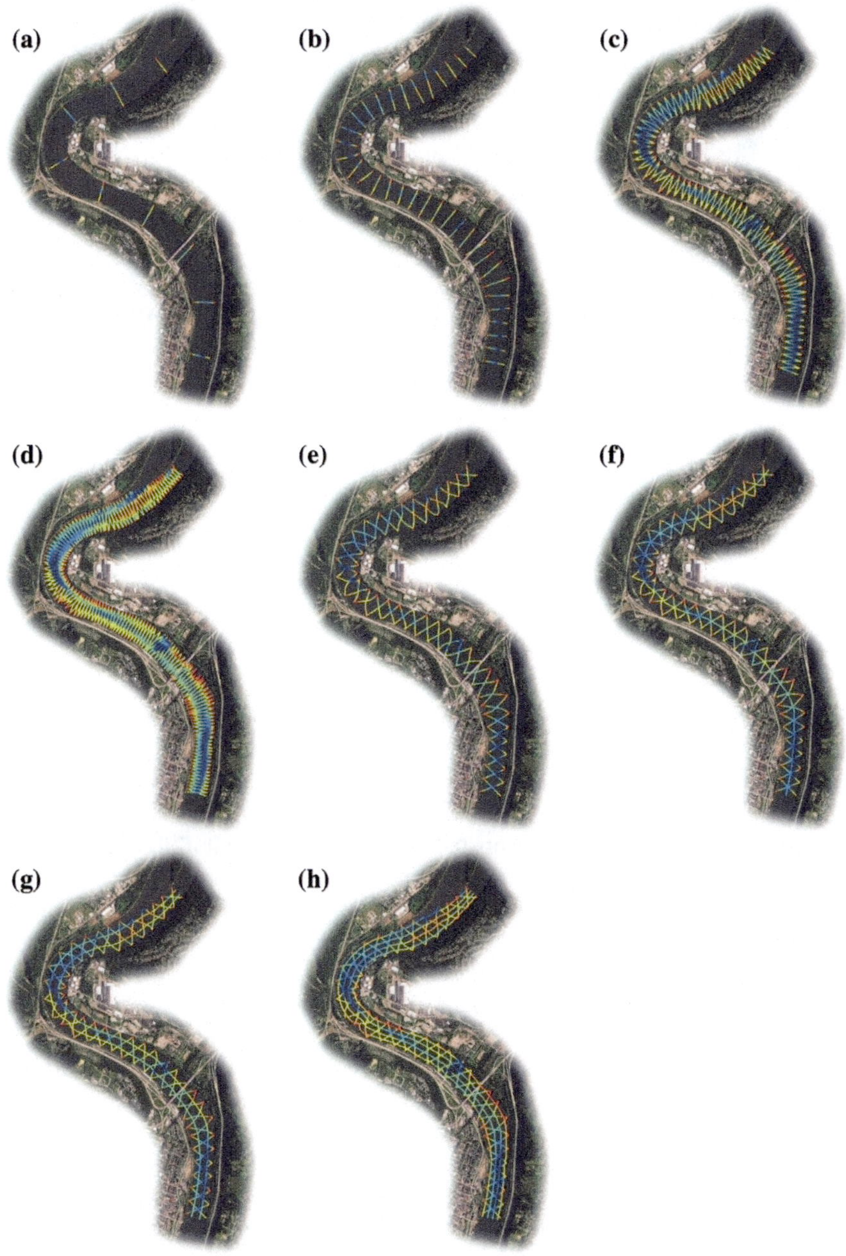

Fig. 4 Examples of different acquisition strategies: cross-sections (**a**, **b**), zig-zag (once) (**c**), zig-zag (twice) (**d**, **e**), zig-zag + centerline (**f**), zig-zag + 2 parallel trajectories (**g**), zig-zag + 4 parallel trajectories (**h**)

2.3 *Spatial Interpolation Methods*

Five interpolation methods are evaluated: Inverse Distance Weighting, Simple Kriging, Ordinary Kriging and Radial Basis Functions. The parameters are slightly changed for each execution, in order to conduct the sensitivity analysis. The purpose is to find the optimal parameters for each of the methods. Further, it is investigated which parameters have the greatest impact on prediction quality. The sensitivity analysis is conducted only for a selection of strategies (Table 1) to stay with a manageable number of interpolations.

2.3.1 Inverse Distance Weighting (IDW)

IDW is a deterministic method where unknown points are approximated by a distance-decayed weighted average of the values of the nearby points (Shepard 1968; Rase 2016). The exponent if the inverse distance weight (p) is the parameter, which defines the significance of the nearby points.

For the IDW algorithm, three parameter settings are tested. First the exponent p of the weighing function is varied. Second, the number of used data points is modified. According to Vande Wiele (2001), beside the number of points also the layout of the points has an influence on the interpolation result. Since, data points evenly surrounding the location of the interpolated point typically lead to better results than using clustered data points, a quadrant search for data points is used, enforcing a more evenly spread point pattern. For each quadrant N points are used. And third, the anisotropy ratio a_r is altered (Merwade et al. 2006). Instead of using a circular search neighborhood, an elliptical one with the axes aligned to the flow direction, is used. The anisotropy ratio a_r is the ratio of the main and minor axis of the ellipse. The main axis is set parallel to the flow direction (s-coordinate).

2.3.2 Radial Basis Functions (RBFs)

RBFs are also deterministic methods. The resulting surface of interpolated points must fulfill two requirements: Pass through the input data points, and be as smooth as possible. The resulting surface can, in contrast to the IDW method, contain

Table 1 Strategies used for sensitivity analysis

	Acquisition strategy	Points	Density (Pts./10^5 ft^2)
Strategy 1	cross-section – 20 ft	83405	5146
Strategy 2	Zig-Zag (twice) 1280 ft + centerline	11760	726
Strategy 3	cross-section – 160 ft	10488	647
Strategy 4	Zig-Zag (twice) 5120 ft + 4 parallel trajectories	9550	589
Strategy 5	cross-section – 1280 ft	1221	75

values which are smaller or bigger than the minima and maxima of the input points. The distance dependent weights are obtained by solving a system of n linear equations.

There exists a number of RBFs—for this study, four RBFs provided by the ArcGIS Geostatistical Analyst are used (ESRI 2001; Rase 2016): the Multiquadric Function; the Inverse Multiquadric Function; the Spline With Tension Function and the Completely Regular Spline Function.

For all four functions, a sensitivity analysis is conducted to compare the functions to determine the influence of two parameters. Like for the IDW method, a quadrant search neighborhood is used and the number of points N is varied. For the interpolations in the s, n-coordinate system the anisotropy ratio a_r is ascertained according to Merwade et al. (2006) by computing the variance ratio of the variance of z_i across the flow and along the flow. For the four subsets, a_r results to $a_r = 3$.

2.3.3 Ordinary Kriging (OK)

Kriging and its variants (Krige 1966) are geostatistical methods, which take the spatial correlation between the measured points into account, when calculating the weights for the points used in the interpolation. The spatial correlation is quantified by a semivariogram model as a function of the distance (lag) between a pair of data points.

Different Kriging variants have been developed in the past. In this study Ordinary and Simple Kriging are used. Ordinary Kriging assumes an unknown, but constant mean in the data and therefore focuses on the spatial component. Simple Kriging assumes that the mean is constant and known. More detailed information on Kriging can be found in Cressie (1993).

Kriging methods are difficult to automate, as a suitable model for the data must be chosen. The ArcGIS implementation supports the user in finding the right parameters for the model, but the final choice for the right model is up to the user. As the ArcGIS Geostatistical Analyst is including 11 model types, an analysis for these in combination with a sensitivity analysis is not feasible. Thus, the different models are analyzed and models with Nugget-parameter not close to zero are removed, which is preferable for Digital Elevation Models without gaps. The visual selections are quantified by cross validation. The spherical and circular models produce a slightly worse RMSE score compared to the tetraspherical and pentaspherical models, but they are less computational expensive and are thus selected for sensitivity analysis. Once suitable models are chosen, the Kriging process can be partly automated in ArcGIS.

Two parameters have a strong impact on the variogram—size and count of lags. The product of count and size of the lags should be at maximum around 50% of the greatest distance in the dataset (ESRI 2001): Subset 2 has the shortest length of about 13500 ft, hence 100 lags of 50 ft width are chosen.

For the Kriging algorithm, the number of points in the search neighborhood are varied and the quadrant search neighborhood is used. In contrast to the other algorithms, a_r is determined by ArcGIS from the data (direction and both axis). Again, this parameter is only used in the *s, n*-coordinate system. Further it is examined if removing the spatial trend in the data (linear in flow direction, non-linear perpendicular to the direction of flow) prior to the interpolation has an effect on the interpolation result.

2.3.4 Simple Kriging (SK)

For Simple Kriging, the main difference compared to the workflow for OK, is that the data is standardized after the trend removal through a Normal Score Transformation. Afterwards, all steps are similar to the interpolation with the OK algorithm. Accordingly, the influence of the four parameters N, a_r, *trend removal* and *variogram model* is examined. A cross-validation for subset 2 yields the same qualitative order of the models as for OK. Due to the longer processing times compared to OK, the two models with the highest RMSE in SK (circular and spherical) are not used.

2.3.5 TopoGrid Algorithm

The TopoGrid algorithm is first described by Hutchinson (1989) and is a variation of the Thin-Plate-Spline method to create hydrologic correct DEMs. This objective is fulfilled by integrating a drainage-system into the interpolation. The Topogrid algorithm does not need any parameters to be set. Some parameters can be varied by the user, but have standard settings depending on the input data. In this study, the standard settings are used, and thus no sensitivity analysis is conducted. The Topogrid algorithm is taken as a benchmark to compare an approach not requiring parameter settings with the algorithms above, which typically all require parameter settings.

3 Assessing the Bathymetry Interpolation Methods Behaviors

To assess the quality of the interpolated riverbed surfaces, different accuracy measures are calculated. The most widely used accuracy measure for the quality of Digital Elevation models is the Root Mean Square Error (RMSE) (Aguilar et al. 2005). As the RMSE is sensitive to large outliners the Mean Average Error (MAE) is used in this study as well. For both RMSE and MAE 0 indicates no error. The third measure used is the Pearson Correlation coefficient (R), which measures the linear correlation for the predicted and the observed values between 1 (best) and 0 (worst). Thus, it evaluates the spatial fit of the prediction.

For the following sensitivity analysis, the RMSE is calculated for each set of parameters. As the parameters are tested for five different acquisition strategies, the RMSE is normalized for each of them. Subsequently, those normalized RMSEs are averaged for each parameter set. Finally, the parameter sets with the lowest average RMSE are selected for the comparison of the interpolation and acquisition strategies in the next chapter.

3.1 Assessing Inverse Distance Weighting (IDW)

For the IDW method, three parameters (N, p, a_r) are varied. Table 2 shows the ten selected parameter sets for the main analysis, as well as the RMSE for the five tested strategies. The sets with the lowest RMSE for each of the tested strategies are chosen. Further, the parameter sets with the lowest averaged RMSE for the five strategies are selected.

As can be seen in the table, the strategies with the highest point densities yield the lowest RMSE. Further can be seen, that that the RMSE increases as the point density decreases. Figure 5 shows that using $a_r > 1$ (considering anisotropy) and $p < 4$ yields a lower RMSE unless the point density is very high. Figure 6 shows that a high number of points N produces the lowest RMSE for low point densities and low values for N yield the lowest RMSE for high point densities. Figure 7 shows that the RMSE is very sensitive to variations of N for $p \leq 2$ and high point densities.

3.2 Assessing Radial Basis Functions (RBFs)

For the RBF methods, five RBFs are tested and two parameters (N, a_r) are varied. Table 3 shows the three selected parameter sets and functions for the main analysis, as well as the RMSE for the five tested strategies.

Table 2 Selected parameter sets for IDW

Parameter			RMSE (ft)					
N	a_r	p	Strategy 1	Strategy 2	Strategy. 3	Strategy 4	Strategy 5	$RMSE_{norm}$
2	2	2	0.363	2.269	2.023	3.412	4.446	0.9217
8	2	3	0.403	2.258	2.026	3.346	4.274	0.9234
6	2	3	0.394	2.262	2.026	3.362	4.318	0.9234
10	2	3	0.409	2.263	2.028	3.336	4.251	0.9244
8	3	3	0.430	2.288	2.022	3.184	4.269	0.9251
4	3	2	0.479	2.249	1.997	3.181	4.310	0.9386
2	1	2	0.356	2.319	2.029	3.684	4.609	0.9476
4	2	1	0.633	2.211	2.046	3.275	4.194	0.9948
8	5	1	1.113	2.393	2.240	3.039	4.094	1.1801
10	5	1	1.232	2.435	2.309	3.051	4.081	1.2322

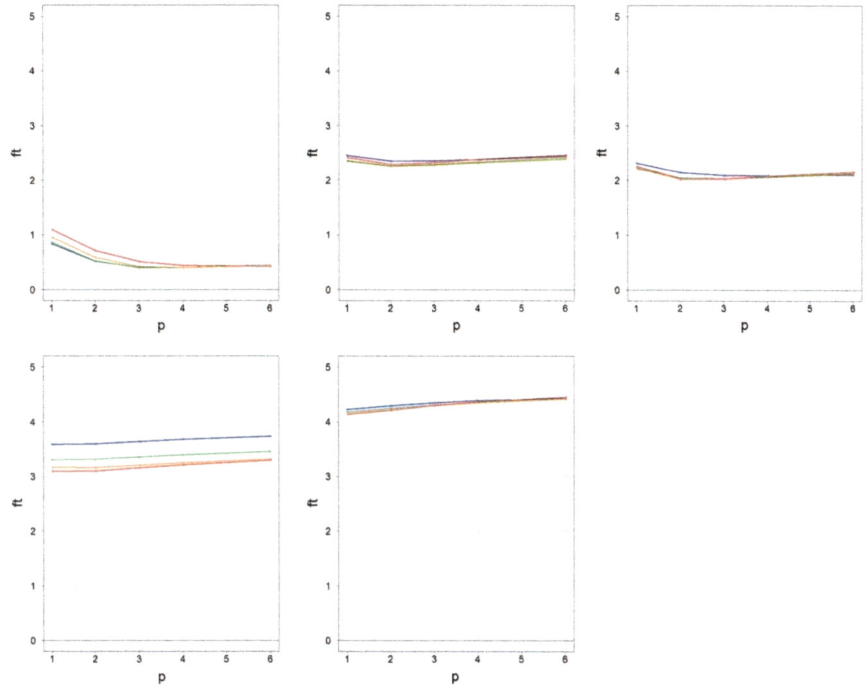

Fig. 5 RMSE for IDW sensitivity analysis: parameters power (p) and anisotropy (a_r): $a_r = 1$ (blue), $a_r = 2$ (green), $a_r = 3$ (orange), $a_r = 5$ (red)

Analogous to the results for the IDW method, the strategies with the highest point densities yield the lowest RMSE. Figure 8 shows, that the differences of the results of different used RBFs are small. Further using an $a_r = 3$ (considering anisotropy) yields the lowest RMSE for each of the tested strategies unless the point density is very high (strategy 1).

3.3 Assessing Ordinary Kriging (OK)

For OK, the most complex sensitivity analysis is conducted. Aside from varying a_r and N, the model of the semivariogram and different trend removal options are tested. Table 4 shows the ten selected parameter sets for the main analysis, as well as the RMSE for the five tested strategies.

Similar to the prior described methods, OK yields the lowest RMSE when using the highest point density. For high point densities, the circular and the spherical models perform worse than Stable, Exponential and the *K-Bessel* models (Fig. 9). Further, using $a_r > 1$ (taking anisotropy into account) yields in higher RMSE for

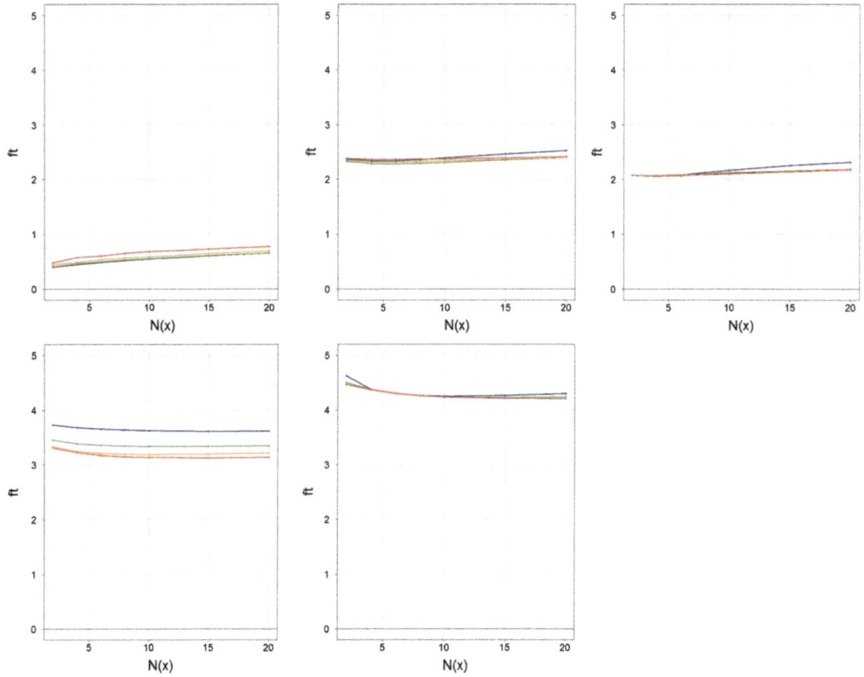

Fig. 6 RMSE for IDW sensitivity analysis: parameters point number (N) and anisotropy (a_r): $a_r = 1$ (blue), $a_r = 2$ (green), $a_r = 3$ (orange), $a_r = 5$ (red)

high point density no matter which model is used. For lower point densities $a_r > 1$ improved the RMSE. While first order trend removal merely improved any of the interpolations, second order trend removal improved the RMSE for medium point densities, but impaired the prediction quality for low point densities. Increasing the number of points in the search neighborhood N yields better results except for strategy 2, where the lowest RMSE is achieved for $N = 4$. Nevertheless, the sensitivity to a variation of N is low.

3.4 Assessing Simple Kriging

For SK the same parameters where varied as for OK. Table 5 shows the seven selected parameter sets for the main analysis, as well as the RMSE for the five tested strategies.

Similar to all other interpolation methods, SK yields the lowest RMSE when using the highest point densities. Also $a_r > 1$ (taking anisotropy into account) yields in higher RMSE for high point densities, no matter which model is used (Fig. 10).

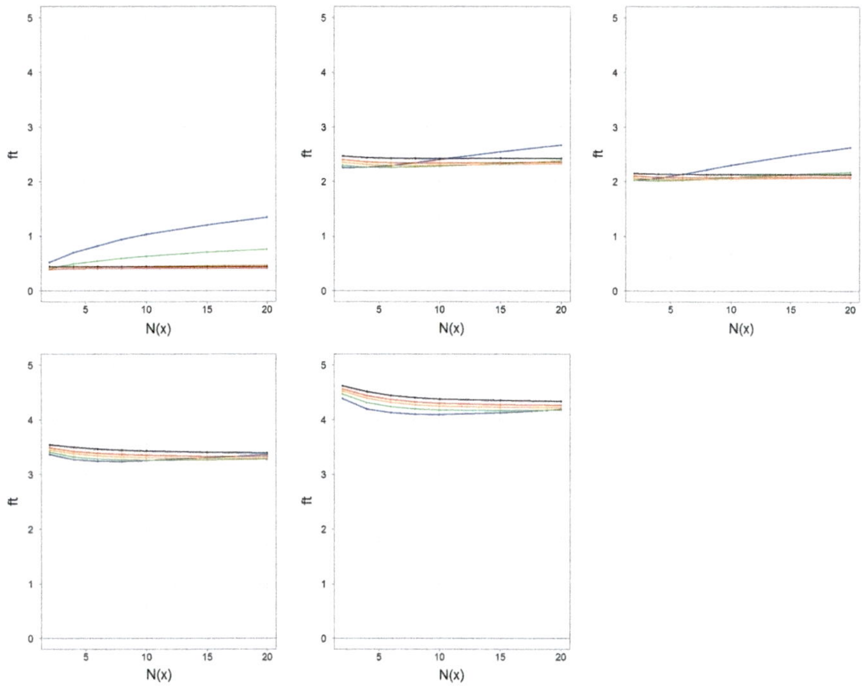

Fig. 7 RMSE for IDW sensitivity analysis: parameters point number N und power (p): $p = 1$ (blue), $p = 2$ (green), $p = 3$ (orange), $p = 4$ (red), $p = 6$ (black)

Table 3 Selected RBF and parameter sets

Parameter			RMSE (ft)					
Type	N	a_r	Strategy 1	Strategy 2	Strategy 3	Strategy 4	Strategy 5	$RMSE_{norm}$
Regular spline	10	3	0.286	2.195	2.013	2.950	4.004	0.9572
Inverse multiquadric	10	3	0.284	2.196	2.036	2.982	3.997	0.9598
Tension spline	8	3	0.278	2.326	2.030	2.950	4.023	0.9657

Apart from the high point densities, where the exponential model performed far worse than K-Bessel and Stable model when using $a_r > 1$, the performance of all models is pretty even. Except for the high point densities, the use of $a_r > 1$ yields a decrease of RMSE. First and second order trend removal improves the RMSE for high point densities when using $a_r = 1$. Nevertheless, the sensitivity to the use of trend removal is small for all models and point densities. Increasing the number of points in the search neighborhood N yields better results except for strategy 2, where the lowest RMSE is achieved for low values of N. The sensitivity to a variation of N is also low.

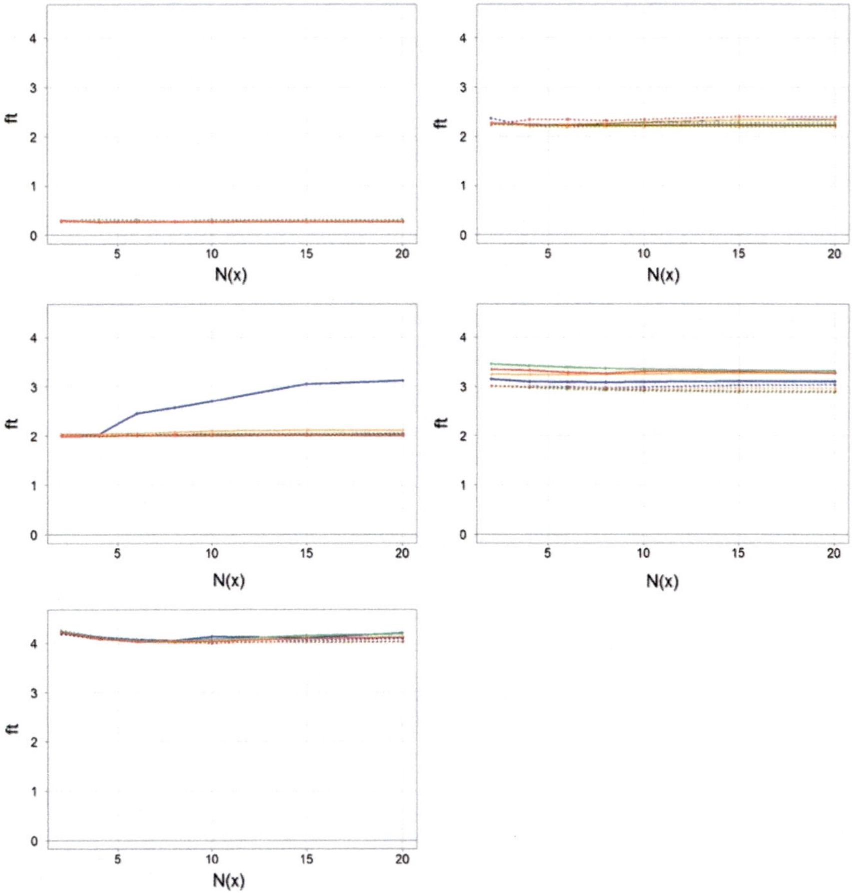

Fig. 8 RMSE for RBFs sensitivity analysis: parameters point number (N), used function and anisotropy (a_r): $a_r = 1$ (solid), $a_r = 3$ (dotted), inverse multi-quadric (blue), multiquadric (green), regularized spline (orange), tension spline (red)

4 Comparisons of the Different Strategies

4.1 Comparing the Different Acquisition Strategies

For the cross-section acquisitions, the influence of the cross-section spacing is examined. The RMSE for the 4 subsets is shown in Fig. 11a. The values for the interpolation methods (IDW, OK, SK and Tension Spline) are averaged to facilitate readability. However, the deviation from the average is smaller than 5%.

Figure 11a shows clearly an increase of the RMSE with increasing spacing of the cross-sections. For the subsets 1, 2 and 4 the chart can be described in three parts. From 10 to 20 ft spacing there is only a small increase of the RMSE (29%),

Table 4 Selected models and parameters for ordinary kriging

Parameter				RMSE (ft)					
Model	*N*	Trend-removal	Anisotropy	Strat. 1	Strat. 2	Strat. 3	Strat. 4	Strat. 5	$RMSE_{norm}$
Stable	20	0	yes	0.326	2.228	2.008	3.056	4.089	0.8624
K-Bessel	20	0	yes	0.342	2.222	2.007	3.057	4.088	0.8679
Stable	8	0	yes	0.326	2.227	2.005	3.142	4.110	0.8680
K-Bessel	20	2	no	0.278	2.230	2.032	3.232	4.371	0.8695
Stable	6	0	yes	0.326	2.226	2.001	3.183	4.123	0.8707
Stable	4	0	yes	0.329	2.230	1.998	3.252	4.138	0.8766
K-Bessel	10	2	no	0.278	2.290	2.021	3.287	4.367	0.8767
Stable	2	0	yes	0.335	2.251	1.999	3.364	4.161	0.8884
Exponential	20	2	yes	0.505	2.235	2.016	2.954	4.105	0.9248
Exponential	4	0	yes	0.645	2.234	1.996	3.270	4.139	0.9951

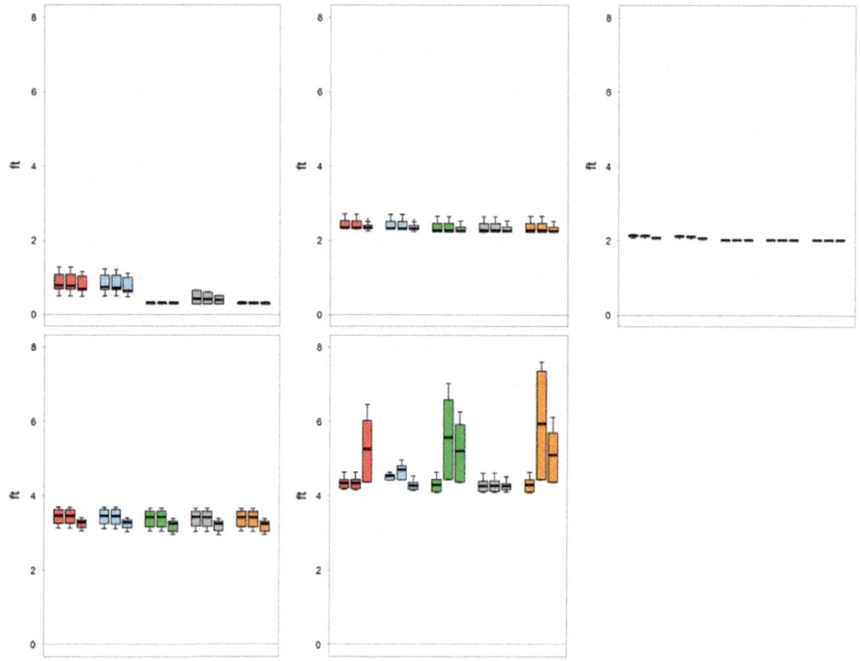

Fig. 9 RMSE for OK sensitivity analysis: parameters model and order of trend removal: groups ordered from left to right: circular, spherical, stable, exponential, K-Bessel and 0, 1, 2—for each model

whereas between 20 and 160 ft the RMSE increases faster (72–108% per step). From 160 ft onwards, the RMSE increase become smaller again (21–30% per step). For Subset 3 (sinuosity = 1.5) the RMSE increase is similar to the other subsets in the range 20–160 ft, but the increase per step stays at a higher level (30–55% per step) than for the straight subsets.

Zig-Zag strategies are simulated for seven wavelengths from 160 to 10240 ft with eight different cruise-scenarios for each wavelength, respectively. Figure 11b shows the averaged RMSE of IDW, OK, SK and Tension Spline, depending on the cruise strategy.

The graph (Fig. 11b) shows the increase of the RMSE with increasing wavelength of the Zig-Zag strategy and asymptotic behavior for high wavelengths. Further, a higher number of flow-parallel trajectories increase the prediction quality. The same is true for the double Zig-Zag strategy in comparison with the single zig-zag strategy. For higher wavelengths (>2000 ft), the latter effect diminishes as the overall amount of input data as there the amount of input data for both strategies almost equals.

An important question to answer is the choice for an appropriate acquisition strategy. To compare cross-sections and cruise strategies, strategies with about the same amount of points (±15% compared to a cross-section) are selected as groups

Table 5 Selected models and parameters for simple kriging

Parameter				RMSE (ft)					
Model	*N*	Trend-removal	Anisotropy	Strat. 1	Strat. 2	Strat. 3	Strat. 4	Strat. 5	$RMSE_{norm}$
K-Bessel	20	2	No	0.280	2.226	2.045	3.162	4.532	0.9422
Stable	20	0	Yes	0.337	2.232	2.014	3.085	4.111	0.9478
K-Bessel	20	0	Yes	0.359	2.225	2.013	3.084	4.116	0.9597
Stable	4	1	Yes	0.331	2.245	2.001	3.277	4.207	0.9606
Exponential	20	2	Yes	0.479	2.239	2.022	2.931	4.118	1.0202
Exponential	10	2	Yes	0.478	2.237	2.020	2.997	4.132	1.0238
Exponential	8	2	Yes	0.479	2.237	2.018	3.023	4.136	1.0257

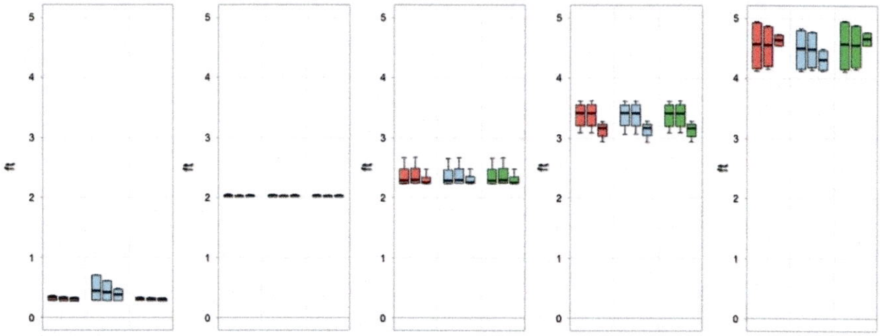

Fig. 10 RMSE for SK sensitivity analysis: parameter model and order of trend removal: groups ordered from left to right: K-Bessel, exponential, stable and 0, 1, 2—for each model

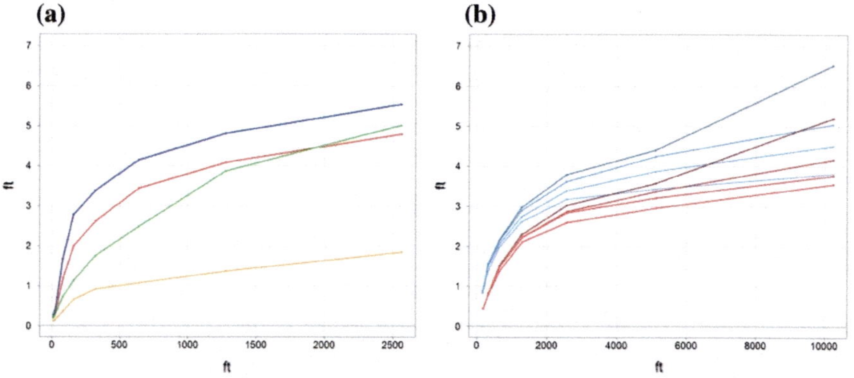

Fig. 11 **a** RMSE in relation to cross section spacing for subsets 1–4 (red, orange, green, blue) **b** RMSE in relation to the wavelength of the zig-zag trajectory, single zig-zag trajectory (blue), double zig-zag trajectory (red), number of additional lengthwise trajectories (0, 1, 2, 4) shown by saturation of color (dark to bright)

to compare. In average, cruise strategies yield a 44% higher RMSE compared to the associated cross-section.

4.2 Comparing the Different Interpolation Methods

To compare the different interpolated methods, a ranking system similar to a method described by Hofstra et al. (2008) is used. A number of skill scores is chosen to rank the interpolation methods for each skill respectively. Here RMSE, MAE, and R are taken as skill scores. As the scale of these scores differ, the achieved rank for each interpolation method is averaged from the three score ranks. The average skill scores for each subset and interpolation method is used to

Table 6 Ranking of interpolation methods

Interpolation method	Mean rank	RMSE (ft)	Rank	MAE (ft)	Rank	R	Rank
Simple kriging	1.5	2.0073	1.3	0.0960	2.0	0.9049	1.3
Ordinary kriging	2.1	2.0291	2.0	0.1003	2.3	0.9010	2.0
Inverse distance weighting	3.0	2.0643	3.3	0.1404	2.8	0.9003	3.0
Tension spline	4.3	2.0926	3.5	0.1565	5.5	0.8966	3.8
Completely reg. spline	5.0	2.1210	5.0	0.1548	5.0	0.8961	5.0
Inverse multiquadric function	5.2	2.2541	6.0	0.1433	3.5	0.8881	6.0
Topogrid	7.0	2.5959	7.0	0.3976	7.0	0.8375	7.0

subsequently rank the interpolation methods. The result of the ranking system is shown in Table 6.

Simple Kriging has the lowest rank for each of the three skill scores and thus the best average rank of all interpolation methods. The RMSE for the first five methods is quite close ($\Delta = 0.114$ ft ~6%), whereas in terms of the MAE Simple and Ordinary Kriging perform much better (~40%) than the other methods. For the correlation factor, all methods except Topogrid, are on about the same level. The parameter free Topogrid method consistently gave worse results than all other methods.

5 Discussion and Conclusion

Starting with a reflection on the data used as reference for this study, the measurement accuracy needs to be critically considered. For the study it is supposed that the measurement accuracy is significantly higher than the actual height changes of the channel profile. Only then, the accuracy of the interpolation results also reflects a valid evaluation measure for the quality of the interpolated channel profile. Closely related is the impact of the sampling rate for the used reference data set. Thus, differences in the depth of the water, which are higher-frequency than the resolution of the reference data set, could only be modeled inadequately or not at all. This effect might result in an (artificial) smoothing and may lead to underestimations of river depth variances.

For the comparative analysis conducted in this study, the reference data uses the flow-related s, n-coordinate system (Sect. 2.1) as input data to the interpolations. The applied transformation ensures that the anisotropic characteristics of the channel profile are taken into account. However, the transformation requires information about the anisotropy, which is typically not given. The river center-line or the possibly existing thalweg are also only proxies for the river anisotropy.

As both, the river sinuosity and the depth variance are relatively small; it is assumed that for the chosen study the influence of this effect is negligible.

The presented study is limited to interpolation methods that are available in the current ArcGIS environment. As it is assumed, that these methods reach a wider user group, the study focusses on the stability and automation potential of these interpolation methods. Comprehensive assessments regarding the effect of different interpolation methods, parameterizations and driving strategies would require that analyzes are actually performed in all possible combinations. To stay with a manageable amount of interpolation scenarios and results, this approach was rejected in favor of the presented two-stage analysis process. The analysis of the driving strategies is preceded by a sensitivity analysis of the parameter settings. This still leads to a total of 10010 different interpolations and it is assumed, that this forms a sound basis for the presented evaluation.

The evaluation of the interpolation results builds on different established statistical parameters. The comparative statistical analysis could be expanded and complemented by further statistical tests. The current measures are global measures, and thus only reflect the global fit of the predicted surface. However, Fig. 12 shows that errors the interpolation results are unevenly distributed. The analysis of the spatial distribution of errors is not part of this study, but will be further investigated.

The presented results confirm the findings of Glenn et al. (2016) and Heritage et al. (2009), that the choice of the data acquisition strategy is more important, than the choice of the interpolation method. Several Interpolation methods have found to be on about the same accuracy level, although the Kriging variants consistently gave the best results.

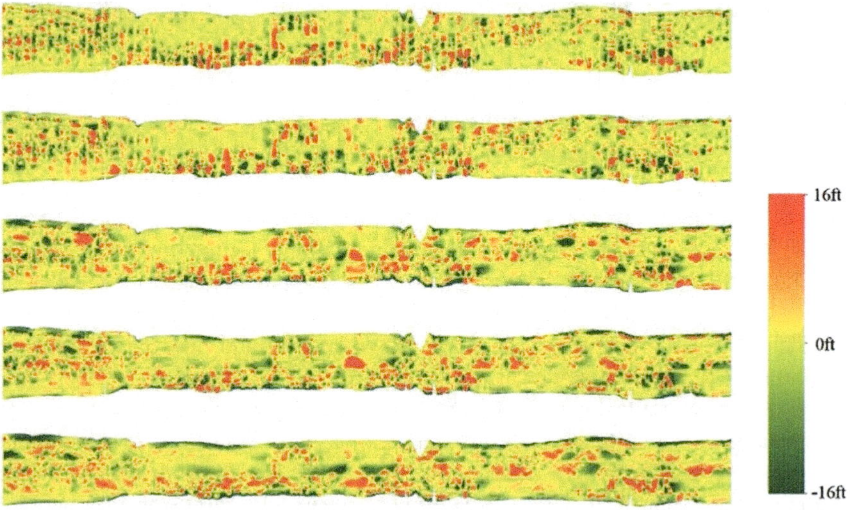

Fig. 12 Overview on error distributions based on the OK residuals comparing five acquisition strategies with same number of data points (±10%) and a RMSE between 2.01 and 3.06 ft

The study results further confirm the findings of Santillan et al. (2016), that the cross-section strategy produces better results compared to the cruise strategies when using the same amount of data points. However, obtaining data from a vessel moving alongside the river might be quicker and therefore more cost effective than measuring cross-sections at a high density. In this context, Fig. 13 indicates the factor of the RMSE decrease compared to the factor of workload (i.e. number of points to acquire) for selected cruising strategies. The figure helps in choosing the best cruising strategy for the accuracy desired.

Moreover, the question arises how far the characteristic of the input data allows general conclusions regarding the interpolation results. This aspect is addressed by considering different river subsets, with different characteristics and by considering several acquisition scenarios. Whether the results can be (easily) transferred to other rivers can only be confirmed by further analyses using different (types of) rivers. Nevertheless, the presented results give a good impression on how different interpolation methods, parameterizations and driving strategies work and what influence they have in an individual use case. The results of this study can in general support judging which point densities and driving strategies are necessary to achieve a defined level of accuracy. Therefore, the results can also assist in the planning of measurement campaigns for other rivers and transferability of the results to other study areas is partially given.

With regard to the requirements of the European Water Framework Directive the results not only allow the determination of river depths and their variances, but also the extraction of longitudinal and cross profiles. Future studies will combine the method toolbox from this study with bathymetric survey campaigns using low-cost technology (e.g. single-beam echo sounder) on unmanned boats (see Fig. 1), to further evaluate and develop strategies for time and cost-efficient monitoring of river landscapes.

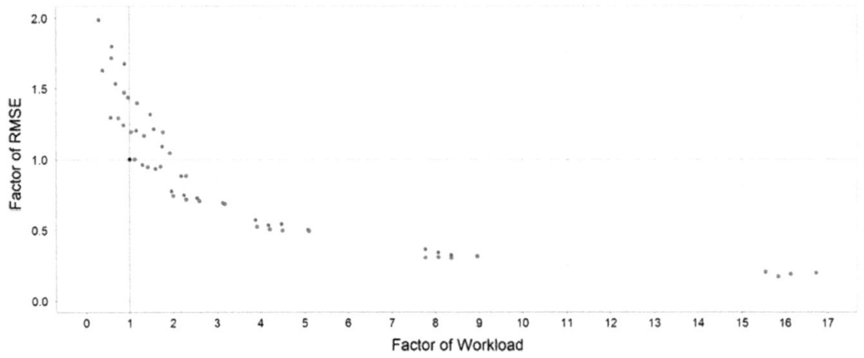

Fig. 13 RMSE reduction related to the factor of workload for different acquisition strategies. For the basic cruise strategy (single zig-zag trajectory with a wavelength of 1280 ft which equals the channel width) the RMSE is set to 100%. Considered are single zig-zag trajectories (blue) and double zig-zag trajectories (red)

Acknowledgements This work was supported by the European Social Fund [grant number 100270097, project "Extruso"] and by the Federal Ministry of Education and Research of Germany [grant number 033W039A, project "Boot monitoring"].

References

Aguilar FJ, Agüera F, Aguilar Ma, Carvajal F (2005) Effects of terrain morphology, sampling density, and interpolation methods on grid DEM accuracy. Photogram Eng Remote Sens 71:805–816. https://doi.org/10.14358/PERS.71.7.805

Casas A, Benito G, Thorndycraft V, Rico M (2006) The topographic data source of digital terrain models as a key element in the accuracy of hydraulic flood modelling. Earth Surf Proc Land 31:444–456. https://doi.org/10.1002/esp.1278

Conner JT, Tonina D (2014) Effect of cross-section interpolated bathymetry on 2D hydrodynamic model results in a large river. Earth Surf Proc Land 39:463–475. https://doi.org/10.1002/esp.3458

Cressie NAC (1993) Statistics for spatial data (revised edition). New York

ESRI (2001) Using ArcGIS geostatistical analyst. ESRI, Redlands

European Union (2000) Directive 2000/60/EC of the European parliament and of the council. Off J Eur Communities 43:1–73

Glenn J, Tonina D, Morehead MD et al (2016) Effect of transect location, transect spacing and interpolation methods on river bathymetry accuracy. Earth Surf Proc Land 41:1185–1198. https://doi.org/10.1002/esp.3891

Heritage GL, Milan DJ, Large ARG, Fuller IC (2009) Influence of survey strategy and interpolation model on DEM quality. Geomorphology 112:334–344. https://doi.org/10.1016/j.geomorph.2009.06.024

Hilldale RC, Raff D (2008) Assessing the ability of airborne LiDAR to map river bathymetry. Earth Surf Proc Land 33:773–783. https://doi.org/10.1002/esp.1575

Hofstra N, Haylock M, New M, et al (2008) Comparison of six methods for the interpolation of daily, European climate data. J Geophys Res Atmos 113. https://doi.org/10.1029/2008jd010100

Hutchinson MF (1989) A new procedure for gridding elevation and stream line data with automatic removal of spurious pits. J Hydrol 106:211–232. https://doi.org/10.1016/0022-1694(89)90073-5

Krige DG (1966) Two-dimensional weighted moving average trend surfaces for ore-evaluation. J South African Inst Min Metallurgy 66:13–38

Merwade VM, Cook A, Coonrod J (2008) GIS techniques for creating river terrain models for hydrodynamic modeling and flood inundation mapping. Environ Model Softw 23:1300–1311. https://doi.org/10.1016/j.envsoft.2008.03.005

Merwade VM, Maidment DR, Goff JA (2006) Anisotropic considerations while interpolating river channel bathymetry. J Hydrol 331:731–741. https://doi.org/10.1016/j.jhydrol.2006.06.018

Merwade VM, Maidment DR, Hodges BR (2005) Geospatial representation of river channels. J Hydrol Eng 10:243–251. https://doi.org/10.1061/(ASCE)1084-0699(2005)10:3(243)

Milne Ja, Sear Da (1997) Modelling river channel topography using GIS. Int J Geogr Inf Sci 11:499–519. https://doi.org/10.1080/136588197242275

Pebesma EJ (2004) Multivariable geostatistics in S: the gstat package. Comput Geosci 30:683–691. https://doi.org/10.1016/j.cageo.2004.03.012

Rase W-D (2016) Kartographische Oberflächen. Books on Demand

Santillan JR, Serviano JL, Makinano-Santillan M, Marqueso JT (2016) Influence of river bed elevation survey configurations and interpolation methods on the accuracy of lidar Dtm-based river flow simulations. ISPRS Int Arch Photogram Remote Sens Spat Inf Sci **XLII**-4/W1:225–235. https://doi.org/10.5194/isprs-archives-xlii-4-w1-225-2016

Shepard D (1968) A two-dimensional interpolation function for irregularly-spaced data. In: Proceedings of the 23rd ACM National Conference, pp. 517–524. https://doi.org/10.1145/800186.810616

US Army Corps of Engineers: Hydrographic survey data. http://www.lrp.usace.army.mil/Missions/Navigation/Navigation-Charts/HydrographicSurveyData/. Accessed 18 Sep 2016

Vande Wiele, T.: Mapping with multibeam data : are there ideal model settings ? In: International Cartographic Conference, Beijing, China (2001)

PLANTING: Computing High Spatio-temporal Resolutions of Photovoltaic Potential of 3D City Models

Syed Monjur Murshed, Amy Lindsay, Solène Picard and Alexander Simons

Abstract Photovoltaic (PV) production from the sun significantly contributes to the sustainable generation of energy from renewable resources. With the availability of detailed 3D city models across many cities in the world, accurate calculation of PV energy production can be performed. The goal of this paper is to introduce and describe PLANTING, a numerical model to estimate the solar irradiance and PV potential at the resolution of individual building surfaces and hourly time steps, using 3D city models. It considers the shading of neighboring buildings and terrains to perform techno-economic PV potential assessment with indicators such as installed power, produced electrical energy, levelized cost of electricity on the horizontal, vertical and tilted surfaces of buildings in a city or district. It is developed within an open-source architecture using mostly non-proprietary data formats, software and tools. The model has been tested on many cities in Europe and as a case study, the results obtained on the city of Lyon in France are explained in this paper. PLANTING is flexible enough to allow the users to choose PV installation settings, based on which solar irradiance and energy production calculations are performed. The results can also be aggregated at coarser spatial (building, district) and temporal (daily, monthly, annual) resolutions or visualized

---

Electronic supplementary material The online version of this chapter (https://doi.org/10.1007/978-3-319-78208-9_2) contains supplementary material, which is available to authorized users.

S. M. Murshed (✉) · A. Simons
European Institute for Energy Research, Emmy-Noether Str. 11, 76131 Karlsruhe, Germany
e-mail: murshed@eifer.org

A. Simons
e-mail: simons@eifer.org

A. Lindsay
EDF Inc. Innovation Lab, 4300 El Camino Real, Los Altos, CA 94022, USA
e-mail: amy.lindsay@edf-inc.com

S. Picard
École Supérieure D'Électricité, 3 Rue Joliot Curie, 91190 Gif-sur-Yvette, France
e-mail: solene.picard@supelec.fr

A. Mansourian et al. (eds.), *Geospatial Technologies for All*,
Lecture Notes in Geoinformation and Cartography,
https://doi.org/10.1007/978-3-319-78208-9_2

in 3D maps. Therefore, it can be used as a planning tool for decision makers or utility companies to optimally design the energy supply infrastructure in a district or city.

Keywords Solar irradiance · Photovoltaic potential · Building surfaces 3D city model · CityGML · Python

1 Introduction

1.1 Background

With an increasing willingness to improve energy efficiency and reduce greenhouse gas emissions, there is a clear trend among scientists, policy makers and energy producers to search for alternative renewable energy options. In many European cities, the decision makers encourage citizens to opt for local renewable energy through economic incentives and binding legislative acts. For example, the European Commission (EU) establishes different measures to reach multiple goals regarding energy efficiency and use of renewables by 2020 e.g., 20% cut in greenhouse gas emissions (from 1990 levels), 20% of EU energy from renewables, and 20% improvement in energy efficiency (European Commission 2017). According to Commission data, buildings in the EU are responsible for 40% of the total energy consumption and 36% of total CO_2 emissions (European Commission 2011). In this regard, research on energy transition with a focus on buildings is rapidly expanding and many projects and initiatives have emerged recently to achieve the EU targets.

Central and local governments also formulate short and long-term energy master plans with special consideration of the utilization of local renewable energy resources. Several tools and services are developed to assist citizens in identifying the potential generation of renewable energies e.g., photovoltaic (PV) or geothermal on their buildings and plots. Energy generation from PV is most promising when buildings are concerned. Different horizontal and vertical building surfaces are exposed to the sun, which can be utilized to generate energy by efficient design and installation of the PV panels. By assessing techno-economic PV potentials on these surfaces in a comparative manner and then aggregating the results at the building, district or city scale, the location of solar installations or the energy supply infrastructure can be optimized.

1.2 Literature Review

Several tools and methods have been developed to calculate solar irradiance and PV potential at different extents and scales—ranging from a simplified assessment on rooftops, with or without considering the shading from neighboring buildings, to the potential of a whole country, or a very detailed analysis of PV potential on individual buildings. A comparative review of the different approaches was performed by Mainzer et al. (2014), Freitas et al. (2015) and Santos et al. (2014). Different sets of vector and raster datasets, such as LIDAR, DSM, 2D or 3D have been used to calculate solar irradiance using both proprietary and open-source tools (Sarralde et al. 2015; Šúri and Hofierka 2004; Hofierka and Zlocha 2012; Redweik et al. 2013; Catita et al. 2014; Buffat 2016; Huld 2017; Li and Liu 2017; Bahu et al. 2014; Lee and Zlatanova 2009; Gueymard 2012). Most of these studies perform calculation at coarser spatial (e.g., building) and temporal (e.g., annual) resolution. Some other studies also performed shadow analyses on the PV installations due to trees, terrain or buildings, with or without direct applications to PV potential analyses (Palmer et al. 2015; Cole et al. 2016; Alam et al. 2012; Jaillot et al. 2017). Some web-based tools, such as Cythelia,[1] InSunWeTrust,[2] Mapdwell[3] and ProjectSunroof[4] can calculate solar radiation and to some degree PV potential on the roofs only.

With the emergence of semantic 3D city models (such as CityGML) of different levels of details (OGC 2012) and their availability for different cities,[5] several applications are possible (Biljecki et al. 2015; Murshed et al. 2017). Using such data of detailed building geometry and semantic relationships, PV potential on horizontal and vertical surfaces of buildings can be accurately calculated for the extent of a district or a city. Several attempts have been made to calculate solar irradiance on building surfaces of 3D city models (Wieland et al. 2015; Strzalka et al. 2012; Wate and Coors 2015; Chaturvedi et al. 2017; Murshed et al. 2018). Some limitations are observed in the modelling of PV potential at the district level. For example, no combined calculation of solar irradiance and techno-economic PV potential were performed at the district or city scale. The results originate at a coarser spatio-temporal resolution, which are not possible to aggregate at other extents (e.g., districts). Moreover, the irradiance and PV potential were calculated on the building surfaces, not directly on the PV installation (tilted) surfaces and without consideration of shading from neighboring buildings or terrains. Some studies used proprietary software and tools, which limit the application cases.

[1]http://www.cythelia.fr/en/renewable-energies/expertise/solar-cadastre/.

[2]https://www.insunwetrust.solar.

[3]https://www.mapdwell.com/en/solar.

[4]https://www.google.com/get/sunroof#p=0.

[5]http://www.citygmlwiki.org/index.php?title=Open_Data_Initiatives.

No decision support tool was developed to allow the users (e.g., decision makers or utility companies) to optimize the panel setting and therefore the energy production and economic benefits.

1.3 Objectives

The main objectives of this paper are to overcome the research gaps explained earlier through the development of the PLANTING (Photovoltaic Potential Based on Three Dimensional Building) model. It calculates both solar irradiance and techno-economic PV potential on the horizontal, vertical and tilted (i.e., PV installation) surfaces of the buildings, considering the 3D city models and the shading due to neighboring buildings or terrains.

It is a coupling between a simplified backward ray-tracing approach (defined number of hemisphere points and meshing of the surfaces) and a PV production model. The ray-tracing approach considers 3D city models to calculate shading (due to terrain and neighboring buildings), sky view factor and sun position of the points on the meshed building surfaces. The PV production model takes into account solar radiation transposition, optical losses from reflection, thermal heat exchanges within the PV module and impact on photo conversion efficiency. These models have been developed through an extensive literature research and implemented in an open-source software architecture. The model can be used as a planning tool for the decision makers and utility companies to generate numerical results about solar irradiance and techno-economic PV potential and visualize them on the roofs and facades of the buildings. It will assist them in optimizing the city energy supply infrastructure.

The rest of this paper is structured as follows: Sect. 2 describes the overall research approach—including a short description of required data, software architecture and methodological steps. Section 3 explains the 3D analyses e.g., sky view factor, shading, sun position. The methods to calculate solar irradiation and PV potential are described in Sect. 4. Then in Sect. 5, the results of the PLANTING model are explained for the city of Lyon in France. Afterwards, the validation approach and associated results are discussed in Sect. 6. Finally, the conclusion is drawn by highlighting the limitations and further improvements of this research.

2 Methodological Overview

2.1 Input Data

Several input datasets are required in PLANTING. For example, semantically and topologically correct 3D city model of the CityGML format with levels of details LoD1 or LoD2, hourly weather data on wind speed, temperature and horizontal

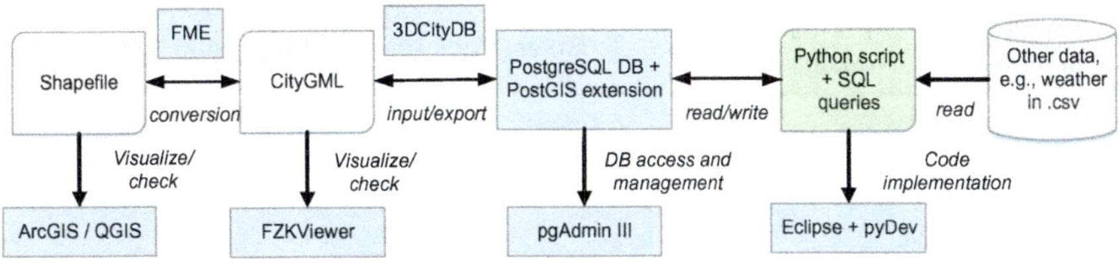

Fig. 1 Description of the software architecture and related technologies deployed in PLANTING model

radiation in TMY3 format (Wilcox and Marion 2008), as well as the techno-economic data on the panels. Assumptions concerning techno-economic values are made according to the state of the art literature review (see Sect. 4.1).

2.2 *Software Architecture*

Three main software e.g., 3DCityDB, PgAdminIII and Eclipse are deployed in PLANTING (see Fig. 1). The appropriate CityGML data of LoD1 and LoD2 format are imported in a PostgreSQL database provided with PostGIS extension in order to be analyzed and run the codes. PostGIS extension facilitates treating spatial objects by creating a special structure (or schema) in the database. That is why the original GML file is modified into the specific relational schema—through 3DCityDB, which reorganizes the data into specific tables, and then imports the modified file in the PostgreSQL database.

The algorithms are written in Python scripts, using the Eclipse Integrated Development Environment (IDE) of PyDev and some libraries (Psycopg2, Numpy, ORG, etc.). The DB can be retrieved using the Python codes and SQL queries. They perform data treatment, algorithm implementation, etc. The model outputs are saved as tables in the DB or can be exported as csv or shape files to visualize in 2D or 3D environments.

2.3 *Methodological Steps*

The PLANTING model is divided into 3 main modules and several sub-modules (see Fig. 2). They are programmatically structured into seven packages, where numerous scripts and functions perform specific tasks. The 3D analysis module includes the preparation of data, checking of the semantics of the 3D city models, and performance of the necessary geometric calculations such as shading, sky view factor and sun position.

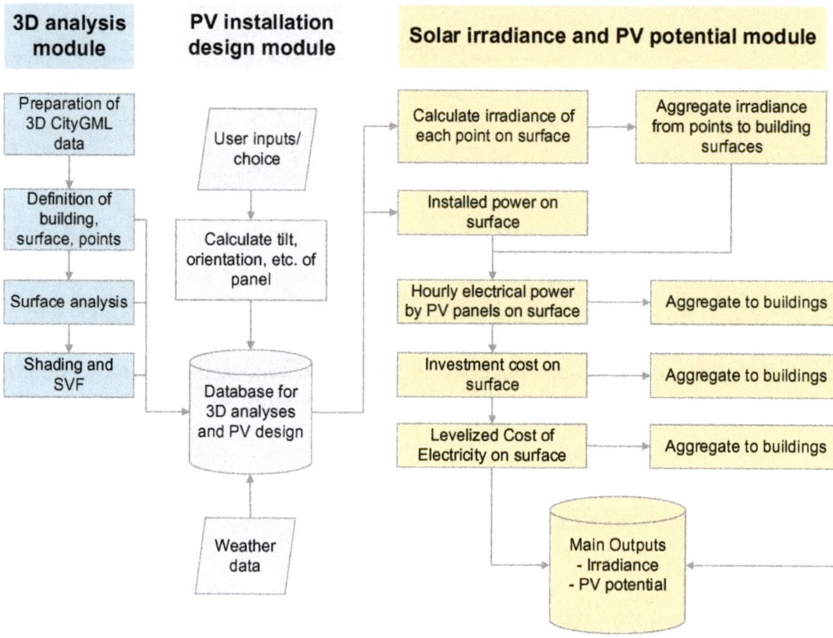

Fig. 2 General overview of the different modules, submodules and their relationships in the PLANTING model

The PV installation design module considers the users' inputs. The user needs to choose the type of technology (3 technologies), orientation (landscape/portrait), direction (N, S, E, W), and tilt of the PV panels on rooftops (slope < 20°). Moreover, the share of surfaces to be dedicated for PV on rooftops and on vertical surfaces (slope > 20°) must be given. Basically, for surfaces with a slope of >20°, the solar panels are considered to be building integrated, meaning the tilt and azimuth of the solar installation are dictated by that of the surface. For surfaces with a slope of <20°, the user can choose to have the solar panels in sheds on the surface, so he/she can choose the azimuth of the installation as well as the tilt, which must be higher than the tilt of the surface. Based on these inputs, the tilt, orientation of the PV installation are calculated and stored in the DB.

Finally, the solar irradiance and PV potential module calculates the hourly solar irradiances and then estimates the technical and economical PV potential. Users can generate outputs, such as solar irradiance (W/m^2), installed power (kWp), produced electrical energy (MWh), producible (kWh/kWp), CAPEX (capital expenditure in €), LCOE (Levelized cost of Electricity in €/kWh), for the horizontal, vertical and tilted PV installation surfaces. They are also saved in the database as tables for further analyses and visualization.

A detailed description of these three modules and associated assumptions is given in Sects. 3–5.

3 3D Analysis

This module is adapted after Wieland et al. (2015), Murshed et al. (2018), Šúri and Hofierka (2004) and Quaschning (2011) to analyze the shading characteristics of the grid points on the surfaces for each hour throughout the year. The methodological flow chart is illustrated in Fig. 3.

The first step is to define and analyze the geometrical objects (buildings, surfaces) of the 3D city model (LoD1 or LoD2) within the PostgreSQL DB. Then, a surface type (wall, roof or floor) is assigned to each surface based upon which the grids of the surface points (except for ground surfaces) will be defined. Each surface point is subsequently tested to identify whether it is shared with another surface. Different attributes (e.g., area, orientation, and azimuth) of the geometrical objects are calculated as well.

For each non-shared surface point, shading and sky view factor are calculated. First, a hemisphere of points is created according to the horizontal and vertical intervals chosen before the model starts. Each surface point gets a list of the hemisphere points that are visible from its location without being intersected by any

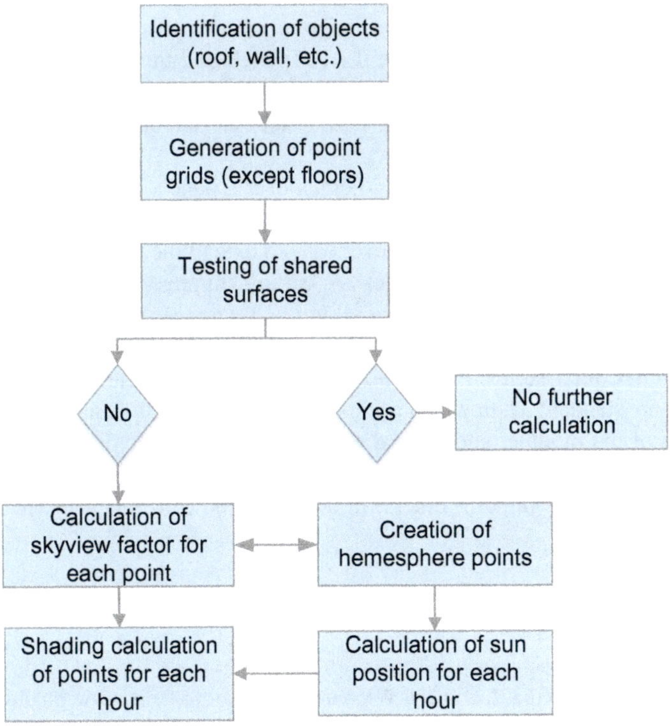

Fig. 3 Different steps involved in 3D analysis module

obstacle. The sky view factor for each surface point is calculated by dividing the number of visible hemisphere points by the total number of hemisphere points.

Afterwards, the sun position of each hour of the year is calculated considering the latitude and longitude of the location as well as local time. The closest hemisphere points replace these hourly positions (8760 in total). Finally, according to the list of visible hemisphere points, each surface point gets a list of 8760 Boolean values that corresponds to the information on whether the particular point is shaded or not for each hour of the year.

Based on the 3D analyses, the hourly solar radiation and energy production is calculated. PLANTING is intended to be used as a planning tool at a district level, therefore, an hourly resolution is reasonable (similar to PVsyst[6] for PV simulations, or Energy Toolbase[7] for building energy). It is not intended to be used for short-term energy production forecasting or for grid integration studies, where a higher time resolution is required.

4 PV Installation Design

4.1 Techno-Economic Assumptions

All building surfaces are supposed rectangular in order to simplify the geometrical considerations to define the length of the PV panels and the spacing between rows. Users can decide how much of the surface areas is to be dedicated to solar panels on the horizontal and vertical surfaces. Users can also choose either the portrait or landscape setting of the panels.

Three of the most dominant technologies of PV panels installed worldwide are considered (VDMA 2016). Assumptions on PV technology are made based on a review carried out in 2016. It is however easy to extend the model with more recent and efficient panels and technologies. A certain number of global variables are defined for the 3 PV technologies by taking as references the following modules: PANDA 60-cell 270Wp by Yingli for mono-Si,[8] 60-cell 260Wp by Jinko for poly-Si,[9] First Solar series 4 110Wp for thin-film.[10] The variables considered are: efficiency in standard testing conditions, temperature coefficient of Pmax, length and width of the module, and normal operating conditions temperature. Moreover, in order to quantify the losses due to resistive losses in the cables and conversion losses in the inverter (among others), a default value of 0.1 is selected for LBOS

[6]http://www.pvsyst.com/en/.

[7]https://www.energytoolbase.com/.

[8]PANDA 60 CELL SERIES 2, 260-280 Wc, Yingli Solar, available at www.yinglisolar.com.

[9]JKM270P-60-V, 255-270 W, Jinko Solar, available at www.jinkosolar.com.

[10]First Solar Series 4™ PV Module, available at www.firstsolar.com.

(losses due to balance of system), assuming that most inverters now have EU efficiencies around 92–98%.

The economic assumptions on the cost of the installation, Weight Averaged Cost of Capital (WACC) and Operation and Maintenance costs (O&M) are based on the report from the French Environment and Energy Efficiency Agency (ADEME 2015).[11] Costs in €/Wp for different installation sizes (0–3; 3–9; 9–36; 36–100; 100–250 kWp) depend on whether the PV is building-integrated (BIPV) or not (not-BIPV). The WACC depends on the actor (e.g., individual, company or utility) making the investment. The annual O&M costs are supposed to be a percentage of the total CAPEX spent each year. Some parameters (e.g., WACC) are fixed for the current year. Economic calculations (investment) are performed for current year but over a 20 years of lifetime.

4.2 PV Installation Design

The python script applies some basic rules for the design of PV installations. In this regard, the surface characteristics calculated in the previous step e.g., surface ID, building ID, represented area [m^2], slope of surface [°], aspect of surface [0° as South, 90° as West, 180° as North and −90° as East] are considered. The model enables the choice of each PV installation at a surface level, considering two cases: if the slope of the surface is below 20° or greater than 20°.

If the slope is below 20°, the PV installation is considered as not being building-integrated (BIPV), and the tilt is set equal to the minimum between the latitude of the location and a maximum tilt value defined by the user. It will also constrain the choice of orientation based on the aspect of one of the walls of the same building, with the underlying hypothesis that the roof is rectangular to simplify the geometrical considerations. The user can choose one of four directions for the PV installation, although it is suggested that the azimuth be as close as possible to South if the location is in the northern hemisphere and as close as possible to North, if it is in the southern hemisphere. This helps to find out the orientation that would maximize PV yield (see Fig. 4).

If the slope of the surface is above 20°, the PV installation is considered as being BIPV, and the tilt and orientation of the PV installation are equal to the slope and aspect of the surface, respectively.

[11]The economic assumptions can be improved with more recent data on affordable PV panels.

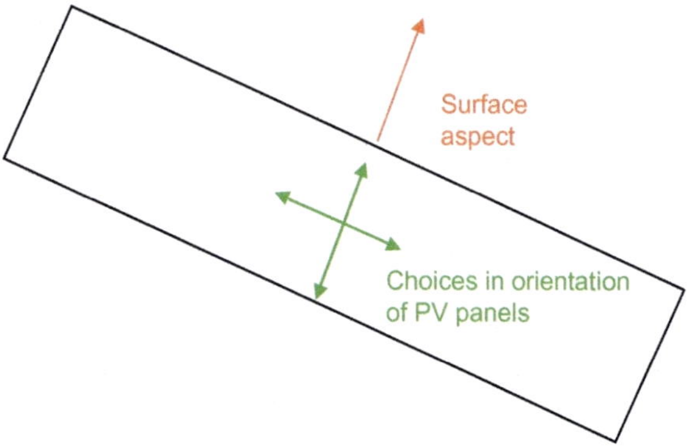

Fig. 4 Orientation of the PV installation depends on the aspect of the surface (in the case the slope of the surface is below 20°)

5 Solar Irradiance and PV Potential Calculation

5.1 Solar Irradiance

Solar irradiance computation requires a TMY3 weather file to determine different components of irradiances on each surface point for each hour of the year. The spatial and temporal aggregation is done to produce results at a surface, building and district scales, as well as at hourly, monthly and yearly resolutions.

Hypothesis: Partial shadings between rows (i.e., shading from other panels) are neglected and the sun position is calculated depending on coordinates and time. Only direct and diffuse radiation are calculated.

Steps: Solar irradiance is calculated in four steps:

1. First of all, a certain number of solar position calculations are made based on the equations explained in Duffie and Beckman (2006), pp. 90–93: declination, equation of time, real solar time, hour angle, sun position vector (the first component of the sun position vector, CosDir [0] gives the sinus of the solar elevation angle).
2. Calculate Direct Normal Irradiance: it can be computed as:

$$DIRN = \frac{GLOH - DIFH}{sinh} \tag{1}$$

   Where *sinh* is the sinus of the solar elevation angle, *GLOH* and *DIFH* are global horizontal and diffuse horizontal irradiances (W/m^2) that can be found in weather data.
3. Calculate incidence angle of beam irradiance on the PV surface: The cosine of the beam incidence angle is then calculated based on the sun position vector CosDir and the normal vector to the plane of array (l, m, n) as:

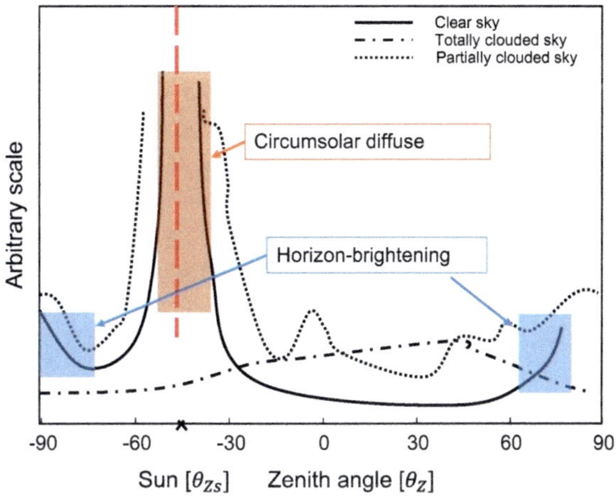

Fig. 5 Consideration of circumsolar diffuse irradiance and horizon brightening, modified after Duffie and Beckman (2006)

$$\cos i = l * CosDir[0] + m * CosDir[1] + n * CosDir[2] \tag{2}$$

4. Calculate the different components of Plane Of Array irradiance: To take into account circumsolar diffuse irradiance and horizon-brightening illustrated in Fig. 5, an anisotropic sky model is used, known as the Hay-Davies-Klucher-Reindl model (Duffie and Beckman 2006).

   The different components of the plane-of-array irradiance are:

$$DIRECT = bool_{shadow} \cdot \left(\cos i \cdot DIRN + \frac{\cos i}{\sin h} \cdot AI \cdot DIFH\right) \tag{3}$$

$$DIFFUSE = (1 - AI) \cdot VF_{sky} \cdot \left(1 + f \cdot \sin\left(\frac{tilt}{2}\right)^3\right) \cdot DIFH \tag{4}$$

with,

$$\begin{gathered} AI = \frac{DIFH}{GLOH_{ext}}, f = \sqrt{\frac{DIRH}{GLOH}} \\ GLOH_{ext} = I_{sc} \cdot \left(1 + 0.033 \cdot \cos\left(\frac{360 \cdot n}{365}\right)\right) \cdot \sin h \end{gathered} \tag{5}$$

where,

i	*the incidence angle of beam irradiance on plane of array (°)*
h	*sun elevation (°)*
tilt	*tilt of the panels (°)*
AI	*an anisotropy index*

f	*corrective factor accounting for horizon-brightening*
I_sc	*the solar constant, 1367* W/m^2
n	*the day of the year*
VF_{sky}	*sky view factors (Computed in 3D analysis)*
$Bool_{shadow}$	*1 if the point ID considered is not in the shade at this given time-step and equals 0 if the point ID considered is in the shade at this given time-step (direct irradiance does not reach this point)*
DIRN, DIRH, DIFH, GLOH, $GLOH_{ext}$	*direct normal, direct horizontal, diffuse horizontal, global horizontal and extraterrestrial global irradiances* (W/m^2), *respectively*

5.2 PV Potential

The techno-economic PV potential is calculated in six main steps:

Nominal Power of PV Installed on Each Surface. Step 1 determines the nominal power of PV installed for each surface, depending on available surface, whether the PV is building-integrated or not, the tilt of the PV panels, the choice of PV technology (which has an impact on the dimensions of the PV modules), the orientation of the PV panels (landscape or portrait) and the share of surface used for the PV installation.

1. Calculate the number of panels that can be installed on a given Surface with a certain share dedicated to PV ($Share_{PV}$) (see Fig. 6).

   The limit angle is by default considered as being equal to 20°.

- If the PV is building-integrated (BIPV = 1), the number of panels can be calculated very simply by the following formula:

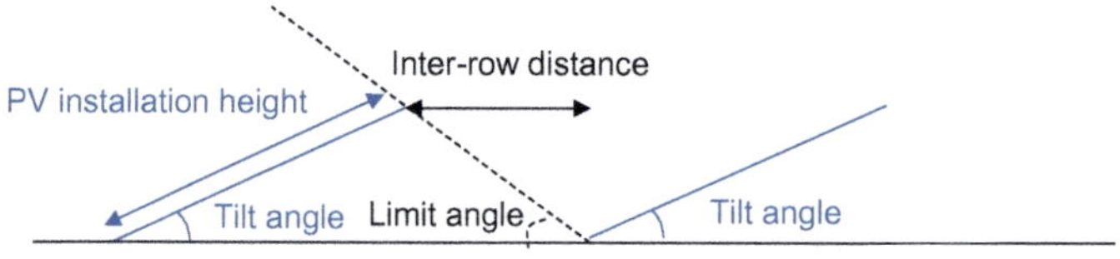

Fig. 6 Calculate number of panels on a given surface

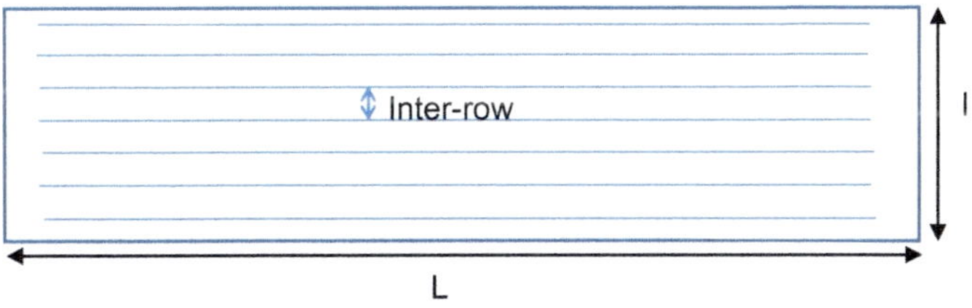

Fig. 7 Calculation of inter-row distance and number of panels to be installed on a building surface

$$nb_{panels} = \frac{Surface * Share_{PV}}{length_{module} * width_{module}} \tag{6}$$

- If the PV is not building-integrated (BIPV = 0), the inter-row distance can first be calculated as follows:

$$interrow = \frac{height_{PVinstallation} * \sin(tilt)}{\tan(limit_{angle})} \tag{7}$$

The variable $height_{PVinstallation}$ will either be equal to the length of the PV panel ($length_{module}$) if the panels are displayed in portrait or to the width of the PV panel ($width_{module}$) if the panels are set-up in landscape.

Once the inter-row is calculated, the number of panels is basically the number of panels we can fit in along one of the sides of the rectangular surface (i.e., along one row) multiplied by the number of rows that can be fit into a perpendicular side of the rectangular surface (see Fig. 7).

$$nb_{panels} = \frac{l * L * Share_{PV}}{effectivewidth_{PVpanel} * (interrow + height_{PVinstallation} * \cos(tilt))} \tag{8}$$

$$= \frac{Surface * Share_{PV}}{effectivewidth_{PVpanel} * (interrow + height_{PVinstallation} * \cos(tilt))} \tag{9}$$

where the *effective width of the PV panel* equals the width of the PV panel if the PV is installed in portrait, and is equal to the length of the PV panel if it is installed in landscape.

2. Calculate the installed power on a given surface in kWp: to get the installed power per surface in kWp, the number of panels is then multiplied by the nominal power of one PV panel:

$$installed_{power} = nb_{panels} * \eta_{STC} * width_{module} * length_{module} \tag{10}$$

The efficiency η_{STC}, width and length of module depend on the technology choice made.

Optical Loss at the Air-Glass Interface. Step 2 defines analytical functions to compute reflection optical losses at the air-glass interface for the two components of solar irradiance: beam and diffuse. It is an improvement of the ASHRAE model, which is not valid for incidence angles above 80° (Luque and Hegedus 2011, pp. 934–936) and (Martin and Ruiz 2001).

1. Calculate transmission factor for direct irradiance: The transmission factor for direct irradiance is given by:

$$FT_B = 1 - \frac{e^{\frac{-\cos i}{a_r}} - e^{\frac{-1}{a_r}}}{1 - e^{\frac{-1}{a_r}}} \tag{11}$$

2. Calculate transmission factor for diffuse irradiance: The transmission factor for diffuse irradiance is given by:

$$FT_D = 1 - e^{\frac{-1}{a_r} \cdot \left(c_1 \cdot \left(\sin(tilt) + \frac{\pi - \frac{tilt.\pi}{180} - \sin(tilt)}{1 + \cos(tilt)} \right) + c_2 \cdot \left(\sin(tilt) + \frac{\pi - \frac{tilt.\pi}{180} - \sin(tilt)}{1 + \cos(tilt)} \right)^2 \right)} \tag{12}$$

where,

c_1	*coefficient equal to* $\frac{4}{3\pi}$
c_2 *and* a_r	*coefficients that depend on the soiling level of the PV panels. We will suppose that the PV panels are clean, leading to* $c_2 = -0.069$ *and* $a_r = 0.17$

Temperature of the PV Cells. Step 3 calculates the temperature of the PV cells, which is based on the technical characteristics of the PV module and the following equation (Duffie and Beckman 2006, p. 760):

$$\frac{PV_{temp} - T_{ext}}{T_{NOCT} - 20°C} = \frac{transmitted_{irr}}{800\,W/m^2} * \frac{U_{NOCT}}{U} * \left(1 - \frac{\eta}{n_{inc}}\right) \tag{13}$$

where,

PV_{temp}	*temperature of the PV cells in °C*
T_{NOCT}	*Normal Operating Conditions Temperature (NOCT) of the specified PV panel, established at an irradiance of 800* W/m^2*, an ambient temperature of 20 °C and a wind speed of 1 m/s*
$Transmitted_{irr}$	*irradiance that is actually received by the PV cells*

η	*efficiency of the panel*
n_{inc}	*optical transmittance at normal incidence*
U	*heat transfer coefficient for convective and radiative heat exchanges with the environment,* $U = 8.55 + 2.56 * V$ *with V wind speed in m/s* (Test et al. 1981, pp. 262–267)

Electrical Power of the PV Installation. Step 4 calculates the electrical power delivered by the PV installation at each given time step. It is defined based upon the irradiances received.

1. Calculate the photo conversion efficiency based on module temperature: The efficiency of the PV panel decreases with increasing temperatures of the PV cells (Duffie and Beckman 2006, p. 757), and this behavior is characterized by μ, the temperature coefficient on Pmax (% °C), given in the module datasheet:

$$\eta(T) = \eta_{STC} \cdot \left(1 + \mu \cdot \left(PV_{temp} - 25°C\right)\right) \tag{14}$$

2. Calculate the electrical power delivered by the PV installation: This efficiency is then multiplied by the irradiances received and scaled to the installed power of the installation, also accounting for conversion losses in the inverters, cables and other power electronics by using the discount coefficient LBOS (global parameter).

Total Investment Cost. Step 5 calculates the total investment costs in € for a given installed power. The unitary costs (€/Wp) decrease with the size of the installation, due to scale effects. The costs are also higher for Building Integrated PV (BIPV) than for standard shed installations (not-BIPV). The default values for economic assumptions are based on literature (ADEME 2015).

Levelized Cost of Electricity. Step 6 calculates the Levelized Cost of Electricity (LCOE) i.e., the cost of the electricity generated by a PV installation over its lifetime (e.g., 20 years), considering the following equation:

$$LCOE = \frac{CAPEX + \sum_{t=1}^{t=N} \frac{O\&M_t}{(1+i)^t}}{\sum_{t=1}^{t=N} \frac{E_t}{(1+i)^t}} \tag{15}$$

where,

$O\&M_t$	*the operation and maintenance costs for year t, generally taken as a percentage of initial investment costs (CAPEX)*
i	*discount rate*
E_t	*energy generated in year t*

6 Implementation

The PLANTING model has been implemented for several cities with a varying number of buildings (in LoD1 and LoD2) in order to test the model performance and validation of results. This paper explains the results obtained in the district of Gerland in the city of Lyon in France. In this regard, the consideration of techno-economic input data and associated assumptions are made in the French context (Sect. 4). The CityGML data of LoD2 format, describing 2750 buildings (with around 100000 wall and roof surfaces) and the weather data of TMY3 format for the city are used. Several scenarios can be prepared by varying the five user inputs. However, as an example, we choose the Mono-crystalline Silicon PV module, dedicate 50% of surfaces to PV installation, and set the panels in portrait mode, with a maximum tilt of 45° in the South direction. Afterwards, the model computes the incoming solar irradiance and other techno-economic potentials on all of the building surfaces.

On Fig. 8, the total annual energy production from solar panels can be seen for each building surface. This is an interesting visualization that considers both available solar irradiance on the building surfaces, but also building size. The surfaces that strike out in red are those that have few neighbouring buildings obstructing sunlight but also large available surfaces. Generally, these are rooftops and this visualization quickly identifies the rooftops that have the best sun exposure. When the users intend to build just a few big solar projects that have a meaningful impact in terms of energy, this planning tool can be very useful.

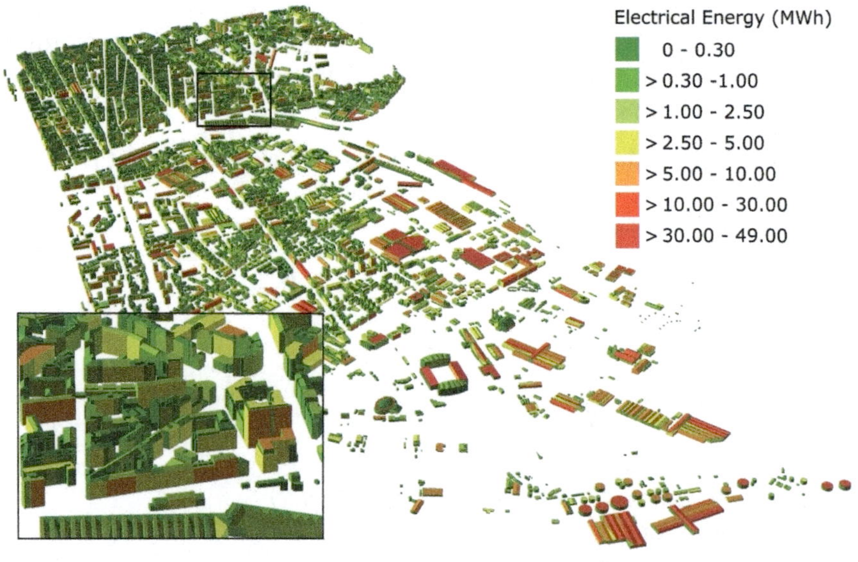

Fig. 8 Total annual energy production (MWh) on the vertical and horizontal surfaces is calculated considering the solar irradiance and building size

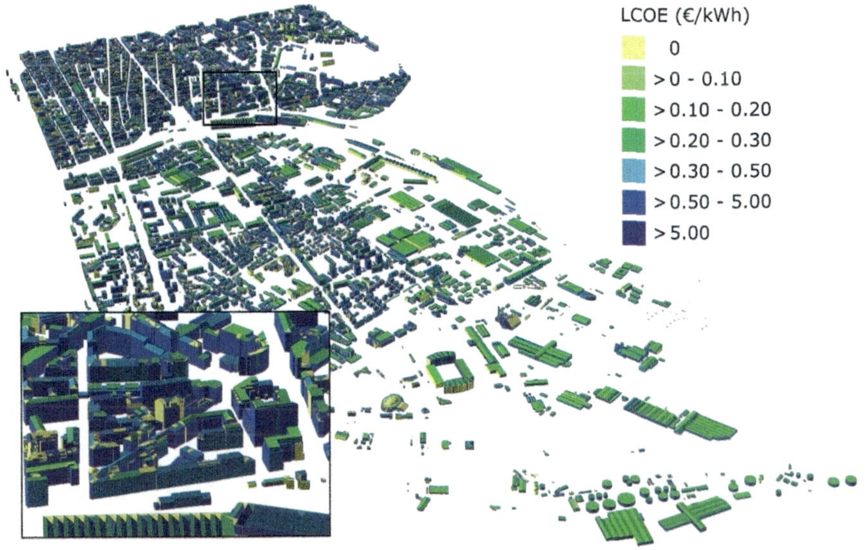

Fig. 9 Levelized Cost of Electricity (LCOE in €/kWh) calculated in each of the building surfaces (considering 20 years of lifetime of PV panels)

On Fig. 9, the Levelized Cost of Electricity is displayed on all building surfaces of this district in Lyon. High potential areas can be noticed with LCOEs beneath 20 c€/kWh, but even achieving costs under 30 c€/kWh in a dense urban area is quite remarkable. This type of display can help decision makers identify areas where solar energy development makes the most sense, as well as quantify the impact of new building projects on available sunlight.

Figure 10 displays the specific yield of solar panels for all surfaces, meaning the production is normalized to the installed capacity (kWh/kWp). This does not account for impact of size of the installation on costs, which is reflected on the LCOE map in Fig. 9. Specific yield identifies the surfaces that have the best sun exposure, without taking into account available surface for PV installations.

These techno-economic indicators can also be analysed at further spatio-temporal resolutions. For example, the hourly energy production results of the surfaces are aggregated at IRIS[12] statistical units to estimate monthly average energy production density (monthly energy production/number of buildings) of these units (see Fig. 11). Each IRIS unit consists of different numbers, forms and shapes of building, which result in varying energy density. Such analyses will help the decision makers and utility companies to estimate possible energy production and plan an efficient energy supply infrastructure.

[12]https://www.insee.fr/en/metadonnees/definition/c1523.

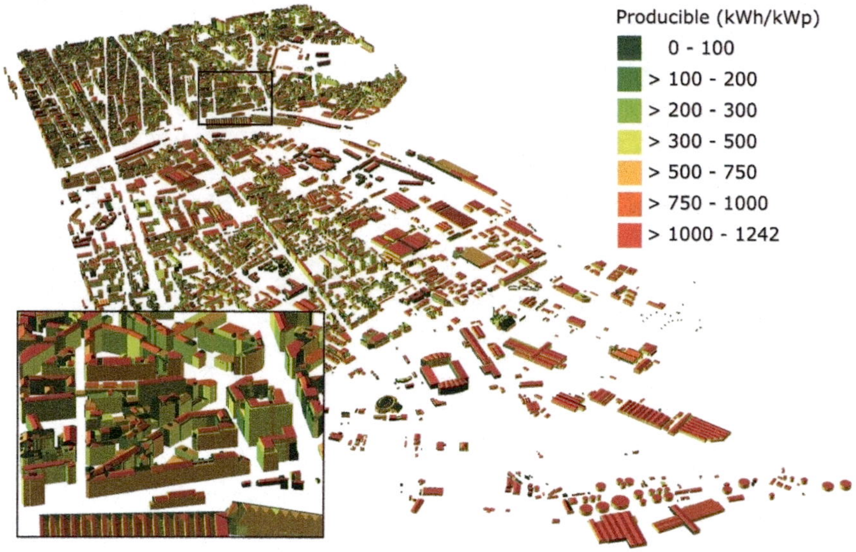

Fig. 10 Producible (kWh/kWp) on the vertical and horizontal surfaces of the buildings displays normalizing yields with regard to PV installed

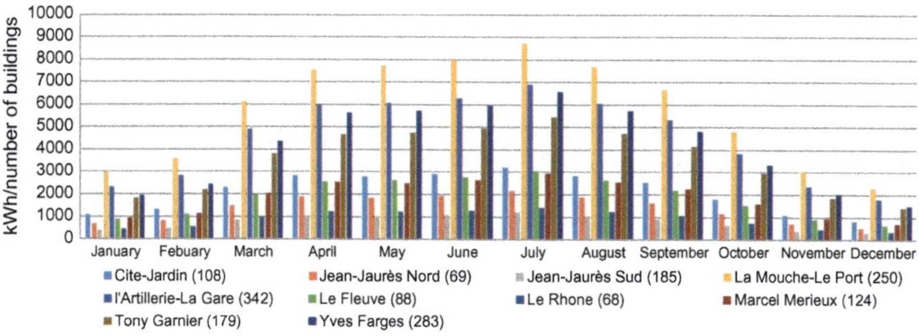

Fig. 11 Monthly PV energy production density in the 10 IRIS statistical units (and total number of buildings) in the district Gerland in Lyon, France

7 Validation and Model Performance

7.1 Validation

Approach. An extensive description of the validation of each module of the PLANTING model is out of the scope of this paper. 3D analyses, calculation of irradiances and techno-economic analysis of PV potential are performed based on literature reviews, previous studies, assumptions on the users' settings of parameters as well as techno-economic data on panels. Therefore, this section mainly explains the validation of irradiance and energy produced on the building surfaces.

Validation of PLANTING results is performed with two software modules, developed at EDF R&D: (a) a ray-tracing tool that performs a detailed computation of irradiance on ground mounted or roof based PV installations and (b) Dymola/BuildSysPro that conducts computation of PV production based on a thermo-electrical model of PV modules. The modelling and validation of these modules have been performed by simulating the PV production in the UK and Turkey (Jourdier et al. 2016).

A representative building in a typical European city is chosen to validate the annual irradiance and PV production on both sides of its saddle roof (globally West and East oriented). Two cases have been chosen (a) an isolated building, without any obstacles and (b) the same building surrounded by neighboring buildings (see Fig. 12). In order to ensure comparative analyses, the input parameters and other assumptions are considered identical in both models (see Table 1).

In both validation cases, the LoD2 buildings of the CityGML format have been translated into a Sketch Up format and then imported into the ray-tracing tool. The Meteonorm weather data is considered as an input to compute the irradiance on both sides of the roof (West and East). Afterwards, the irradiance data computed by the ray-tracing tool is used as an input to BuildSysPro to compute the annual energy production of the building. An overview of the validation approach is given in Fig. 13.

Discussion of Results. A satisfactory general agreement is obtained between the numerical computations done by ray-tracing and PLANTING for irradiance calculations (see Fig. 14). The maximum relative discrepancy of +3.24% is obtained for the case of West-oriented solar panels on the surrounded building (El Hajje and Boyere 2017).

In terms of annual energy yield calculations, a maximum relative discrepancy of 10.92% is obtained for the case of the West-oriented surrounded building (Fig. 15).

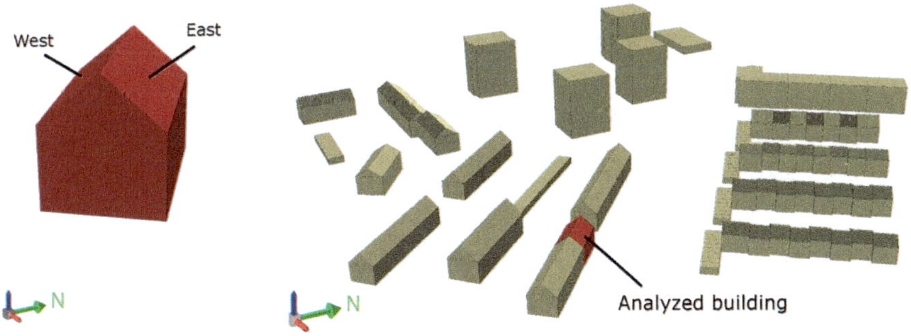

Fig. 12 Validation of results with an isolated building (left) and the same building surrounded by other buildings (right)

Table 1 Assumptions and inputs considered in validating the irradiance and energy production

Input parameters	Roof surface West	Roof surface East
Tilt of the roof	40.35°	40.33°
Azimuth of the roof	230.3° (West)	50.264° (East)
Surface of the roof	84.78 m^2	84.71 m^2
Module type	Mono-crystalline Silicon	
Max tilt	45°	
Orientation of panels	Landscape	
Surface dedicated for PV panels	80%	

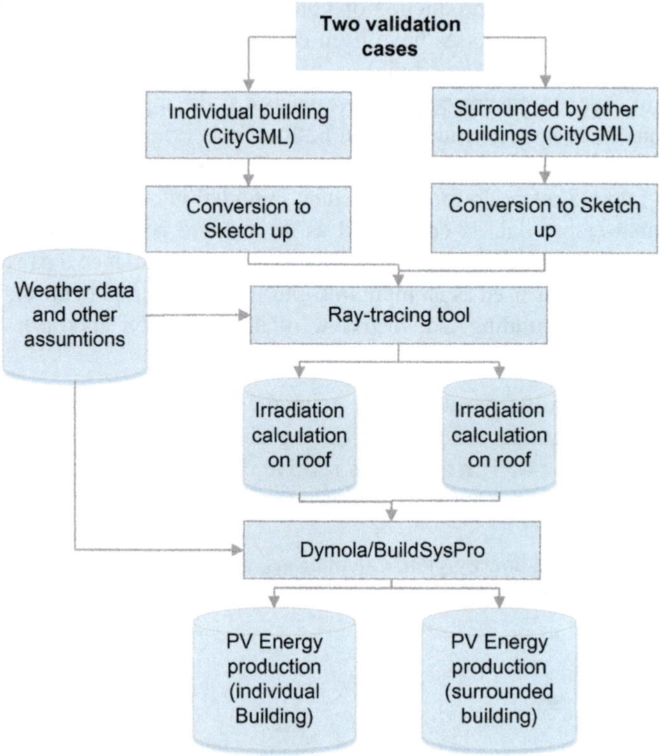

Fig. 13 Validation approach for irradiance and PV energy production for an individual building without obstacles and the same building surrounded by other buildings

This can be explained by the respective irradiance discrepancies and mostly the electrical calculation models of each numerical tool. By simple subtractions between these discrepancies observed in the West and East oriented surfaces of the isolated and surrounded buildings, we obtain the following values (7.55, 7.67, 7.68, and 7.55%). Therefore, we can almost safely conclude that the electrical yield

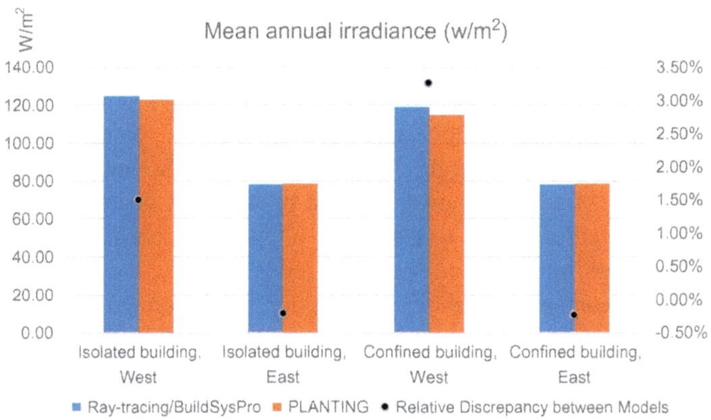

Fig. 14 Comparison of annual irradiance calculation results from the two models

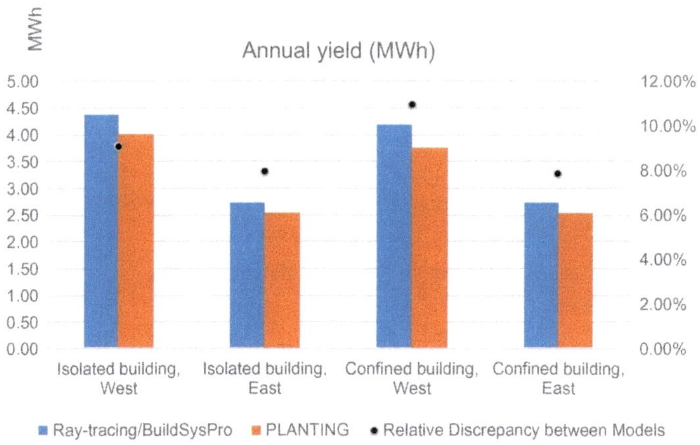

Fig. 15 Comparison of annual energy yields calculation results from the two models

calculations performed by both models solely induce a consistent relative discrepancy of around 7.6% in the global discrepancy calculations, regardless of the considered case study (orientation and/or shading). Therefore, in order to better understand how to minimize this 7.6% relative difference, both electrical models need to be compared and analyzed in detail. It can be concluded that the obtained relative mismatches are roof orientation-dependent. A greater discrepancy values for the West-oriented rooftop calculations are observed.

7.2 Evaluation of Run Time

The PLANTING model is run on virtual machines with the following configuration: (a) Linux Server for Python with 64 GB Ram, 12 GB used, 10 cores, HDD 8.5 GB free and (b) PostgreSQL Server for the DBs with 16 GB Ram, 3.5 GB used, 10 cores, HDD 44 GB free. The model is tested on different sets of LoD1 and LoD2 data across many cities in the world. Each of them are constituted of a varying number of buildings, surfaces and points. In order to have a comparative overview, all relevant parameters (e.g., 5 m * 5 m grid point resolution, 96 hemisphere points) are considered identical.

Table 2 explains the computing time required for running different modules on the LoD1 and LoD2 datasets. We observe that the model run time increases with the increasing number of surfaces. In general, due to the lesser complexity of LoD1 city models, PLANTING requires less time than for LoD2 models. In evaluating the most time consuming component, we observe that the irradiance and PV calculation part takes about 50% of the total model run time. However, if the CityGML data contains terrain information and if the shading effect due to the terrain is calculated (e.g., LoD2 data in Lyon), the PLANTING model requires significantly more time.

Table 2 Comparative evaluation of run time of different modules in PLANTING with multiple LoD1 and LoD2 3D city models

Location	#Buildings #Surfaces #Points	Computing time for different modules				
		3D analysis	Internal DB calculation	Irradiance and PV (surface)	Irradiance and PV (building)	Total
Hong Kong (LoD1)	1085 16548 213933	1 h 40 min 44 s	37 min 46 s	3 h 20 min 4 s	10 min 19 s	5 h 48 min
Abu Dhabi (LoD1)	280 2624 44623	13 min 51 s	8 min 12 s	45 min	13 s	1 h 7 min 16 s
Kuwait (LoD1)	567 5595 44462	37 min 15 s	18 min 54 s	1 h 45 min 45 s	2 min 17 s	2 h 42 min 11 s
Karlsruhe (LoD2)	95 745 2756	53 s	30 s	3 min 20 s	1 s	4 min 44 s
Karlsruhe (LoD2)	12035 95672 441689	6 h 4 min 45 s	1 h 17 min 58 s	7 h 54 min 55 s	15 s	15 h 17 min 54 s
Lyon (LoD2) with terrain	2750 99558 501459	9 h 30 min 34 s	4 h 2 min 19 s	5 h 38 min 14 s	4 s	19 h 11 min 13 s

8 Conclusion and Future Research

8.1 Summary

This paper introduces and describes the numerical model, PLANTING that estimates and evaluates the solar irradiance (W/m^2) and techno-economic photovoltaic production in terms of installed power (kWp), produced electrical energy (MWh), producible (kWh/kWp), CAPEX (capital expenditure in €) and LCOE (Levelized cost of Electricity in €/kWh) on the horizontal, vertical and tilted PV installation surfaces of the buildings in a city or district. It is built within an open-source architecture using mostly non-proprietary data formats, software and tools. The model is divided into three main modules. First, it analyzes the 3D city models to calculate the shading and sky view factor, and then considers user choices (e.g., type of PV technology, orientation, tilt, etc.) to determine the optimal setting of PV panels. Finally, solar irradiance and PV potential are calculated on the basis of scientific literature and country specific techno-economic assumptions.

The model has been applied to several cities with a varying number of buildings in order to test the model performance and validate the results. Considering different climatic and topographic conditions, the model also proves robust. This paper explains the results obtained in the city of Lyon in France. In this regard, the CityGML data of LoD2 format of 2750 buildings in the district of Gerland in Lyon is used.

The model is also validated with EDF R&D's own software modules: ray-tracing and Dymola/BuildSysPro, considering two different case studies: (a) an individual residential building with West and East-oriented roofs (b) the same building surrounded by other buildings. Based on the final irradiance and PV energy calculation results, we obtain a satisfactory general agreement.

PLANTING is flexible enough to allow the users to choose different PV installation settings, based on which solar irradiance and energy production calculation is performed. The results can also be aggregated at coarser spatial (building, district) and temporal (daily, monthly, annual) resolutions or visualized in 3D maps. Therefore, it can be used as a planning tool for decision makers or utility companies to optimally design the energy supply infrastructure in a district or city.

8.2 Limitations

PLANTING requires semantically and topologically correct CityGML data of LoD1 or LoD2 format. The data should be of appropriate standard and imported correctly into the DB, otherwise 3D analysis, solar irradiance and PV potential calculations cannot be performed. We have experienced different types of problems associated with the CityGML data, which had to be corrected. Also, the current version of the model cannot run on LoD3 or LoD4 datasets.

In calculating sky view factor and shadow, accuracy depends on the meshing of the surfaces; here a regular orthonormal grid is defined. The reduction of the meshing leads to diminishing model run time, but reduces the accuracy. Moreover, the hemisphere points determine the sky view factor and the sun positions. The reduction of hemisphere points leads to decreasing model run time, but reduces the accuracy. In calculating solar irradiance, only direct and diffuse radiation are considered. However, reflected radiation generally accounts for a small percent in the global radiation (Šúri and Hofierka 2004).

In designing PV installations, all horizontal and vertical surfaces are assumed rectangular. Tilt angle and azimuth of the PV installation have some constraints depending on the tilt of the surface and on the aspect of the surface. Users have to decide on the portrait or landscape setting of the PV panels, based on which PLANTING calculates the PV potential for the 3D city models. Unfortunately, selection of individual settings for individual buildings is not possible.

Three dominant types of PV technologies are considered, but other technologies can easily be incorporated into the model. No degradation rate is taken into account in the energy yield over the lifetime of the installation. The efficiency of photo-conversion of PV panels decreases linearly with the temperature of the module. In the LCOE, for the moment, no decrease in energy production is considered. However, a change could easily be implemented in the code. Economic calculations (investments) are performed based on the economic hypothesis for the current year and supposing a discount rate for the future years. Some other economic parameters (e.g., WACC) are set to default values fixed for the current year and should be updated regularly.

The model does not suggest the optimal location to install the PV panels on a surface. However, by inspecting the point irradiance values on the surfaces, users can identify the portion of the surfaces that receive maximum solar irradiance.

8.3 Future Studies

The results are currently saved in a PostgreSQL database as tables and are exported to visualize in GIS software or to prepare graphs in spreadsheet programs. However, a browser-based 3D decision support system is currently under development, where users can interactively visualize and explore different techno-economic outputs at the point, surface or building level and at varying temporal resolutions.

The developed model only considers shading due to neighboring buildings and terrains. In an urban environment, shading due to roof protrusions or trees can affect the PV production. Therefore, consideration of detailed tree models and CityGML data of LoD3 or LoD4 formats will help in assessing the shadow effects and enhance the accuracy of solar radiation or energy production results. Thus, in the future, the model will be adapted to incorporate trees and other obstructions.

In the current version, the run time and computational efficiency of handling large 3D city models and terrains have been significantly improved by restructuring and profiling the code, as well as deploying a multi-processing package of Python. However, it can be further improved e.g., by grouping the points that have the same visible hemisphere, avoiding the calculation of nighttime hourly irradiances. These consume a significant amount of computational resources.

Other techno-economic parameters e.g., feed in tariffs, energy prices and avoided of CO_2 emissions can be incorporated in the future. Reflected radiation can also be calculated, by considering the ground view factor and albedo. Finally, it will be useful to perform a sensitivity analyses to understand the impact of uncertain model parameters on the results.

Acknowledgements We are grateful to EDF R&D and Métropole de Lyon for funding this research within the projects "Smart and Low Carbon Cities" and "Modélisation Urbaine Gerland (MUG)", respectively. Our heartiest gratitude to the city of Lyon and Laboratoire LIRIS for providing the 3D city model of Lyon. We would like to thank City of Karlsruhe for the permission of using the CityGML data for evaluating the model. We are very grateful to Gilbert El Hajje and Emmanuel Boyere of EDF R&D for comparison and cross validation of the model. We also acknowledge Fabrice Casciani, Monika Heyder, Pierre Imbert, Omar Benhamid, Alexandru Nichersu, Céline De Pin, Alice Duval, Manfred Wieland and other colleagues for their input. Finally, our sincere gratitude extends to the editors and two anonymous referees for their insightful comments, which helped us to improve the manuscript.

References

Ademe (2015) Filière Photovoltaïque Française: Bilan, perspectives et Stratégie. Agence De l'Environnement et de la Maitrise de l'Energie, Angers

Alam N, Coors V, Zlatanova S, Van Oosterom P. Shadow effect on photovoltaic potentiality analysis using 3D city models. In: XXII congress of the international society for photogrammetry and remote sensing, 25 August–1 September 2012, Melbourne. ISPRS

Bahu J-M, Koch A, Kremers E, Murshed, SM (2014). Towards a 3D spatial urban energy modelling approach. Int J 3-D Informat Model 3:1–16

Biljecki F, Stoter J, Ledoux H, Zlatanova S, Çöltekin A (2015) Applications of 3D city models: state of the art review. ISPRS Int J Geo-Informat 4:2842–2889

Buffat R (2016) Feature-aware surface interpolation of rooftops using low-density lidar data for photovoltaic applications. In: Sarjakoski T, Santos M, Sarjakoski L (eds.) Geospatial data in a changing world. Lecture notes in geoinformation and cartography geospatial data in a changing world. Cham, Springer

Catita C, Redweik P, Pereira J, Brito MC (2014) Extending solar potential analysis in buildings to vertical facades. Comput Geosci 66:1–12

Chaturvedi K, Willenborg B, Sindram M, Kolbe TH (2017) Solar potential analysis and integration of the time-dependent simulation results for semantic 3D city models using dynamizers. ISPRS Ann Photogramm Remote Sens Spatial Inf Sci IV-4/W5:25–32

Cole IR, Palmer D, Betts TR, Gottschalg, R (2016) A fast and effective approach to modelling solar energy potential in complex environments. In: 32nd European photovoltaic solar energy conference and exhibition, 20–24 June 2017. Munich

Duffie JA, Beckman WA (2006) Solar engineering of thermal processes. Wiley, New York

El Hajje G, Boyere E (2017) Comparison and cross validation of PV production models used by EIFER and EDF R&D/TREE. EDF R&D, Moret Sur Loing Cedex

European Commission E (2011) Review of the energy performance of buildings directive 2010/31/ EU. EC Directorate-General of Energy, Brussels

European Commission E (2017) Climate strategies and targets [Online]. http://ec.europa.eu/clima/policies/strategies/2020. Accessed 2017
Freitas S, Catita C, Redweik P, Brito M (2015) Modelling solar potential in the urban environment: State-of-the-art review. Renew Sustain Energy Rev 41:915–931
Gueymard CA (2012) Clear-sky irradiance predictions for solar resource mapping and large-scale applications: Improved validation methodology and detailed performance analysis of 18 broadband radiative models. Sol Energy 86:2145–2169
Hofierka J, Zlocha M (2012) A new 3-D solar radiation model for 3-D city models. Trans GIS 16:681–690
Huld T (2017) PVMAPS: software tools and data for the estimation of solar radiation and photovoltaic module performance over large geographical areas. Sol Energy 142:171–181
Jaillot V, Pedrinis F, Servigne S, Gesquière G (2017) A generic approach for sunlight and shadow impact computation on large city models. In: 25th international conference on computer graphics, visualization and computer vision, May 29–June 2, 2017. Pilsen, Czech Republic
Jourdier B, Hoang T-T-H, Chiodetti M (2016) Reconstitution de la production photovoltaïque horaire en Grande-Bretagne et Turquie sur 58 ans et validation d'un nouveau modèle physique de production PV. EDF R&D, Chatou cedex
Lee J, Zlatanova S (2009) Solar radiation over the urban texture: LIDAR data and image processing techniques for environmental analysis at city scale. In: Lee J, Zlatanova S (eds) 3D Geo-information sciences. Lecture notes in geoinformation and cartography. Springer, Berlin, Heidelberg
Li Y, Liu C (2017) Estimating solar energy potentials on pitched roofs. Energy Build 139:101–107
Luque A, Hegedus S (2011) Handbook of photovoltaic science and engineering Wiley
Mainzer K, Fath K, Mckenna R, Stengel J, Fichtner W, Schultmann F (2014) A high-resolution determination of the technical potential for residential-roof-mounted photovoltaic systems in Germany. Sol Energy 105:715–731
Martin N, Ruiz J (2001) Calculation of the PV modules angular losses under field conditions by means of an analytical model. Sol Energy Mater Sol Cells 70:25–38
Murshed SM, Picard S, Koch A (2017) CityBEM: an open source implementation and validation of monthly heating and cooling energy needs for 3D buildings in cities. ISPRS Ann. Photogramm Remote Sens Spatial Inf Sci. IV-4/W5 83–90
Murshed SM, Simons A, Lindsay A, Picard S, De Pin C (2018) Evaluation of two solar radiation algorithms on 3D city models for calculating photovoltaic potential. In: 4th international conference on geographical information systems theory, applications and management, 17–19 March 2018. Funchal, Madeira, Portugal
Ogc 2012. OGC City Geography Markup Language (CityGML) Encoding Standard 2.0.0. Open Geospatial Consortium
Palmer D, Cole IR, Goss B, Betts TR, Gottschalg R (2015) Detection of roof shading for PV based on LiDAR data using a multi-modal approach, 14–18 September 2015. In: 31st European photovoltaic solar energy conference and exhibition. Hamburg
Quaschning V (2011) Regenerative energiesysteme. München, Carl Hanser Verlag, Technologie-Berechnung-Simulation
Redweik P, Catita C, Brito M (2013) Solar energy potential on roofs and facades in an urban landscape. Sol Energy 97:332–341
Santos T, Gomes N, Freire S, Brito M, Santos L, Tenedório J (2014) Applications of solar mapping in the urban environment. Appl Geogr 51:48–57
Sarralde JJ, Quinn DJ, Wiesmann D, Steemers K (2015) Solar energy and urban morphology: scenarios for increasing the renewable energy potential of neighbourhoods in London. Renew Energy 73:10–17
Strzalka A, Alam N, Duminil E, Coors V, Eicker U (2012) Large scale integration of photovoltaics in cities. Appl Energy 93:413–421
Šúri M, Hofierka J (2004) A new GIS-based solar radiation model and its application to photovoltaic assessments. Trans GIS 8:175–190

Test F, Lessmann R, Johary A (1981) Heat transfer during wind flow over rectangular bodies in the natural environment. J Heat Transf 103:262–267

Vdma (2016) International technology roadmap for photovoltaic (ITRPV) 2015 results., 7th edn. VDMA, Frankfurt am Main

Wate P, Coors V (2015) 3D data models for urban energy simulation. Energy Procedia 78:3372–3377

Wieland M, Nichersu A, Murshed SM, Wendel J (2015) Computing solar radiation on CityGML building data. In: 18th AGILE international conference on geographic information science, June 9–12. Lisbon

Wilcox S, Marion W (2008) Users manual for TMY3 data sets. Colorado, National Renewable Energy Laboratory (NREL)

Interpolation of Rainfall Through Polynomial Regression in the Marche Region (Central Italy)

Matteo Gentilucci, Carlo Bisci, Peter Burt, Massimilano Fazzini and Carmela Vaccaro

Abstract Notwithstanding its small size (less than 10,000 km^2), because of its varied topography, ranging from the Apennines Range (up to more than 2000 m amsl) to coastal environments, the Marche Region (the Adriatic side of Central Italy), is characterized by many different types of climate. In this region there are no fully satisfactory models to interpolate and generalize rainfall data from the 111 available meteorological recording stations; however, in this study an innovative way to interpret data linking precipitation to many topographic parameters is introduced. Based on those considerations, statistical analyses were carried out on rainfall historical series in order to assess significantly variations during the last 60 years and to create a model capable of explaining rainfall distribution based on geographical and topographic parameters. The model highlighted a significant decrease of rainfall from 1961–1990 to 1991–2016, over the whole period, in the hilly and mountainous sectors (100–200 mm), while closer to the coast the difference is slight (about 0–100 mm). The new model also highlights the presence of some outliers in the rainfall values, which may lead to a better comprehension of climatic dynamics in this area.

M. Gentilucci (✉) · C. Bisci · M. Fazzini
School of Sciences and Technologies, University of Camerino, Camerino, Italy
e-mail: matteo.gentilucci@unicam.it

C. Bisci
e-mail: carlo.bisci@unicam.it

M. Fazzini
e-mail: fzzmsm@unife.it

M. Gentilucci
V.le Indipendenza, 180, 62100 Macerata, Italy

P. Burt
Department of Agriculture, Health and Environment, Natural Resources Institute, University of Greenwich at Medway, Chatham, Kent ME4 4TB, UK
e-mail: p.j.a.burt@greenwich.ac.uk

C. Vaccaro
Department of Physics and Earth Sciences, University of Ferrara, Ferrara, Italy
e-mail: vcr@unife.it

A. Mansourian et al. (eds.), *Geospatial Technologies for All*,
Lecture Notes in Geoinformation and Cartography,
https://doi.org/10.1007/978-3-319-78208-9_3

Keywords Rainfall · Marche region · Multiple regressions · GIS Climate

1 Introduction

1.1 Aim of the Study

The Marche Region has various environments and climate types that influence precipitation. The present study about precipitation of Marche Region was performed in a period from 1961 to 2016, divided into two sub-periods of 30 years, in order to make it comparable with other climate analysis in the rest of the world, following the protocol of the World Meteorological Organization.

This research has two aims:

- to investigate the spatial distribution of precipitation variations in the Marche Region from 1961–1990, in order to assess possible climate change in the last period 1991–2016;
- to assess how local precipitation is correlated with geographical and topographic parameters, in order to develop a predictive computer program for the creation of an acceptable mathematical model of prediction.

1.2 Geography of the Area

Since systematic statistical analyses of the influence of geographical and topographic features on precipitation are lacking for the Marche Region (Fig. 1), data recorded by more than 100 rain gauges in the Marche Region and in its neighbours (Emilia Romagna, Tuscany and Lazio, Fig. 1) have been analyzed.

The study area stretches over 10,000 km^2 and is located on the Adriatic side of Central Italy. With the exception of a small sub-basin draining to the Tyrrhenian Sea (the Nera River sub-basin of the Tiber River), the Region is characterized by elevations progressively decreasing eastwards (from the Apennines to the Adriatic Sea). Within the Region, all rivers follow the regional altitudinal gradient, flowing almost perpendicularly to the coastline. No lakes are present, even though there are many reservoirs of different size. The area is mostly hilly (ca. 69%) and subordinately mountainous (ca. 31%, to the west): the maximum elevation is Mt Vettore (2476 m a.s.l.), located at the Region's SW boundary. Alluvial plains are small and narrow. There are four main climate types (Köppen 1900; Geiger 1954; Spina et al. 2002) in Marche Region:

- Cfa—temperate climate with sufficient rainfall in all months and the hottest month above 22 °C; it is present up to 30–40 km inland from the coast;

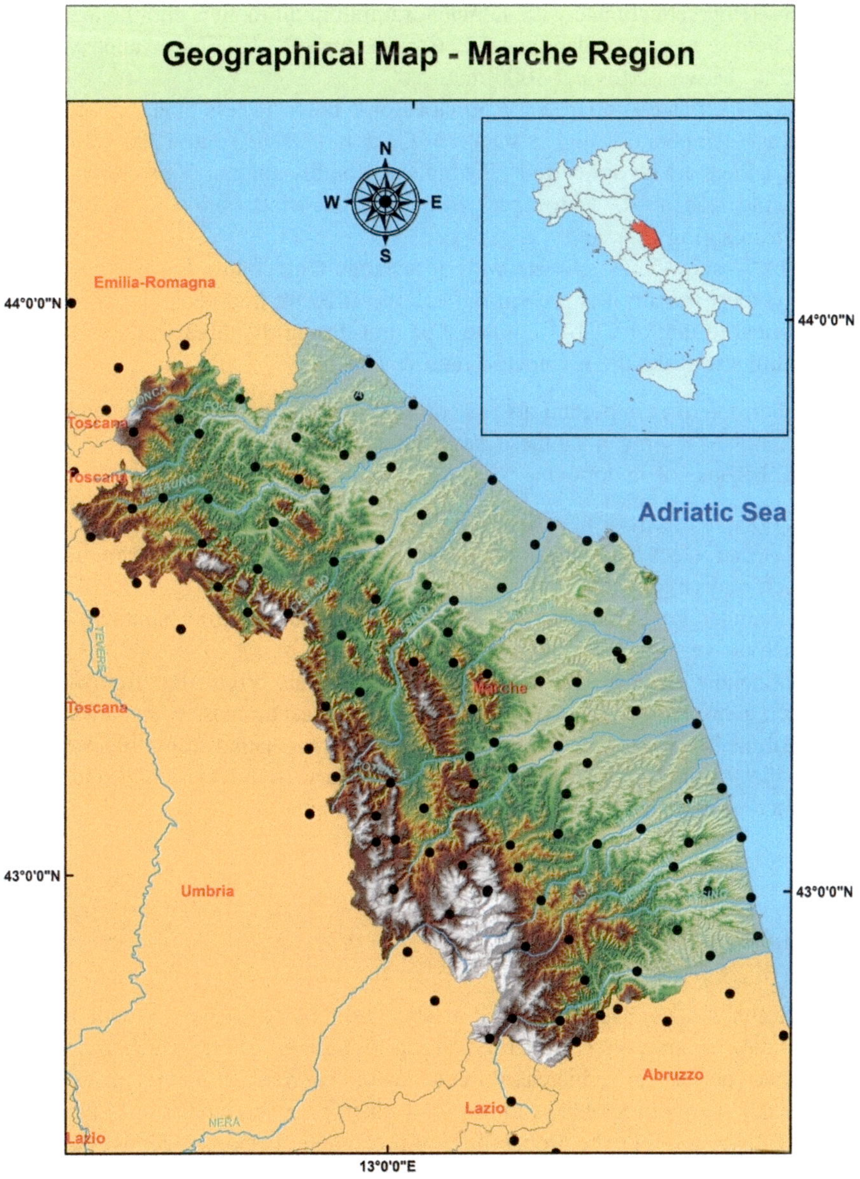

Fig. 1 Geographic map of the Marche region; the 111 rain gauges taken into account are marked by circles

- Cfb—temperate climate with sufficient rainfall in all months and the average of the hottest month colder than 22 °C; it is typical of altitudes approximately ranging between 500 and 1000 m;
- Cfsbx″—climate similar to the previous one, but with less than 4 months with average temperature higher than 10 °C, it is present from 1200 to 1800 m, where there are no dry periods, the highest monthly amount of precipitation is in the cold season (fs), with a peak in the autumn-winter and a secondary maximum in spring (x″), and
- Dfsbx″—snow-forest climate with an average temperature lower than 3 °C in the coldest month; it is present only above 1800 m; the precipitation regime is identical to that of Cfsbx″ climate type, but during the three winter months on the highest peaks the nivometric ratio reaches 90%.

Furthermore it is important also to consider air masses. Italy is affected by 8 air masses, however only 4 of these have a considerable influence for the Marche Region, because of its topography:

1. Continental Arctic cold. This originates from North Russia, above all Siberia and it can affect Italy from the end of October to April. This is the coolest air which may involve the Italian nation;
2. Continental Polar cold. Cold and dry air originating from southern Russia, which arrives over Italy from the Balkan area;
3. Continental Tropical warm. Hot and very dry air originating from arid and desert areas, its source is from south. It is the hottest air mass which affects Italy.
4. Maritime Tropical warm. This kind of air mass originating from the South-West (in the Atlantic Ocean near Azores and Canary Islands), is mild and wet in winter, while becoming hot and muggy in summer.

2 State of the Art

Climate analyses are usually focused on standard themes, such as extreme monthly precipitation (Gutowski et al. 2008; Wang and Zhou 2005), which has a great importance because of an increased frequency of occurrence of natural disasters; the relationship between global warming and changes in precipitation (Chou et al. 2009; Trenberth 2011); the probabilities of precipitation changes (Jones et al. 1995; Tebaldi et al. 2004) and in order to provide predictions for the future climate scenarios, and precipitation analysis (Serrano et al. 1999; Partal and Khaya, 2006) applied to other fields of interest (e.g. geomorphology, hydrogeology, ecology).

Spatial and temporal distributions of precipitation over central Italy during the last century have shown a moderate, irregular reduction of annual and seasonal rates equating to a precipitation decrease between 5 and 10% during the 20th Century (Brunetti et al. 2000a, b; Brunetti et al. 2006a, b; Colombo et al. 2007). Winter is the season characterized by a heavier reduction in rainfall over central Italy as a whole. However, only a few studies have taken into account the relationships between local features and precipitation to obtain a better interpolation of sparse data; most studies consider elevation as the only dependent variable at a regional scale (Brunsdon et al. 2001). The scarcity of this type of investigations can be caused by local differences in rainfall, depending on the climatological and topographical conditions of the investigated region, which need a detailed analysis obtained through GIS software.

Only a few analyses have been carried out on precipitation in the Marche Region and there is no recent work in this area. Some of these take into account the standard precipitation index or analyze the relation between precipitation and other geographic features in a graphical way (Bisci et al. 1994, 1996, 2001, 2002; Rossetti et al. 1997; Fazzini et al. 2002). Furthermore, there is a study that dealing with precipitation and altitude in order to observe the ratio, but show the result as example only for August and December from 1948 to 1981 (Bordi et al. 2001). Some other studies have to be considered as reports of historical climatic records (Biondi et al. 1991; Amici and Spina 2002; Bisci and Fazzini 2002).

The most recent report for the Marche Region is from Amici and Spina (2002), whose work summarizes the amount of precipitation in this area and highlights a trend of decreasing precipitation, due to climate change. However, there is an analysis for the standard precipitation index from 1948 to 1981, which takes into account the altitude in order to have a better interpolation of precipitation in the area (Bordi et al. 2001).

3 Methods

3.1 Quality Control

Rainfall data were collected for 111 rain gauges located in the regional territory and in its neighbours (Fig. 1) for 1961–2016. The dataset was submitted to a quality control test composed of two parts (Aguilar et al. 2003; World Meteorological Organization (WMO) 2011):

1. Gross error checking—all the digitization errors and the strongly anomalous values were removed from the "raw" data after an investigation, which highlighted any negative or the clearly wrong values (too high for each climatic zone in the world), with the "conditional formatting" (Microsoft® Excel©).
2. Internal consistency check—a threshold has been set (from 0 to 800 mm of rain per month, threshold that exceeds of about 100 mm the highest values recorded by a rain gauges in this area) through the data validation tool of Microsoft® Excel©.

Furthermore, for each rain gauge only complete monthly records were taken into account and also the annual means were calculated when all the months were complete. At the end of this validation procedure, 1102 values have been deleted (1.67% of the source data).

Finally, the Craddock test (Craddock 1979) for the identification of inhomogeneity of the time series confirmed an acceptable homogeneity of the dataset, while the Mann-Kendall test (Salmi et al. 2013) showed, for all the monthly series, an absence of significance of their trend, with p-values around 0.05 and 0.10.

3.2 Climate Analysis Method

In order to identify climate changes, the rainfall data set has been split into two different periods, from 1961 to 1990 and from 1991 to 2016. These two periods have been chosen because they represent different climatological standard normals.

The data were examined to see if there was a statistically significant difference in precipitation between the two periods. If so, climate change could be one possible explanation or other reasons, such as temporary fluctuations, might be plausible.

This led to the construction of a multivariate model adopting several geographical and topographic features as independent variables influencing precipitation.

Thus it was possible to identify the most important independent variables for the Marche Region through a multiple regression analysis. This analysis was chosen to consider many topographic parameters in one model. Furthermore a second order polynomial regression was chosen as this allows better results of fitting to be obtained compared to, for example, the standard OLS (ordinary least squares) regression. The presence of outliers was also investigated in the analysis: some rain gauges are named outliers if their data don't follow the most common ratio with the independent variables, but show an abnormal distance from other values.

The first technical operation to edit maps was to create a DEM (Digital Elevation Model). All the sheets of the CTR (Carta Tecnica Regionale—technical map of the Region; 1:10000 nominal scale, provided by the Marche Region local government as AutoCAD files) were merged, extracting all the features relevant to determine relief (such as contour lines, elevation points, hydrographic network) and using them, in addition to elevation data of all available weather stations, to create a TIN (Triangulated Irregular Network). The latter was then corrected and optimized, checking suspect elevation values and adding new elevation points and polylines taken from geographical and topographical maps after their georeferencing in the software, in order to provide a better representation for the morphology of the area. Finally, the resulting TIN has been transformed into a detailed raster DEM with a cell size of 15 m (Fig. 1). All the geographical and topographic variables adopted for the analyses were obtained starting from the Digital Elevation Model.

4 Results and Discussion

4.1 Rainfall Analysis

The analysis of rainfall of Marche Region starts from the report of standard parameters preparatory for the central part of the research. The calculated mean precipitation is 943 mm for the first period (1961–1990) and 915 mm for the second one (1991–2016), even though there is a similar geographical distribution, observed graphically through the ArcGis tool "Cluster and outlier Analysis" that identifies statistically significant distributions in the rain gauge data, using the Anselin Local Moran's I statistic (Anselin 1995). Standard deviations highlight a higher dispersion of data for 1961–1990 (209.46) than for 1991–2016 (177.95). In order to improve the evaluation of precipitation, symmetric percentiles have been considered (Table 1); 50% of the data ranges between 791 and 1032 mm for the period 1961–1990, while it ranges from 785 to 1018 mm for the period 1991–2016.

Figures 2 and 3, created using the "Geostatistical analyst" extension of ArcGis© and selecting an IDW (Inverse Distance Weighted) method (Johnston et al. 2001; Wong and Lee 2005), depict the distribution of precipitation in the study area for the two periods (1961–1990; 1991–2016). The IDW has been optimized finding the optimal power (control the weight of the measured values in relation to the distance,

Table 1 List of the adopted independent variables

Time interval	Independent variables					
	First	Second	Third	Fourth	Fifth	
1961–1990	Year	Distance from sea	Latitude	Local relief	Elevation	Distance from divide
	Spring	Distance from sea	Local relief	Latitude	Elevation	Distance river
	Summer	Distance from sea	Latitude	Elevation	Distance river	Distance from divide
	Autumn	Distance from sea	Local relief	Latitude	Elevation	Distance river
	Winter	Distance from sea	Local relief	Elevation	Elevation	Distance from divide
1991–2016	Year	Distance from sea	Latitude	Elevation	Distance river	Local relief
	Spring	Distance from sea	Elevation	Distance river	Latitude	Local relief
	Summer	Distance from sea	Elevation	Latitude	Local relief	Distance river
	Autumn	Distance from sea	Latitude	Elevation	Distance river	Local relief
	Winter	Distance from sea	Local relief	Latitude	Elevation	Distance river

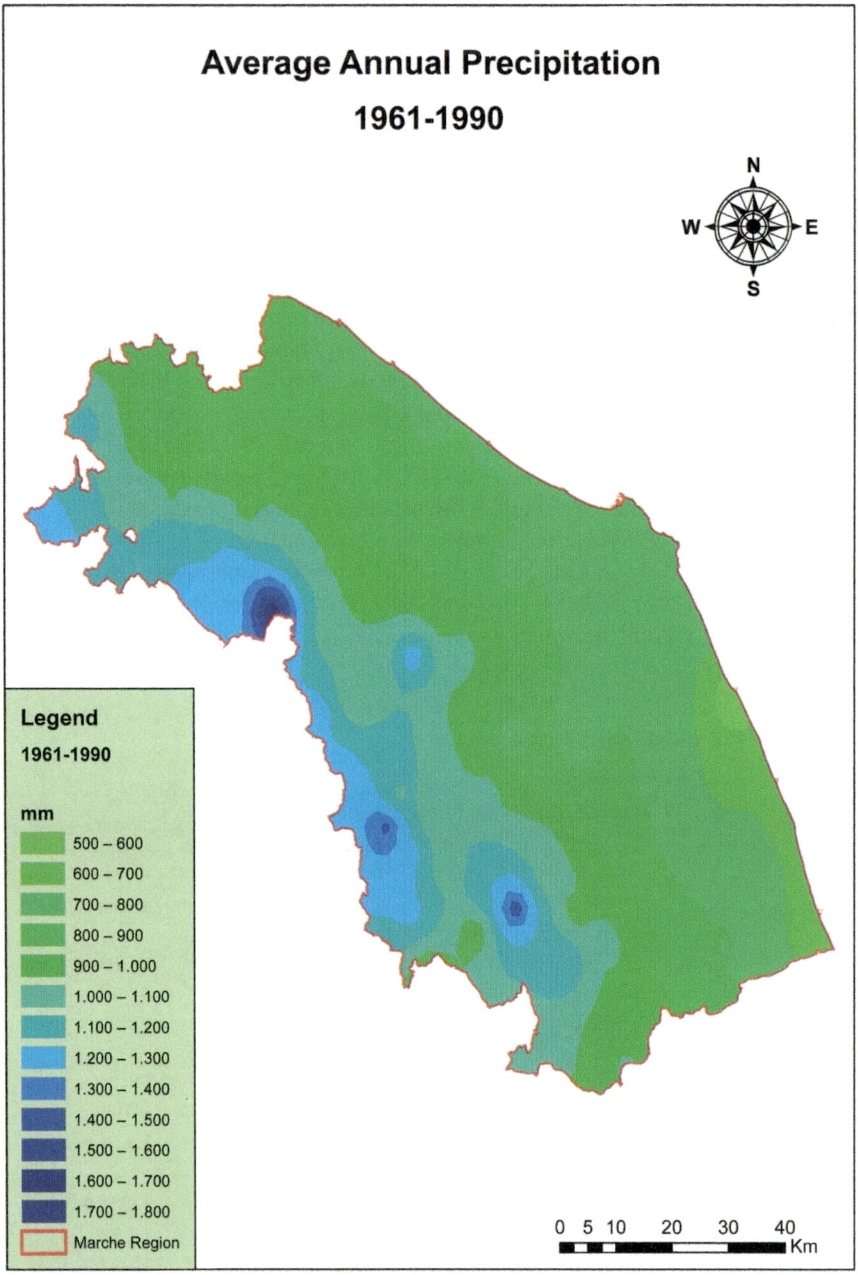

Fig. 2 Average annual precipitation 1961–1990

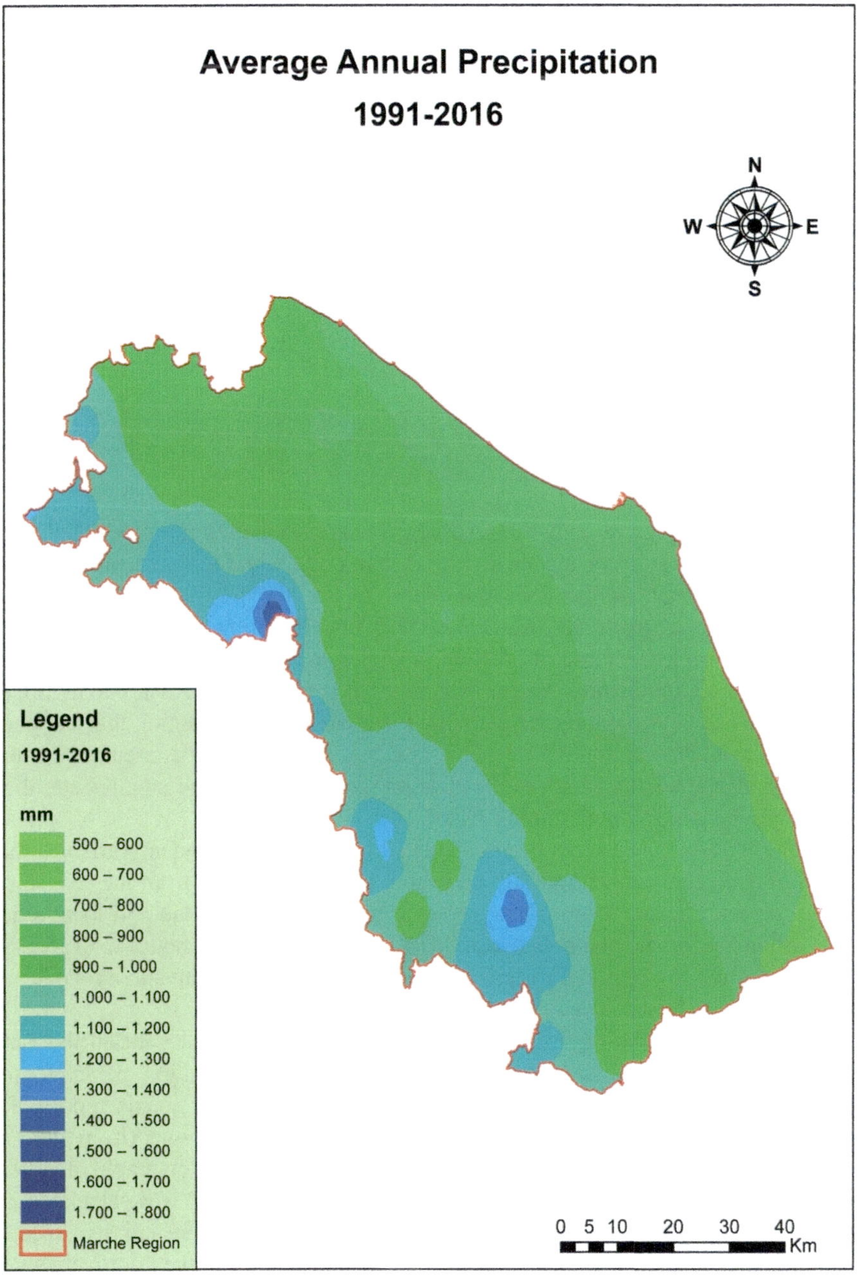

Fig. 3 Average annual precipitation 1991–2016

in the interpolation) for both periods. In fact the power that minimizes the root mean square error was chosen for the interpolation, through a cross validation that removes each data location one at a time and predict the value in the same location:

$$\widehat{Z}(s_0) = \sum_{i=1}^{N} \lambda_i Z(s_i)$$

where $\widehat{Z}(s_0)$ is the value predicting for s_0 location; N is the number of measured values; λ_i are the weights; $Z(s_i)$ is the observed value at s_i location. The comparison of Figs. 2 and 3 highlights the differences between the two periods, in which there is a clear reduction of precipitation from 1961–1990 to 1991–2016.

4.2 Regression Analysis

In the Marche Region in winter the Adriatic Sea is too small to mitigate the climate in a sensitive manner, as happens on the western coast with the Tyrrhenian Sea and the Atlantic Ocean. In fact the Apennines impede the free circulation of mild western air masses from Atlantic Ocean. This situation allows the incursion of air from the east and north-east, especially in winter, while continental tropical warm air is present predominantly present in summer. It is therefore necessary to highlight the importance of geography for the climate variability of this region that stretches gradually from the sea to the mountains. In fact, if the atmospheric circulation is the same for all the territory under investigation, both for large and small scale, then geography becomes a determining factor.

Multiple regressions have been calculated using six independent variables that could have a relation with rainfall: elevation, latitude, distance from the sea, slope angle, aspect, distance from rivers, distance from the main divide and local relief (i.e. difference of elevation between local divide and valley bottom) (Table 1). These variables are important because they are related to the climatic dynamics of this study area (Basist et al. 1994).

The first step in the regression analysis was to relate rainfall with all the variables, in order to estimate the best variable to use as the first parameter of regression. Data were analyzed using scatter plots, where the y-axis reports the dependent variable (rainfall) and the x-axis the independent one (from the six geographic or topographic parameters). For each of those scatter graphs, the best fitting polynomial curve of the second order ($y = ax^2 + bx + c$) was calculated, as well as the co-efficient of determination R^2 obtained. A polynomial curve of second order was chosen because it fitted the data better than a line and the improvement of R^2 is negligible increasing the order. The whole process was required to find a model that could explain the analyzed data. Figure 4, where the x-axis represents the distance from the sea, shows an example of good fitting.

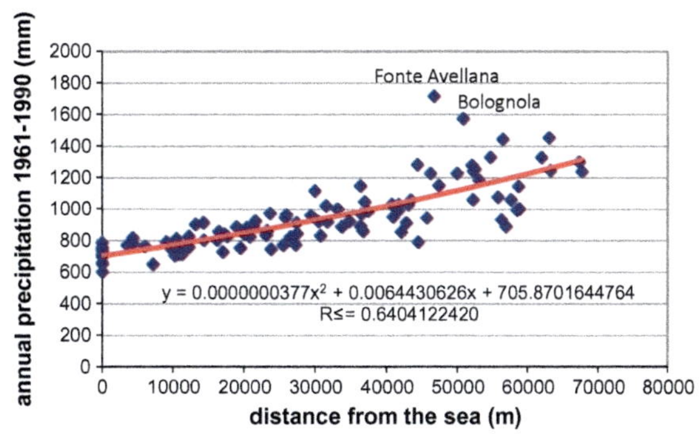

Fig. 4 Relationship between average annual precipitation 1961–1990 and distance from the sea

The representation by scatter graphs also allows assessment of the presence of outliers (stations showing anomalous values), related to particular environments or atmospheric dynamics (Alexander et al. 2006; Hijmans et al. 2008; Van den Brink and Können 2008). In Fig. 4 two outliers are present (Bolognola and Fonte Avellana), showing a precipitation significantly higher than it could have been hypothesized; it is evident that the improvement of the fitting curve without the outliers in the second, where the co-efficient of determination (is a ratio between data variability and correctness of the model) increases from 0.64 to 0.73.

Based on this first level of regression, for each environmental parameter and for each period, the expected rainfall has been calculated using the resulting polynomial equation: this value has then been subtracted from the recorded value to obtain the residual unexplained value. The analysis then continues through the comparison of residuals with the environmental features left, thus individuating the second best fitting independent variable, once more on the basis of the best coefficient of determination: continuing in this way, it was possible to calculate further new residuals and individuate independent variables, until the end of the procedure.

There are five levels in this analysis and the value of the coefficient of determination at the end of each level of regression, is added to the previous value of R^2, up to the final value that represents the amount of weather comprehension by the model.

The following is a practical example of a complete regression, in this case for the period 1991–2016 and the annual average:

first parameter of regression (distance from the sea): $R^2 = 0.625$
second parameter of regression (latitude): $R^2 = 0.041$
third parameter of regression (altitude): $R^2 = 0.038$
fourth parameter of regression (distance from the river): $R^2 = 0.012$
fifth parameter of regression (local relief): $R^2 = 0.010$
TOTAL: $R^2 = 0.728$

At the end of multiple regression analysis, it is useful to make a summary with percentage of comprehension of data that express the relation between the dependent variable (precipitation) and the independent variables (distance from the sea, latitude, elevation) (Table 1) to evaluate the best for each time series period and interval of time.

It is shown that there is a noteworthy increase of the co-efficient of determination when outliers are excluded from calculations: the increase for the period 1961–1990 is higher than in the last one and it reaches values of greater reliability (from 3.7% in summer to 11.4% in autumn) (Table 2).

Figures 5 and 6 describe the differences between real and predicted values for the average annual rainfall during the periods: 1961–1990 (Fig. 5), 1991–2016 (Fig. 6).

Predicted values have been calculated on the basis of the resulting model from the 5 levels of regression analysis, while the real values are the measured ones. For both maps the best results have been obtained for the low elevation sector close to the Adriatic coast, while the highest differences (both positive and negative) are located close to the Apennines Range. In particular, there is a strong overestimation in the inner part of the province of Macerata (Centre-West part of the Region), while in the north-west part of the Region there is a substantial underestimation. The ratio between expected and real values is similar for both the adopted time intervals (Fig. 7). To assess the differences in a map between the models of the two periods, a raster subtraction (1961/1990–1991–2016) has been performed. Figure 7 shows that there is a higher difference especially in those area affected by outliers (300/350 mm of negative difference and 250/200 mm of positive).

The outliers highlighted more frequently during this analysis are the values for Fonte Avellana and Bolognola.

From the study it is evident that rainfall increases moving westward from the coast to the mountain range for both periods investigated, even if this relationship can be due prevailing to the topography of the Marche Region eastward from the Apennines and exposed to the same air masses. The minimum precipitation is always recorded in the coastal and hilly area located in the south-central part of the Region. During spring, this dry zone widens up to Ancona in the Central part of the

Table 2 Percentage of variance explained by the regression

	1961–1990 (%)	1991–2016 (%)
All stations		
Year	74.3	72.8
Spring	70.1	70.6
Summer	62.2	45.8
Autumn	66.4	74.5
Winter	77.0	71.3
Without outliers		
Year	81.7	75.2
Spring	78.5	72.8
Summer	65.9	45.7
Autumn	77.8	75.8
Winter	84.6	74.0

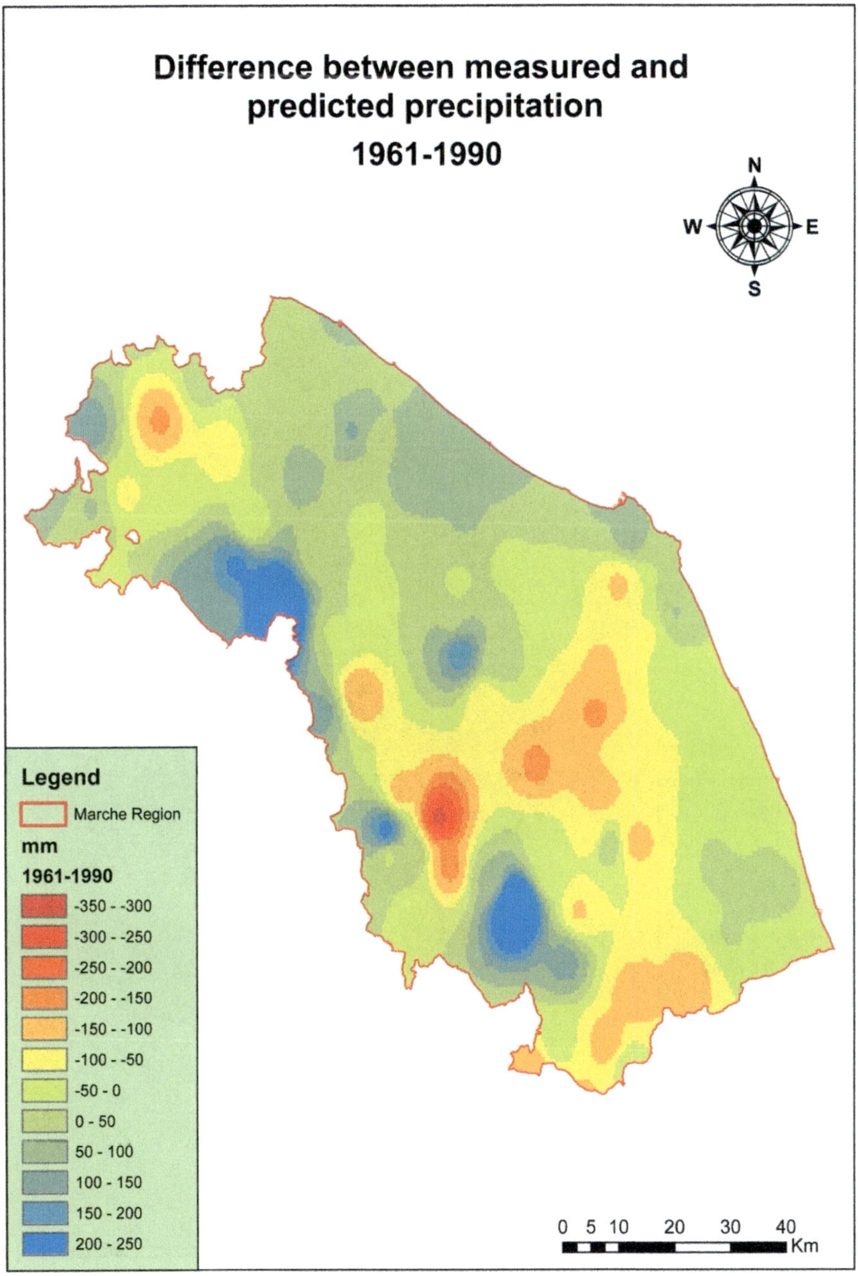

Fig. 5 Map of difference between predicted and observed values in 1961–1990

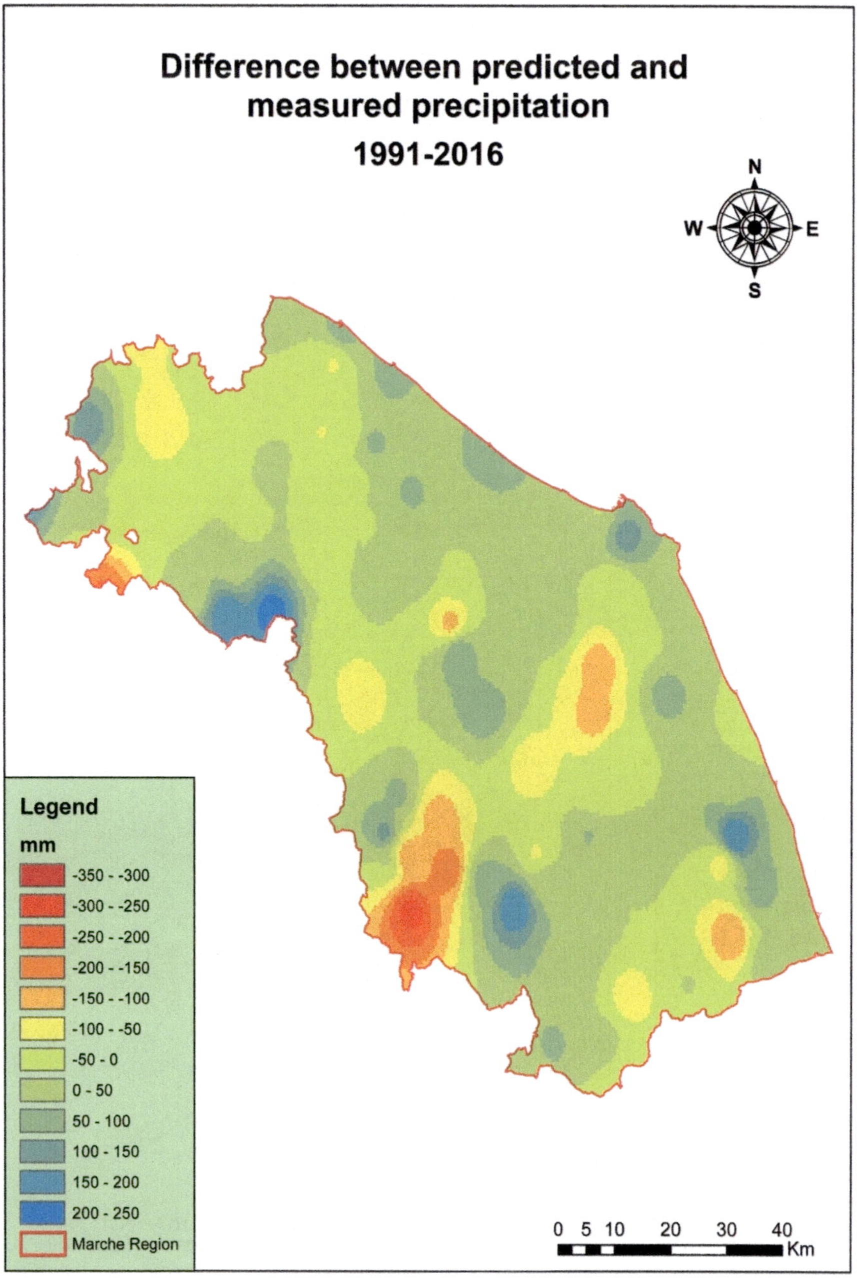

Fig. 6 Map of difference between predicted and observed values in 1991–2016

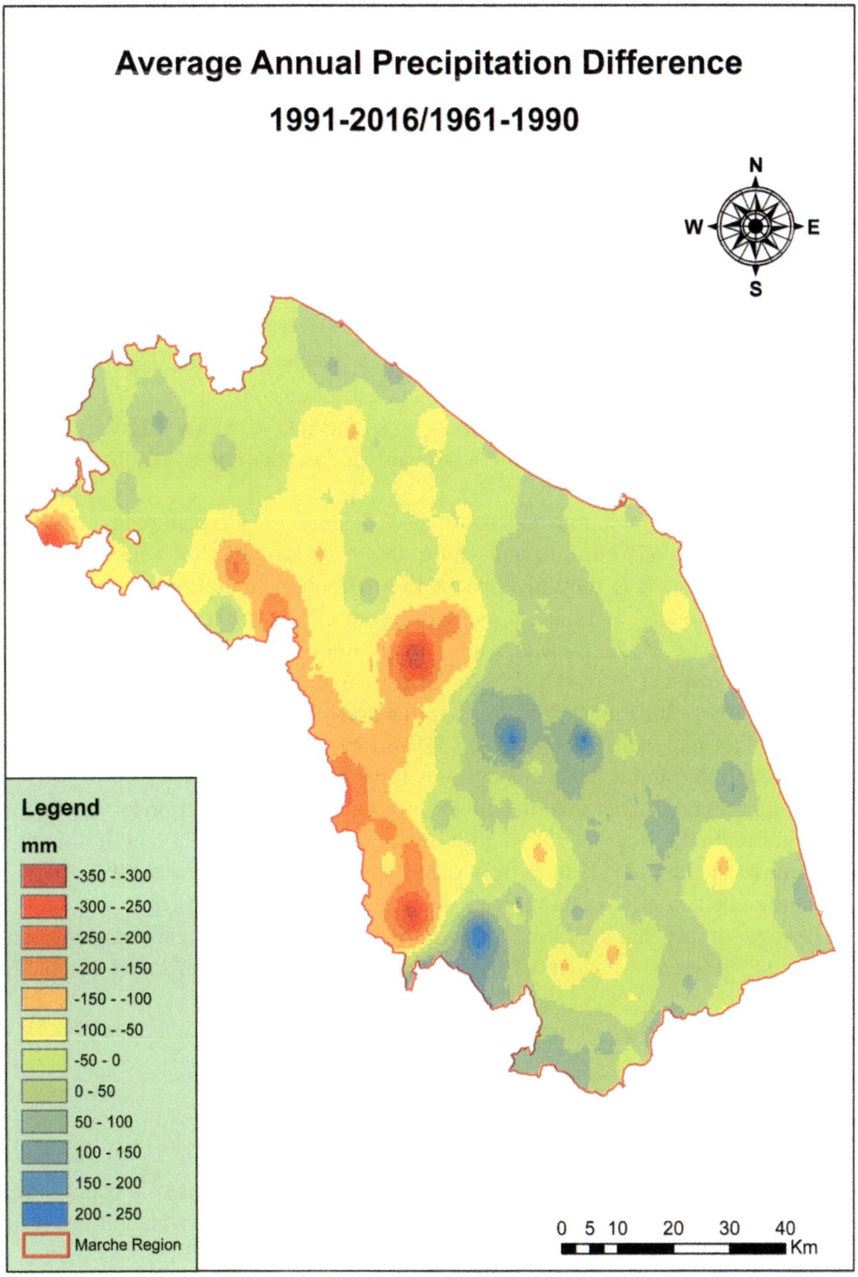

Fig. 7 Map of model performance, subtraction between the differences real predicted values 1991–2016 and 1961–1990

regional coast. However, these results are affected by a strong underestimation of precipitation for the mountain rain gauges, probably due to the strong winds which don't allow a correct measurement of rainfall (Crisciotti and Preziosi 1996; Andermann et al. 2001). This problem is further amplified by the scarcity of rain gauges located above 1000 m (only three) and the absence of reliable gauges above 1400 m (this latter problem obliged an extrapolation of values at higher altitudes, with severe reduction of precision and reliability). This lack of reliable data in the Apennine Mountains probably leads to the creation of the outliers highlighted above. In fact it is a possible explanation of the strongest outliers detected in the graphs through the observation of their topographical features or the analysis of atmospheric dynamics.

Fonte Avellana is located at an elevation of 689 m along the eastern slope of high relief connected to the main watershed (Acuto Mountain, 1475 m). In instances of advection air masses coming from the east it is affected by an intense *stau* effect (cold air build-up on the windward side of the mountain); therefore, the model underestimates precipitation by up to 27%.

Rainfall at Bolognola is underestimated despite its high altitude (1070 m), because it is relatively close to the sea (around 50.8 km) and has a high mountain range to the west. It is the only one of the measurement stations in the Marche Region to be affected by both the meteorological effects of the passage of Atlantic disturbances and Mediterranean ones, when linked to cyclogenes is over the Balkans, the lower Tyrrhenian (Ponza Low) or over the Ionian Sea.

Some further rain gauges behave as outliers only in one particular season: this is caused by local dynamics, often caused by topographical reasons, that in some seasons can represent a cover for the prevailing air masses.

The elimination of the major outliers has led to an improved understanding of precipitation in the Region, above all in the period 1961–1990. In fact, in this time the value of the total R^2 changes from 0.743 to 0.818, compared with 1991–2016 in which the coefficient of determination increases from 0.728 to 0.752.

This may imply that interpolation gives better results with longer time series and highlights the great influence of the anomalous rain gauges on the result of the model.

5 Conclusions

This analysis can be considered an innovation because take into account many topographic parameters for each period, in order to choose those which fit better the precipitation data. The results achieved can be summarized in 3 main points:

1. Precipitation decreases moving eastward, with a minimum to the south-east. This trend is in accordance with topography (it seems to be plausible that it can cause orographic precipitations proportional to local elevation, that produce a result of R^2 as first independent variable slightly less high than distance from the

sea) and distance from the sea (a value of about 0.6 as R^2 for both period probably due to the mitigation effect increasing close to the coast).

2. A downward trend for precipitation (of about 80 mm) in the last period has been detected, except for in autumn (when precipitation remains constant) and summer (which showed a highly unpredictable behaviour without any prevailing trend: it can be assumed that rainfall may derive both from cold oceanic air advection and from ascending currents caused by surface warming resulting in local short but heavy thunderstorms). There is a significant decrease of rainfall from 1961–1990 to 1991–2016 in the hilly and mountainous sectors (100–200 mm), while in the area close to the coast the difference is slight (about 0–100 mm).
3. Outliers of rainfall amount have been identified and interpreted. For a better explanation of their behaviour, an analysis of the seasonal distribution of wind direction is needed, but wind records are scarce and too poorly distributed in the area to accomplish such a task. It would be interesting to know also the trend of temperature in relation to precipitation, as well as solar radiation and air moisture, since all these parameters together could generate a complex model that could significantly improve the comprehension of outliers. Unfortunately, once more only a few recording stations furnish these observations.

The good continuity of rainfall data has allowed greater comprehension of rainfall space distribution and variation in the Marche Region than previously. This can constitute a relevant step toward the creation of a complete climate model that, in turn, may lead to an even more accurate interpolation and to a better explanation and characterization of outliers. Furthermore this study could be a tool to investigate slope stability, as well as both to characterize water reserves and to carry out agrometeorological studies.

The scarcity of recording stations in the mountain sector severely limits the accuracy of interpolation at elevations higher than 1000 m. In the future this limitation will be less relevant, since some more rain gauges have been installed in the Sibillini Mountains starting from year 2000 at altitudes ranging between 1400 and 1900 m.

Finally, in future it would be interesting to test geostatistical methods such as co-kriging that could give good results with topographic variables reducing the estimation errors.

References

Aguilar E, Auer I, Brunet M, Peterson TC, Wieringa J (2003) Guidelines on climate metadata and homogenization, WMO/TD No. 1186, WCDMP No. 53; WMO, Geneve, CH

Alexander LV, Zhang X, Peterson TC, Caesar J, Gleason B, Klein Tank AMG, Griffiths G (2006) Global observed changes in daily climate extremes of temperature and precipitation. J Geophys Res 111. https://doi.org/10.1029/2005jd006290

Amici M, Spina R (2002) Campo medio della precipitazione annuale e stagionale sulle Marche per il periodo 1950–2000. Macerata, IT, Centro di Ecologia e Climatologia - Osservatorio Geofisico Sperimentale

Andermann C, Bonnet S, Gloaguen R (2001) Evaluation of precipitation data sets along Himalayan front. Geochem Geophys Geosyst

Anselin L (1995) Local indicators of spatial association—LISA. Geograp Anal 27:93–115. https://doi.org/10.1111/j.1538-4632.1995.tb00338.x

Basist A, Bell GD, Meentemeyer V (1994) Statistical relationships between topography and precipitation patterns. J Clim 7:1305–1315

Biondi E, Baldoni MA, Talamonti MC (1991) Il fitoclima delle Marche. Atti Conv, Salvaguardia e Gestione dei Beni ambientali nelle Marche, Ancona, IT

Bisci C, Dramis F, Fazzini M, AltobelloL Dorigato S (2001) Analyse des trends termo-pluviometriques du versant Adriatique compris entre la lagune de Venice et le Cap de Santa Maria di Leuca (Italie orientale). Actes XIV Congr Ass Intern Climatologie, Seville, ESP, Climat et environnement

Bisci C, Farabollini P, Fazzini M, FolchiVici C, Viglione F (1996) Variations récents des précipitations en la Région Marche (Italie Centrale). Coll Assoc Intern deClimatologie, Strasbourg, FR

Bisci C, Fazzini M (2002) Climatic features of the central southern Marches (Central Italy). In: Proceedings of "Natural hazard on built-up areas" CERG—Camerino, 45–47

Bisci C, Fazzini M, Coccia N (2002) Analyse spatio-temporelle des séries des températures dans l'Apennin centre-méridionale italien par rapport aux paramètres topo-géographiques. Applications de la climatologie aux echelles fines. Actes XV Congr Ass Intern Climatologie, Besançon, FR

Bisci C, Fazzini M, Folchi Vici C, Viglione F (1994) Multivariateanalysis of time trend of rainfall in the Marche area (Central Italy). In: I.G.U. Commission on Climatology, Contemporary Climatology, Brno, CZ

Bordi I, Frigio S, Parenti P, Speranza A, Sutera A (2001) The analysis of the standardized precipitation index in the Mediterranean area: regional patterns. Ann Geof 44:979–993

Brunetti M, Buffoni L, Mangianti F, Maugeri M, Nanni T (2006a) Temperature, precipitation and extreme events during the last century in Italy. Glob Planet Change 40:141–149

Brunetti M, Maugeri M, Nanni T (2000a) Variations of temperature and precipitation in Italy from 1866 to 1995. Theor Appl Climatol 65:165–174

Brunetti M, Maugeri M, Nanni T (2000b) Trends of minimum and maximum daily temperatures in Italy from 1865 to 1996. Theor Appl Climatol 66:49–60

Brunetti M, Maugeri M, Monti F, Nanni T (2006b) Temperature and precipitation variability in Italy in the last two centuries from homogenised instrumental time series. Int J Climatol 26:345–381. https://doi.org/10.1002/joc.1251

Brunsdon C, McClatchey J, Unwin DJ (2001) Spatial variations in the average rainfall–altitude relationship in Great Britain: an approach using geographically weighted regression. Int J Climatol 21. https://doi.org/10.1002/joc.614

Chou C, Neelin JD, Chen CA, Tu JY (2009) Evaluating the "Rich-Get-Richer" mechanism in tropical precipitation change under global warming. J Clim 22:1982–2005. https://doi.org/10.1175/2008JCLI2471.1

Colombo T, Pelino V, Vergari S, Cristofanelli P, Bonasoni P (2007) Study of temperature and precipitation variations in Italy based on surface instrumental observations. Glob Planet Change 57(3):308–318

Craddock JM (1979) Methods of comparing annual rainfall records for climatic porposes. Weather 34

Crisciotti C, Preziosi E (1996) Analisi della variabilità spaziale della precipitazione nel bacino del Fiume Nera (Italia centrale): primi risultati. V Conv. Naz, Giovani Ricercatori in Geologia Applicata, Cagliari, IT

Fazzini M, Bisci C, Dramis F, Altobello L, Dorigato S, Fubelli G, Molin P (2002) Statistic analisys of thermometric and pluviometric trends along the Adriatic side of the Italian peninsula. Proc. IAG Intern. Symp, Addis Ababa, ETH
Geiger R (1954) Landolt-Börnstein – Zahlenwerte und FunktionenausPhysik, Chemie, Astronomie, Geophysik und Technik. alteSerie, vol 3, Ch. Klassifikation der Klimatenach W. Köppen, Springer, pp 603–607
Gutowski WJJR, Raymond WA, Kawazoe S, Flory DM, Takle ES, Biner S, Snyder MA (2008) Regional extreme monthly precipitation simulated by NARCCAP RCMs. J Hydrometeor 11
Hijmans RJ, Susan E, Cameron SE, Parra JL, Jones PG, Jarvis A (2008) Very high resolution interpolated climate surfaces for global land areas. Int J Climatol 25:1965–1978. https://doi.org/10.1002/joc.1276
Johnston K, VerHoef JM, Krivoruchko K, Lucas N (2001) Using ArcGis geostatistical analyst. Redlands, USA, ESRI
Jones RG, Murphy JM, Noguer M (1995) Simulation of climate change over Europe using a nested regional-climate model. I: assessment of control climate, including sensitivity to location of lateral boundaries. Q J R Meteorol Soc 121:1413–1449. https://doi.org/10.1002/qj.49712152610
Köppen W (1900) VersucheinerKlassifikation der Klimate, vorzugsweisenachihren Beziehungen-zur Pflanzenwelt. Geogr Zeitschr 6(593–611):657–679
Partal T, Kahya E (2006) Trend analysis in Turkish precipitation data. Hydrol Process 20, 2011–2026. http://dx.doi.org/10.1002/hyp.5993
Rossetti R, Bisci C, Dramis F, Fazzini M, Speranza A (1997) Etude de la distribution des precipitations en fonction des caracteres geographiques et morphometriques de la règion Marche (Italie centrale, cote adriatique). Coll Assoc Intern de Climatologie, Quebec, CAN
Salmi T, Määttä A, Anttila P, Amnell T (2013) Makesens 1.0.xls, Meteorological Finnish Institute
Serrano A, Matos VL, Garcia JA (1999) Trend analysis of monthly precipitation over the iberian peninsula for the period 1921–1995. Phys Chem Earth Pt B 24:85–90
Spina R, Stortini S, Fusari R, Scuterini C, Di Marino M (2002) Caratterizzazione climatologica delle Marche: campo medio della temperatura per il periodo 1950-200. Centro di Ecologia e Climatologia - Osservatorio Geofisico Sperimentale, Macerata, IT
Tebaldi CL, Mearns O, Nychka D, Smith RL (2004) Regional probabilities of precipitation change: a Bayesian analysis of multimodel simulations. Geophys. Res 31. https://doi.org/10.1029/2004gl021276
Trenberth KE (2011) Changes in precipitation with climate change. Clim Res 47:123–138. https://doi.org/10.3354/cr00953
Van den Brink HW, Können GP (2008) The statistical distribution of meteorological outliers. Geophys Res 35. https://doi.org/10.1029/2008gl035967
Wang Y, Zhou L (2005) Observed trends in extreme precipitation events in China during 1961–2001 and the associated changes in large-scale circulation, Geophys Res 32. https://doi.org/10.1029/2005gl022574
World Meteorological Organization (2011) Guide to climatological practices. WMO No. 100, WMO, Geneve, CH
Wong WSD, Lee J (2005) Statistical analysis of geographic information with arcview GIS and ArcGIS. Wiley, Hoboken, USA

An Artificial Stream Network and Its Application on Exploring the Effect of DEM Resolution on Hydrological Parameters

Haicheng Liu

Abstract Digital elevation models (DEM) are widely used in various distributed hydrological models. The stream network can be extracted from it so that runoff routing can be calculated. With the advent of remote sensing and computing technologies, the computation based on DEM with high resolution becomes possible. However, there still exist regions with poor resolution, particularly in developing countries. Previous work only conducted comparisons between results by implementing hydrological models for specific basins in the real world and resolutions were only assigned to several fixed values, such as 30 and 90 m. So, the results derived were thus not in a general sense. To roughly understand how DEM resolution influences the hydrologic response, in this paper, first an artificial stream network of which the principle is originated from fractal theory is constructed. Then by implementing calculation on such artificial networks in an iterative way and performing aggregation, the influence of DEM resolution on several hydrological parameters, namely, the number of basins, drainage density of all basins, total stream length, average stream slope and average topographic index used to assess the spatial distribution of soil saturation of the largest basin can thus be acquired. It is found that DEMs of low resolution would reduce drainage density, total stream length and average stream slope, but would increase topographic index. But the effect is insignificant regarding the number of basins. In the end, the results of the simulation as well as the quality of the fractal terrain are validated by referencing field data.

Keywords Fractal terrain · DEM · Stream network · Hydrological parameter

H. Liu (✉)
Faculty of Architecture, Delft University of Technology, Delft, The Netherlands
e-mail: H.Liu-6@tudelft.nl

A. Mansourian et al. (eds.), *Geospatial Technologies for All*,
Lecture Notes in Geoinformation and Cartography,
https://doi.org/10.1007/978-3-319-78208-9_4

1 Introduction

DEM as a common approach to express the topographic information is used as the basis for many fields and applications, for example (Li et al. 2004): determination of hydrological terrain parameters, highway and railway design, orthoimage generation, wind models for environmental study and so on. Widely used distributed hydrological models like TOPMODEL (Beven and Kirkby 1979), SWAT (Arnold et al. 1994) and DHSVM (Wigmosta et al. 1994) are all based on DEMs to calculate stream flow. The effect of DEM resolution on these hydrological models or hydrological parameters has been the focus of many researchers in the past several decades. For example, Wolock and Price (1994) found that degradation of the spatial resolution of the topographic data resulted in higher minimum, mean, variance, and skew values of the TOPMODEL topographic index distribution. Wei et al. (2004) indicated that the decrease of DEM resolution would result in the increase of total stream length and the reduction of average slope of stream flow from a perspective of topographic entropy. Furthermore, this might cause the reduction of both simulated flood peaks and hydrological response time. Similarly, by implementing DHSVM, Kenward et al. (2000) also observed lower predicted peaks and additionally higher base flow for DEMs of lower resolutions. However, all these studies including more recent ones (Sørensen and Seibert 2007; Yang et al. 2014; Jain and Sahoo 2017) share the same methodology, that is, demonstrating the law using data in the real world since it is impossible to collect data of all basins throughout the world to prove the discovery. In other words, whether those results can be applied to other watersheds is still under suspicion. The aim of this paper is to propose a new approach in a generic sense to resolve how the resolution affects hydrological parameters extracted from the DEM. To do so, the simulation on terrains is implemented using principles from fractal theory.

Basically, fractal terrain is originated from Mandelbrot and Pignoni (1983) who came up with fractals could be used as a basis for simulating natural scenes and phenomena. Since then, different algorithms (Saupe 1988) for generating fractal terrains have been proposed and implemented mainly in the field of computer graphics. The quality of fractal terrains is judged by their visual analogy to reality. It is true that self-similarity is the core of the fractal terrain, but in order to assign it more physical meaning, Kelley et al. (1988) first proposed a method in which a stream network was first generated using empirical erosion models and then the terrain was developed according to the channel network. Musgrave et al. (1989) adopted physically based models of hydraulic and thermal erosion and sediment movement to simulate the erosion caused by water flow to modify the fractal terrain. Later Stachniak and Stuerzlinger (2005) employed a stochastic local search to identify a sequence of local modifications. This method deformed the fractal terrain to conform to a set of specified constraints. And it served as a general solution to the modification of a fractal terrain. All these approaches make the fractal terrain realistic in terms of physics.

This paper first introduces the generation of a realistic fractal terrain where empirical statistic recognitions of the stream network are enrolled in a filter to select raw fractal terrains. Then, the qualified terrain is used as a platform to derive some commonly referenced hydrological parameters. This is then followed by a discussion about the effect of resolution of DEM on these parameters by adopting aggregation on the digital terrains. Finally, the comparison between results from field data and experiments on virtual channel networks is performed.

2 Artificial Stream Network

The artificial stream network is extracted from a DEM using the D8 algorithm (Tarboton et al. 1991) which assumes the flow direction of one cell can only be one of its eight neighbours. So the DEM, i.e. fractal terrain has to be firstly generated. Here, the diamond-square algorithm which is originated from two-dimensional approximations to fractional Brownian Motion (Fournier et al. 1982) is adopted. The procedure of this algorithm is demonstrated in Fig. 1.

In Fig. 1a, first we choose a common elevation value for four points at the corners on the boundary of the square. But it is also possible to assign random numbers from a normal distribution to the four points. Then, by averaging the elevations of the four points and adding a random number extracted from a normal distribution of which the mean value equals 0 and the standard deviation is a figure which can represent the range of elevation for a certain area, we can get the elevation for the central black point in Fig. 1b. For example, if the minimum height of a particular area is 20 m, and the highest point is 160 m, then it is better to set the standard deviation to 35 m which stands for σ for the normal distribution. The same principle goes for black points in Fig. 1c, but this time the mean is derived from the combination of two points at corners with the central point. After this, the whole square is divided into 4 patches and each of them falls into a similar situation of Fig. 1a. However, when calculating the value for a special black point in Fig. 1d, another normal distribution with a different standard deviation from the one used in Fig. 1b is employed and the relationship between them is expressed as Eq. 1. The mean value for the normal distribution remains unchanged, i.e. 0.

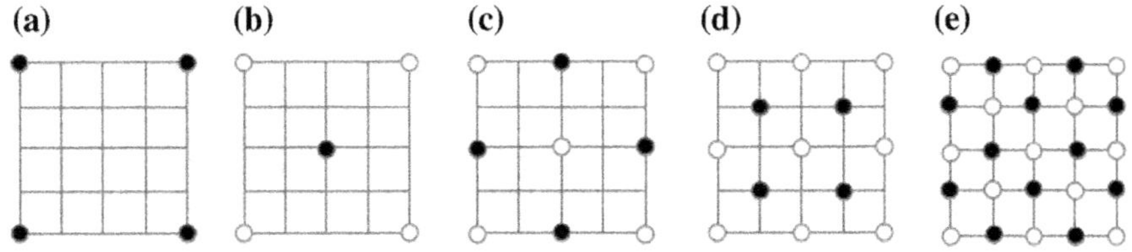

Fig. 1 Generation of a fractal terrain using the diamond-square method

$$\sigma_n = \frac{\sigma_{n-1}}{2^r} \tag{1}$$

where σ is the standard deviation for the normal distribution, n represents the iterative times, and r is basically a factor describing the roughness of the virtual terrain and actually it determines the fractal dimension of the terrain. It ranges from 0 to positive infinity. Visualizations for some specific r values are provided in Fig. 2. As can be seen from the figure, the terrain generated by a higher roughness parameter tends to be more flat. The reason for successive modification of standard deviation is to guarantee the self-similar principle which means by zooming into a specific segment of the whole terrain, the sub-terrain presents a similar pattern as the whole terrain.

By implementing such a procedure in an iterative way, an elevation dataset can be generated in any dimension required. Of course, horizontal coordinates will be later assigned to form a DEM.

As mentioned earlier, terrains produced this way have weak physical meaning. For example, the number of streams in the first strahler order (Strahler 1957) derived from such a terrain may be equal to that in the second order, which contradicts the fact in the nature. Consequently, either modifications to the generation of the fractal terrain should be implemented, such as embedding an erosion model, or some validations should take place to filter the raw fractal terrains to fulfil physical laws. Here, the second option is adopted, and basically these criteria are originated from the statistical knowledge of stream networks though field surveys (Shreve 1966; Smart 1968). They are:

1. The coefficient of variation of drainage density for different basins derived from the DEM should be no more than 0.5.
2. The average bifurcation ratio of the stream network of the largest basin should be between 2.5 and 5.
3. The average stream length ratio for the largest basin can only range from 1 to 3.
4. The average slope of all first order streams should be larger than that of the maximum order streams in the largest basin.

So the procedure to generate an artificial stream network is as follows,

1. Generate a raw DEM dataset with a specific roughness parameter using the diamond-square algorithm.
2. Extract the statistics of stream networks such as the average bifurcation ratio from the raw DEM.

(a) (b) (c)

Fig. 2 Fractal terrains with different roughness values, $r_a = 0.5$, $r_b = 1.0$, $r_c = 2.0$

3. Judge whether the raw DEM follows all the constraints, and if so, enter the next step. Otherwise, return to the first step.
4. Implement the D8 algorithm on the DEM to derive the stream network.
5. Iterate the whole process until a large number of qualified DEMs are produced for experiments.

3 Effect of DEM Resolution

The initial DEM which is acquired by the diamond-square algorithm is assigned 10 m as the resolution. By aggregating the DEM, we can get coarse versions. In addition, as has been explained earlier, the roughness parameter plays a crucial role on the shape of the fractal terrain, so 0.6, 0.8, 1.0 and 1.2 are adopted respectively. Five hydrologic parameters explored are the number of basins of the terrain, drainage density of the terrain, total stream length of the largest basin, average stream slope of the largest basin and average topographic index (Beven and Kirkby 1979) of the largest basin. The specific procedure for the experiment is provided below,

1. Set the initial value of the elevation for the four corner points to 100 and the primary standard deviation to 50. Then 0.6 is chosen as the roughness parameter to generate the fractal terrain in an iterative way for 9 times. The result is an array of the elevation in the size of 513 × 513. This is then followed by the constraints validation. If the constraints are violated, then redo this step, otherwise aggregate the dataset into smaller scales, i.e. 30, 60, 90 m and enter the next step.
2. Derive the stream network from all the DEMs, and then extract the 5 hydrological parameters.
3. Repeat step 1 and 2 1,000 times (Table 1) in order to retrieve the statistical information of the 5 parameters for each resolution, i.e. the distribution of their values.
4. Alter the roughness parameter and repeat step 1 to 3. And after this step, statistics of five hydrological parameters in four terrain shapes can be derived.

Basically, for each parameter, Probability Density Function (PDF) is derived by fitting the data using a specific type of distribution. PDF here is to generalize the

Table 1 Statistics of experiments on fractal terrains

Roughness	Total fractal terrain	Qualified fractal terrain	Success rate (%)
0.6	1,000	507	50.7
0.8	1,000	484	48.4
1.0	1,000	407	40.7
1.2	1,000	262	26.2

quantitative findings, e.g. the mode and deviation, which makes analysis conveniently later. Figure 3 demonstrates this. 1.0 is taken as the roughness while 10 m is assigned to the resolution. Since the fractal terrain is generated in random totally, so the first two parameters, i.e. the number of basins and drainage density fit the normal distribution well which is later adopted as the distribution type for these two parameters. However, when it comes to the last three parameters, the target is changed to the basin with the largest area, which means in order to access the values

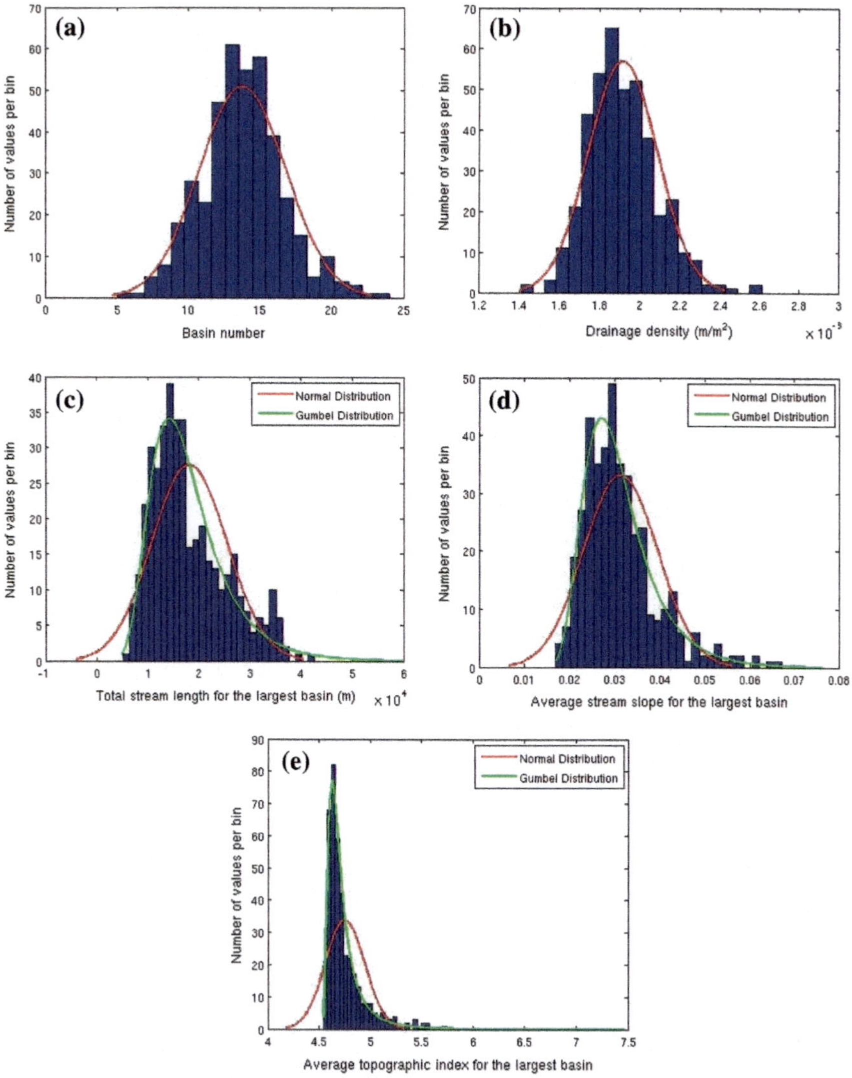

Fig. 3 Determination of probabilistic distributions of five parameters

of these parameters, first the largest basin in the original DEM has to be selected therefore the whole procedure is concerned with the extreme distribution. In Fig. 3c–e, on the one hand, normal distribution is adopted to fit the histogram, on the other hand, Gumbel distribution which belongs to extreme distribution is tried. Figures clearly show that Gumbel distribution leads to better fitting results. Thus it is utilized to analyse the last three parameters.

3.1 The Number of Basins

In Fig. 4, the gaps between the 10 m DEM and the other aggregated DEMs are not large, and the original DEM produces the least basins in general. Actually, in the experiments, there were also results which indicated that aggregation decreased the number of basins.

Additionally, figures also present that the mean value of the number of basins is around 14 and such a large number indicates that the artificial stream network can only be suitable for simulating the head area of rivers where trivial streams are in a large quantity. So the regions of the field data selected later to check the stability of the experiments are also located at source-basin areas of rivers.

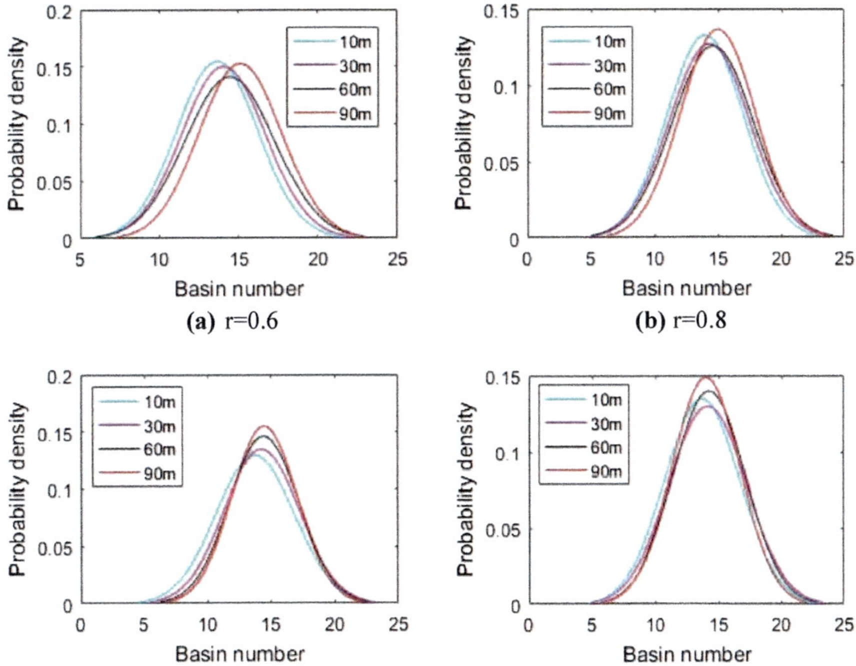

(a) r=0.6
(b) r=0.8
(c) r=1.0
(d) r=1.2

Fig. 4 PDFs for the number of basins where *r* represents the roughness parameter

3.2 *Drainage Density*

Drainage density is one of the most representative parameters to describe a basin. Former studies have shown that several morphological processes have great correlation with drainage density, such as streamflow (Carlston 1963), sediment yield (Hadley and Schumm 1961), flood (Pallard et al. 2009), etc. It is the sum of channel lengths divided by basin area. Horton (1945) indicated that the drainage density tended to be a constant for all basins within a same region because of the common environment.

Figure 5 Shows that the drainage density of the source DEM can be 1.5 times larger than that of the most coarse DEM. The direct cause can be attributed to the decrease of the total stream length after aggregation since the area is nearly the same for all the DEMs. As can be seen in Fig. 6, for the 90 m DEM, some tributaries disappear, which makes the total stream length decline. A further influence on hydrological response can also be deduced. Basically, runoff process is divided into two parts of which one takes place on the slope, and the other occurs in the channel. It is the fact that the water velocity is higher in the channel than on the slope. Thus a high drainage density implies that the flood can be formed in a short time. On the

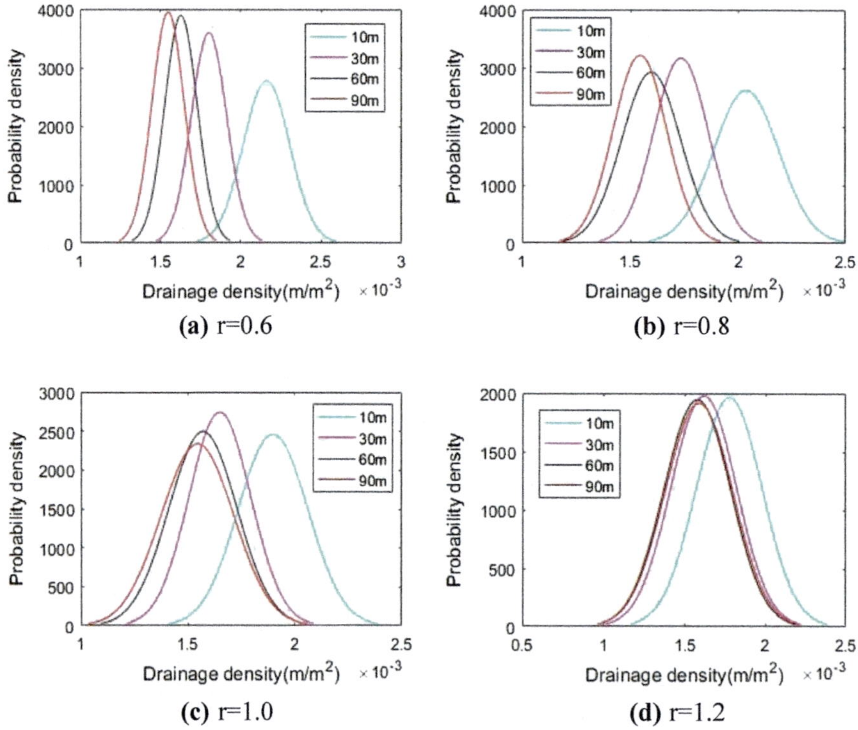

Fig. 5 PDFs for drainage density

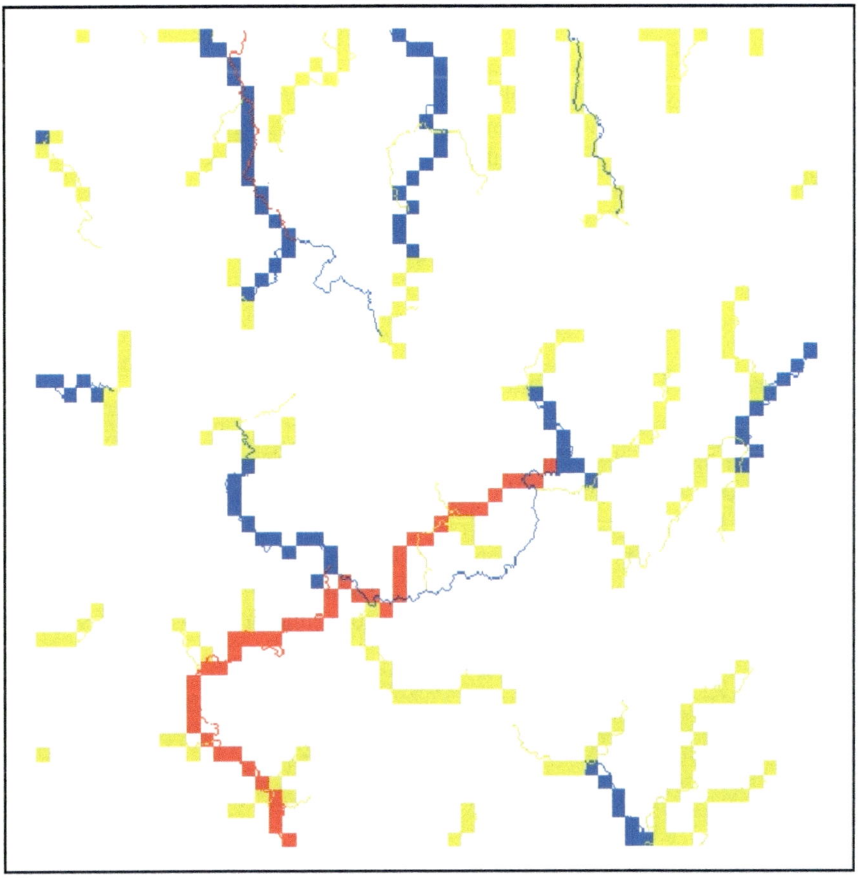

Fig. 6 Stream network for the original DEM represented by thin lines, and the 90 m DEM represented by thick lines. Each colour refers to a distinctive stream order. For instance, yellow streams are all of the first order. The roughness parameter is 0.6

other hand, less water infiltrates into soil on the slope, which can increase the volume of flood. Pallard et al. (2009) also indicate some indirect effects of the drainage density on flood. For instance, drainage density can be regarded as an index of vegetation cover in semiarid areas where bare soil is much likely to be eroded, which results in high drainage density and high runoff production. This further implies large flood peaks and volume.

Another interesting point is that the flatter the terrain is, the weaker the effect of DEM resolution presents. When the roughness parameter is equal to 1.2, the gap between the aggregated DEMs and the original DEM becomes narrow. This is because the average effect of aggregation is not that strong for plane compared with mountain areas where local elevation change can be very serious, hence aggregation significantly affects the basic shape of a stream.

3.3 Total Stream Length of the Largest Basin

Although the effect of DEM resolution on stream networks can be demonstrated in a rectangle place, the widely adopted hydrological research unit is still the basins which is the basic unit for the analysis of hydrological responses.

Analogous to drainage density, with the increase of roughness of the terrain, variations between the original DEM and three coarse DEMs decline. Coarse DEMs have smaller total steam lengths because some tributaries are lost during aggregation (Fig. 6).

3.4 Average Stream Slope of the Largest Basin

In Fig. 7, on the whole, due to the averaging effect of the aggregation, the average stream slope of each coarse DEM is lower than that of the original DEM. More specifically, for the rugged terrain, i.e. r = 0.6, the difference can be as large as 2 to 3 times of the average stream slope derived from coarse DEMs. But when the roughness parameter goes up, the average stream slopes retrieved from different DEMs present little gaps. This can be attributed to that in flat areas, the slope can

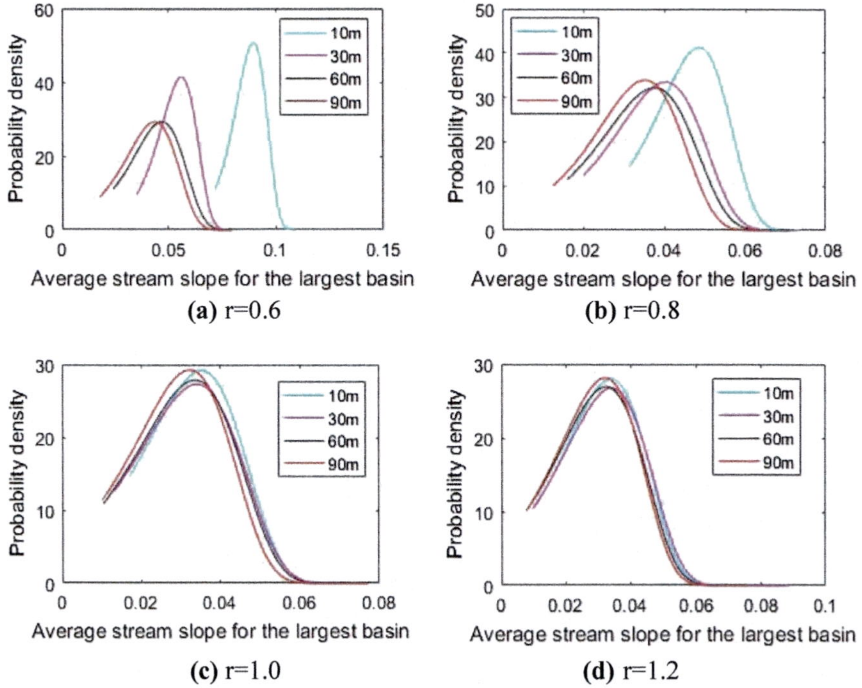

Fig. 7 PDFs for average stream slope of the largest basin

keep a constant for a wide scope, so the aggregation does not influence the value of slope very much. While in mountain areas, the slope can vary severely in a narrow space and averaging elevation will definitely affect the average stream slope.

The average stream slope is also correlated with hydrological processes. Roughly speaking, the speed at which the water flows in the channel would decrease if the slope descended. We usually take the combination of the average stream slope and the total stream length to analyse composite effect on the hydrological process. So taking consideration of the conclusions from previous section, it can be deduced that the arrival of flood would delay and the flood volume would decline if the coarse DEM was adopted to perform related calculation. The focus here is only the effect caused by the change of stream network, but hydrological process is always concerned with several sub processes, like evapotranspiration, infiltration, etc. More accurate results can be got by running a distributed hydrological model, such as DHSVM, which in turn complexes things since more elements are enrolled in the process.

3.5 Average Topographic Index of the Largest Basin

Topographic index, also known as the wetness index is first introduced in TOPMODEL by Beven and Kirkby (1979) and it is used to assess the spatial distribution of soil saturation. A high value of the topographic index indicates the region has high potential to be saturated. It is defined as the logarithm of the ratio of the upslope area and the local slope for a certain pixel. The upslope area means all the area which contributes flow across a particular pixel. Topographic index is a crucial attribute for hydrological responses since soil moisture is closely related to runoff, soil moisture and the depth of ground water.

As can be seen from Fig. 8, the average topographic index is mostly between 4.5 and 5 for the most accurate DEM, while for the coarse DEMs, the index ranges from 5.5 to 8.5. This is mainly due to the decrease of the average slope of the whole basin which can be roughly reflected by the average stream slope from last section. This is based on two assumptions. First, the basic shape of one basin remains the same after aggregation. Second, the stream location in the coarse DEMs would not shift too much from the original DEM and this can be verified by Fig. 6. Basically, the deviation of the slope of a pixel on the one hand, is caused by the change of flow direction, and on the other hand, it results from the change of relative elevation between the pixel and its target neighbour with respect to the flow direction. These two assumptions imply that the flow directions inside the basin stay the same as the original DEM. So the change of the average stream slope is mainly attributed to the averaging effect of the aggregation which works the same for all pixels inside the basin.

Large gaps between topographic indexes derived from DEMs of different resolution indicate the DEM resolution has profound impact on the calculation of the volume of stream flow by TOPMODEL. In fact, the average topographic index

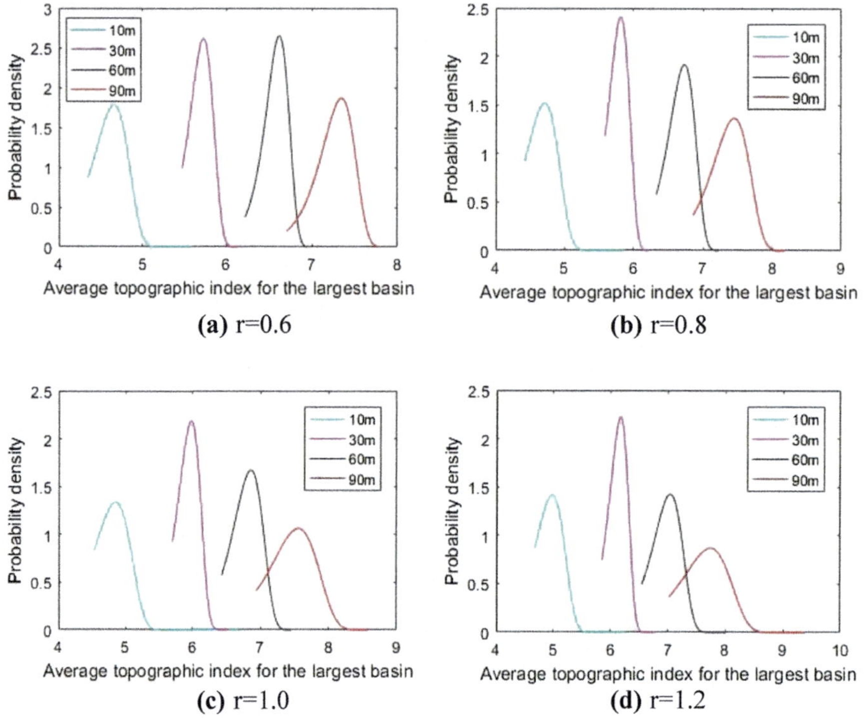

Fig. 8 PDFs for average topographic index of the largest basin

cannot represent the spatial distribution of soil saturation which may differ from pixel to pixel, instead it can only provide a sense of the overall saturation.

4 Validation Against Field Data

4.1 Data Description

The field DEM data all comes from the National Elevation Dataset (NED) of the USA. The NED is a seamless mosaic of best-available elevation data drawn from a variety of sources. Much of the NED is derived from USGS Digital Elevation Models (DEM's) in the 7.5-min series, increasingly large areas are being obtained from active remote sensing technologies, such as LIDAR and IFSAR, and also by digital photogrammetric processes. NED is available in spatial resolutions of 1 arc-s (roughly 30 m), 1/3 arc-s (roughly 10 m), and 1/9 arc-s (roughly 3 m). And the dataset is updated on a two months cycle.

The accuracy of the NED varies spatially because of the variable quality of data sources. An overall vertical accuracy is acquired by comparing it with the geodetic

Table 2 Error statistics (in meters) of the NED versus 13,305 reference geodetic control points (Gesch 2007)

Minimum	Maximum	Mean	Standard deviation	RMSE
−42.64	18.74	−0.32	2.42	2.44

Table 3 Primary metadata of field DEM datasets

Region	Location	Resolution (m)	Minimum value	Maximum value	Rows	Columns	Area (km^2)
Tennessee	−84.0795,35.3331: −84.0207,35.3880	10	678	1,440	593	635	37.66
California	−122.9656,40.3960: −122.9069,40.4506	10	687	1,455	590	635	37.47
Idaho	−115.4908,43.8068: −115.4298,43.8658	10	1,337	2,104	638	659	42.04

control points that the National Geodetic Survey (NGS) uses for gravity and geoid modeling (Smith and Roman 2001; National Geodetic Survey 2003) (Table 2).

The three locations (Table 3) chosen for validation are all source-basin areas of rivers (The reason has been given in Sect. 3.1) and they are square patches from Tennessee, California and Idaho respectively. The resolutions of DEM datasets of these three regions are all 10 m, i.e. 1/3 arc-s and they are all selected at the same area level with our experiments.

4.2 Analysis

DEM aggregation and deviation of hydrological parameters are performed from the field data. The procedure is basically repeating the simulation experiment and all the results are listed in Table 4. The column *mean* from experiment is the average value of the four means with respect to different roughness parameters and the column of standard deviation is analogous. However, more accurate way for their calculation is for each of the field datasets, first trying to determine the corresponding fractal dimension using methods (Brivio and Marini 1993) like box counting, variance analysis, etc. And then construct the fractal terrain of the same scale to derive PDFs. Also, for a same dataset, these methods may return different fractal dimension values (Brivio and Marini 1993) which still need to be selected and thus they are not adopted here.

As is presented in the table, for the number of basins, although all values derived from field data are higher than the mean retrieved from simulations when the resolution is 10 m, still they fall into the one σ range and so do the aggregated DEMs, which implies the simulation can provide reliable number of basins.

As to the drainage density, the simulating result is much larger than values derived from field data, although in order to acquire low drainage density, we can

Table 4 Values of hydrological parameters derived from field data and simulation

Hydrological parameter	Resolution (m)	Tennessee	California	Idaho	Mean from experiments	Standard deviation from experiments
Number of basins	10	16	15	14	13.760	2.903
	30	16	14	13	14.153	2.927
	60	16	14	13	14.405	2.867
	90	17	13	14	14.588	2.675
Dainage density (*10^{-3} m/m^2)	10	1.505	1.310	1.332	1.956	0.166
	30	1.470	1.291	1.331	1.692	0.143
	60	1.441	1.315	1.294	1.590	0.148
	90	1.407	1.312	1.338	1.557	0.149
Total stream length of the largest basin (*10^4 m)	10	2.405	2.073	3.032	2.206	0.823
	30	2.343	2.039	3.017	1.906	0.708
	60	2.294	2.075	2.923	1.803	0.668
	90	2.270	2.080	3.032	1.791	0.672
Average slope of the largest basin	10	0.104	0.177	0.147	0.049	0.012
	30	0.103	0.174	0.159	0.040	0.012
	60	0.113	0.174	0.138	0.037	0.013
	90	0.103	0.168	0.139	0.035	0.012
Average topographic index of largest basin	10	5.010	5.573	5.119	4.807	0.247
	30	5.650	6.034	5.695	5.940	0.161
	60	6.354	6.577	6.325	6.830	0.204
	90	6.936	7.071	6.829	7.540	0.309

increase the threshold used to derive the stream network. But in a way, high drainage density is a characteristic implied in the fractal terrain constructed by diamond-square algorithm. Unless the approach for producing the fractal terrain is modified radically, the drainage density will not decrease in a natural sense. On the other hand, this parameter also differs notably in different regions and Tennessee basins present more dense stream network than the other two regions. So accurately simulating the drainage density needs further research.

Total stream length of the largest basin again presents satisfactory results even though the value derived from experiments seems lower. Actually, the areas covered by the field data are all slightly larger than area of the fractal terrain. And it also shows that the aggregation process does cause the decrease of the total stream length for both field data and virtual data.

When it comes to the average slope of the largest basin, the value of the fractal terrain is much lower than that of the real terrain. This is because as Table 3 shows, the range of elevation selected from different regions are all around 800 m, while the fractal terrain only covers a range of 200 m for elevation. By increasing the initial standard deviation in the first diamond step (Sect. 2), the mismatch issue can be addressed. In addition, for the field data, the aggregation does not always make

Table 5 Tests against constraints on the fractal terrain

	Tennessee	California	Idaho	Constraints
CV of drainage density	0.37	0.81	0.38	<0.5
Bifurcation ratio	3.06	5.00	3.29	2.5–5
Stream length ratio	2.14	3.32	2.22	1–3
$S_1 – S_{-1}$	0.08	0.19	0.17	>0

the average stream slope of the largest basin decline as the data of Tennessee and Idaho indicates. This is possible since after averaging the original DEM, the stream length decreases, while the variation of elevation may not change too much to result in a decreasing slope.

For the average topographic index of the largest basin, the experiment fails to simulate extreme values presented by California with the highest resolution. Another unsatisfactory result is the decreasing rate is faster than that derived from the field data and this can be mostly attributed to the high decreasing rate of average slope. But, on the whole, the topographic index shows an increasing trend for both simulation and field data from aggregation.

Additionally, the rules for filtering the artificial stream network are also validated by the field data (Table 5).

Where S_1 refers to the average slope of streams in the first order while S_{-1} represents the stream slope in the maximum order. In Table 5, stream networks derived from datasets of Tennessee and Idaho satisfy the standard of the artificial stream network while California fails. Its CV of drainage density is above 0.5, which means the drainage density fluctuates a lot within its geographic scope. Besides, its stream length ratio also exceeds the range. In fact, the stream length ratio should not be valued too much because the threshold used to delineate basins and the range of regions selected has significant effect on this value. For example, a lower threshold tends to reduce the stream length in the first order, and also if the geographic range of selected area is enlarged, other tributaries may join into the current stream system, thus disturbing the stream length ratio.

To sum up, the artificial stream network together with the fractal terrain shows some satisfactory results. However, improvements still need to be made. Also if more hydrological parameters were taken into account, situation would become more sophisticated. This research can be seen as an initiative to explore a standard prototype for modeling stream networks and it reveals positive aspects to introduce computer simulations based on fractal theory to discover hydrological laws.

5 Conclusions

In this research, an artificial stream network based on fractal terrain is developed and then several typical hydrological parameters concerned with stream flow are selected to analyse the effect of DEM resolution on hydrological responses roughly.

Actually, it is insufficient to analyse the DEM influence on hydrological responses without considering specific environment such as precipitation and evapotranspiration because a hydrologic system is composed of several hydrological processes. A more comprehensive simulation framework may be coupled with a precipitation model in the future.

The artificial stream network is delineated from a constrained fractal terrain developed by adopting the diamond-square algorithm. And the constraints come from empirical statistical understanding of natural channel networks. These recognitions were proposed around 1960s by Smarts (1968), Shreve (1966) with basic concepts originated from Horton (1945), which might not be advanced. They are only statistical properties like the average stream length ratio and the bifurcation ratio. Yet in reality, their range may not be appropriate (Table 5). But approaches based on the statistical knowledge to define the fractal terrain have the advantage of simplicity. Final simulation results keep consistent with previous studies, i.e. the low resolution of DEM leads to the reduction of the total stream length and the average stream slope (Li et al. 2004) and the increase of the topographic index (Wolock and Price 1994).

A specific characteristic of the artificial stream network described here is that it can only model the situation of stream networks in the source-basin areas. More meaningful work is to extract specific basins from the square fractal terrain and perform researches within a single basin. In order to achieve a large scale of basin, say, 100,000 km^2, much more random points should be generated for the fractal terrain given a high resolution. As current implementation of terrain generation for 1,000 times can cost half a day with a normal laptop, parallel implementation should be developed for the sake of efficiency. Nonetheless, the artificial stream network as a digital platform can serve for other experiments as well, for example, simulating the evolution of stream networks.

References

Arnold JG, Williams JR, Srinivasan R, King KW and Griggs RH (1994) SWAT: soil and water assessment tool. Technical report. US Department of Agriculture, Agricultural Research Service, Grassland, Soil and Water Research Laboratory, Temple, TX

Beven KJ, Kirkby MJ (1979) A physically based, variable contributing area model of basin hydrology/Un modèle à base physique de zone d'appel variable de l'hydrologie du bassin versant. Hydrol Sci J 24(1):43–69

Brivio PA, Marini D (1993) A fractal method for digital elevation model construction and its application to a mountain region. Comput Gr Forum 12(5):297–309

Carlston CW (1963) Drainage density and streamflow. US Govt. Print. Off.

Fournier A, Fussell D, Carpenter L (1982) Computer rendering of stochastic models. Commun ACM 25(6):371–384

Gesch DB (2007) The national elevation dataset. In: Maune D (eds) Digital elevation model technologies and applications—the DEM users manual. American Society for Photogrammetry and Remote Sensing, Bethesda, MD, pp. 99–118

Hadley RF, Schumm SA (1961) Sediment sources and drainage basin characteristics in upper Cheyenne river basin. US Geol Surv Water-Supply Pap 1531:137–196
Horton RE (1945) Erosional development of streams and their drainage basins; hydrophysical approach to quantitative morphology. GSA Bull 56(3):275–370
Jain V, Sahoo R (2017) Sensitivity of drainage morphometry based hydrological response (GIUH) of a river basin to the spatial resolution of DEM data. Comput Geosci 111:78–86
Kelley AD, Malin MC, Nielson GM (1988) Terrain simulation using a model of stream erosion. ACM Siggr Comput Gr 22(4):263–268
Kenward T, Lettenmaier DP, Wood EF, Fielding E (2000) Effects of digital elevation model accuracy on hydrologic predictions. Remote Sens Environ 74(3):432–444
Li Z, Zhu C, Gold C (2004) Digital terrain modelling: principles and methodology. CRC press
Mandelbrot BB, Pignoni R (1983) The fractal geometry of nature. WH freeman, New York
Musgrave FK, Kolb CE, Mace RS (1989) The synthesis and rendering of eroded fractal terrains. ACM Siggr Comput Gr 23(3):41–50
National Geodetic Survey (2003) GPS on bench marks for GEOID03. http://www.ngs.noaa.gov/GEOID/GPSonBM03/index.html. Accessed 27 Jan 2018
Pallard B, Castellarin A, Montanari A (2009) A look at the links between drainage density and flood statistics. Hydrol Earth Syst Sci 13(7):1019–1029
Saupe D (1988) Algorithms for random fractals. In: Peitgen HO, Saupe D (eds) The science of fractal images. Springer, pp. 71–136
Shreve RL (1966) Statistical law of stream numbers. J Geol 74(1):17–37
Smart JS (1968) Statistical properties of stream lengths. Water Resour Res 4(5):1001–1014
Smith DA, Roman DR (2001) GEOID99 and G99SSS: 1-arc-minute geoid models for the United States. J Geodesy 75(9):469–490
Sørensen R, Seibert J (2007) Effects of DEM resolution on the calculation of topographical indices: TWI and its components. J Hydrol 347(1):79–89
Stachniak S, Stuerzlinger W (2005) An algorithm for automated fractal terrain deformation. Comput Gr Artif Intell 1:64–76
Strahler AN (1957) Quantitative analysis of watershed geomorphology. Eos Trans Am Geophys Union 38(6):913–920
Tarboton DG, Bras RL, Rodriguez-Iturbe I (1991) On the extraction of channel networks from digital elevation data. Hydrol Process 5(1):81–100
Wei L, Hao Z, Li L (2004) Information entropy-based assessment of different resolution DEM and its effects on run-off simulation. Hydroelectr Energy 22(4):1–4 (in Chinese)
Wigmosta MS, Vail LW, Lettenmaier DP (1994) A distributed hydrology-vegetation model for complex terrain. Water Resour Res 30(6):1665–1679
Wolock DM, Price CV (1994) Effects of digital elevation model map scale and data resolution on a topography-based watershed model. Water Resour Res 30(11):3041–3052
Yang P, Ames DP, Fonseca A, Anderson D, Shrestha R, Glenn NF, Cao Y (2014) What is the effect of LiDAR-derived DEM resolution on large-scale watershed model results? Environ Model Softw 58:48–57

Spatio-Temporal Analysis of Mangrove Loss in Vulnerable Islands of Sundarban World Heritage Site, India

Biswajit Mondal and Ashis Kumar Saha

Abstract Mangroves are unique ecosystem found mainly in tropical coastal region in saline environment and under tidal influence. It has enormous ecological and economic value to the environment and local people. However, the problems are arising in tropical coastal region like Sundarban, where both natural and ever increasing anthropogenic activities have complicated the growth and development of mangroves. Therefore, spatio-temporal monitoring of mangroves has huge importance for their conservation in Sundarban World Heritage Site, the largest mangrove population in the world. Remote sensing has been proven as an important tool to monitor such ecosystem, but the traditional pixel based approach has several drawbacks. Recently, Object-based Image Analysis (OBIA) approach in remote sensing has helped to overcome such drawbacks. The present study attempts to analyse the status of mangroves over the time period of 40 years (1975–2015) in the study area using Landsat time series images through OBIA. The result reveals that the mangroves are gradually reducing over the last 40 years and about 4% mangrove area has been converted into water. It is a major indication of increase in sea water level, making many islands vulnerable. The time series analysis in some islands, like Bhangaduni, Bulchery, Dalhousie and Halliday shows the land area as well as mangroves have been destroyed more than one-third. If the process continues at the same rate, these islands may soon completely disappear.

Keywords Sundarban · Mangroves · Landsat · OBIA · Sea level rise

B. Mondal · A. K. Saha (✉)
Department of Geography, Delhi School of Economics, University of Delhi,
Delhi 110007, India
e-mail: aksaha@geography.du.ac.in

A. Mansourian et al. (eds.), *Geospatial Technologies for All*,
Lecture Notes in Geoinformation and Cartography,
https://doi.org/10.1007/978-3-319-78208-9_5

1 Introduction

Coastal zones are very complex, dynamic and delicate environments (Mishra 2009). The coastal environment has enormous optimistic and valuable support to the people living near the coast through fishing, mineral resources, forest products etc. However, these regions are under continuous threat due to natural hazards, such as, floods, storm surges, tropical cyclone as well as ever increasing human influence. According to Jusoff (2006), mangrove acts as a living dyke against the effects of the tides along many tropical coasts providing natural support and promoting aquaculture (Nabahungu and Visser 2011). It also maintains water quality by separating sediments and nutrients in polluted coastal areas (Lugo and Snedaker 1974), carbon balance of the coastal zone as well as contribute livelihood to local people (Ha et al. 2014; Wood et al. 2013). Sundarban coastal ecosystem has similar changing characteristics mainly due to both natural hazards and human influences. Cyclonic disturbances are frequent in this region. It is evident from the existing literature that mangrove habitats and as well as total land area have been decreasing at an alarming rate and the remaining areas are under tremendous pressure mainly due to clearing of mangroves, encroachment, hydrological alterations and most importantly climate change (Blasco et al. 2001).

Global warming and associated sea level rise has been a grave matter of concern among the scientific community. The global average sea level has been rising at an alarming rate and also accelerating (Church et al. 2013). The impact of rising sea level can be seen directly in the tropical coastal regions, where large numbers of low altitudinal islands are on the verge of submersion. Interestingly, similar observation on Indian part of Sundarban that alone consist of 102 islands has been reported by some authors (Human Development Report 2009; Ghosh et al. 2015; Hazra et al. 2016). In these circumstances, there is an urgent need to assess the spatio-temporal pattern of change and its dominant causes for efficient conservation planning and management of world's largest mangrove population in Sundarban (Blasco et al. 1988).

In the last few decades, remote sensing has played a very important role in mangrove area mapping and estimation of changes (Emch and Peterson 2006; Davis and Jensen 1998). Mostly, Landsat time series data available from 1972 has been analysed for mangrove mapping through the traditional pixel-based image classification techniques (Malthus and Mumby 2003; Alongi 2008; Field 1999). The pixel based approach can only consider the variation of spectral property of pixels and produce heterogeneous land use/land cover map. However, for demarcating mangrove covered islands and other land covers contextual information like size, shape, texture and position of ground elements have very important role along with spectral properties. To consider contextual elements along with spectral properties, OBIA has been recently introduced to improve the accuracy of the classification (Chen et al. 2012; Huang and Jia 2012; Wang et al. 2004; Conchedda et al. 2008; Navulur 2007).

Therefore, in this research work, an attempt has been made to quantify mangrove cover and land area change in five different time periods (1975, 1989, 1995, 2005 and 2015) using Landsat images in Sundarban World Heritage site and its surrounding region using OBIA approach. Due to complexity of the mangrove ecosystem, it is difficult to generalize the possible causes of these spatial changes (Emch and Peterson 2006; Thom 1984). However, we tried to correlate the possible role of sea level rise in this region on reduction of the mangrove covered land masses.

2 Materials and Methods

2.1 Study Area

Sundarban delta is formed by the deposition of alluvium of three main rivers viz., Ganga, Brahmaputra and Meghna. Mangrove forests in Sundarban is well known for its rich and diverse species composition spread over India and Bangladesh. Sundarban was declared as a "World Heritage Site" in 1987 and "Biosphere Reserve" in 1989 by UNESCO (Ghosh et al. 2015). Sundarban World Heritage has huge significance for its rich biodiversity. The geographic extent of World Heritage Site and its 20 km buffer in Indian part were considered in this study (Fig. 1).

The Sundarban World Heritage Site is situated in a humid region and mainly tropical climate prevails here. Sundarban receives heavy rainfall (144 cm on average), out of which 75% is received during June to September through monsoon (District Statistical Handbook 2009). Cyclonic storm (e.g. Aila in 2009) and floods are frequent in the coastal Sundarban, cause extensive destruction.

In the last few decades, human population who are dependent on forest product (wood, honey, wax, medical plants etc.) collection and aquaculture have increased manifold in the periphery region of Sundarban (Census 2011). It has caused further anthropogenic threat to the mangrove ecosystem.

2.2 Data Base

2.2.1 Remote Sensing Data Base

In this study, the spatio-temporal analysis has been performed using five Landsat images (1975–2015) (Table 1). Cloud free, similar tidal phase and winter season orthorectified Landsat images were downloaded from United States Geological Survey (https://earthexplorer.usgs.gov/) website. GPS assisted field survey have been carried out during December, 2015 to January, 2016 to validate the results.

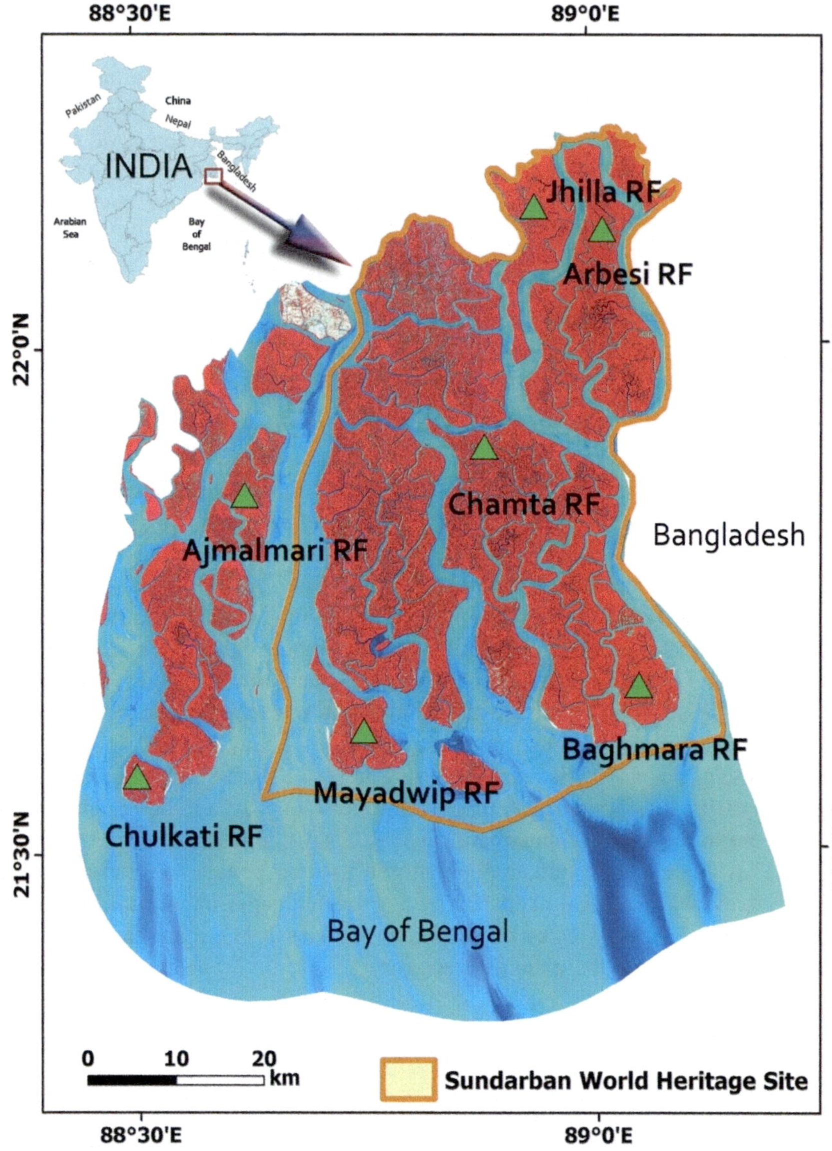

Fig. 1 Location of the Sundarban World Heritage site in the study area. Landsat 8 standard false colour composite image is in the background

2.2.2 Sea Level Data

The Sea level data have been collected from Permanent Service for Mean Sea Level (PSMSL) (www.psmsl.org) website. Five nearest stations have been selected due to

Table 1 Specifications of Landsat images used in the study

Date of acquisition	Satellite/sensor	Product type	Path/row	Pixel size (m)	Spectral bands used
05-12-1975	Landsat 2/MSS	GEOTIFF/6 bit	148/45	57[a]	Band 1 (Green), Band 2 (Red), Band 3 (NIR), Band 4 (NIR)
03-01-1989	Landsat 4/TM	GEOTIFF/8 bit	138/45	30	Band 1 (Blue), Band 2 (Green), Band 3 (Red), Band 4 (NIR), Band 5 (SWIR-1), Band 7 (SWIR-2)
28-01-1995	Landsat 5/TM	GEOTIFF/8 bit	138/45	30	Band 1 (Blue), Band 2 (Green), Band 3 (Red), Band 4 (NIR), Band 5 (SWIR-1), Band 7 (SWIR-2)
07-11-2005	Landsat 5/TM	GEOTIFF/8 bit	138/45	30	Band 1 (Blue), Band 2 (Green), Band 3 (Red), Band 4 (NIR), Band 5 (SWIR-1), Band 7 (SWIR-2)
08-03-2015	Landsat 8/OLI	GEOTIFF/12 bit	138/45	30	Band 2 (Blue), Band 3 (Green), Band 4 (Red), Band 5 (NIR), Band 6 (SWIR-1), Band 7 (SWIR-2)

[a]Resampled to 30 m

unavailability of stations exactly within World Heritage Site, three from India viz. Haldia (22° 1′ 59.98″ N and 88° 5′ 59.99″ E), Diamond Harbour (22° 11′ 59.99″ N and 88° 10′ 0.12″ E) and Gangra (21° 56′ 59.99″ N and 88° 1′ 0.12″ E) and two from Bangladesh side viz. Hiron Point (21° 46′ 59.98″ N and 89° 28′ 0.12″ E) and Khepupara (21° 49′ 59.98″ N and 89° 49′ 59.98″ E).

2.3 *Methodology*

The broad methodology adopted in this study is described in Fig. 2.

2.3.1 Pre-processing of Landsat Images

To minimise the time and scene dependent effects, e.g. atmospheric absorption, scattering, sensor calibration, sensor target illumination geometry, pre-processing of multi date Landsat images were necessary. Therefore, image DN values have been converted to at sensor radiance (Chavez 1988) using COST method (Mahiny and Turner 2007) except for Landsat 8 images. Further, haze is a major problem to get a clear image. The haze correction was done using method given in Chavez (1988). The DN vales of Landsat-8 image have been converted into reflectance using the procedure given in Landsat-8 user guide (2016). All the analytical parts have been carried out in Erdas Imagine software.

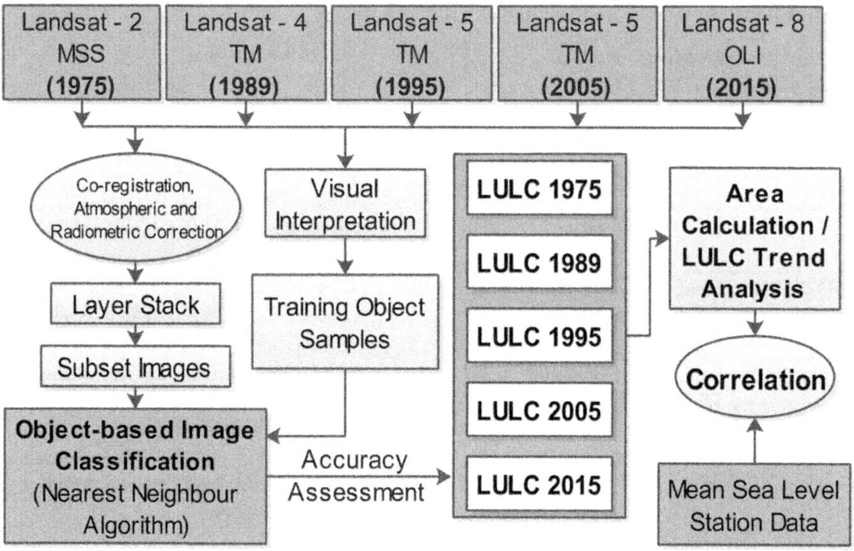

Fig. 2 Methodological framework of spatio-temporal analysis of mangrove distribution and change

2.3.2 Classification of Landsat Images

As mentioned earlier, Object-based Image Analysis (OBIA) is found to be more suitable than traditional pixel based techniques for classification of Landsat image in this study. Image segmentation is the first step of OBIA image analysis approach. Segmentation refers to the process of partitioning digital image into homogenous objects drawing a group of pixels together. The homogeneity is in terms of colour, texture, shape, smoothness, compactness etc. (Navulur 2007; Dornik et al. 2017). There are numbers of image segmentation methods, but none of the methods are universally recognised for image segmentation (Dass et al. 2012). Among different segmentation techniques, multi resolution segmentation is one of the most widely used and well accepted method for Object-based Image Analysis (Conchedda et al. 2008; Kamal et al. 2015; Dass et al. 2012). All the Landsat reflectance images have been segmented by multi-resolution segmentation method with a scale factor of 5, with shape factor 0.1 and compactness factor of 0.5 (Navulur 2007) through trial and error in eCognition Developer software.

Based on visual interpretation of Landsat images and field knowledge five land use and land cover classes were identified in the study area viz. mangroves, water body (river channels and sea water), barren land, sand and agricultural fields. Around 20 randomly distributed objects were selected for each class per Landsat image as training samples for nearest neighbour classification in the OBIA framework (Navulur 2007). The classified images are shown in Fig. 3. As the proportions of barren land, sand and agriculture lands are very less in comparison to

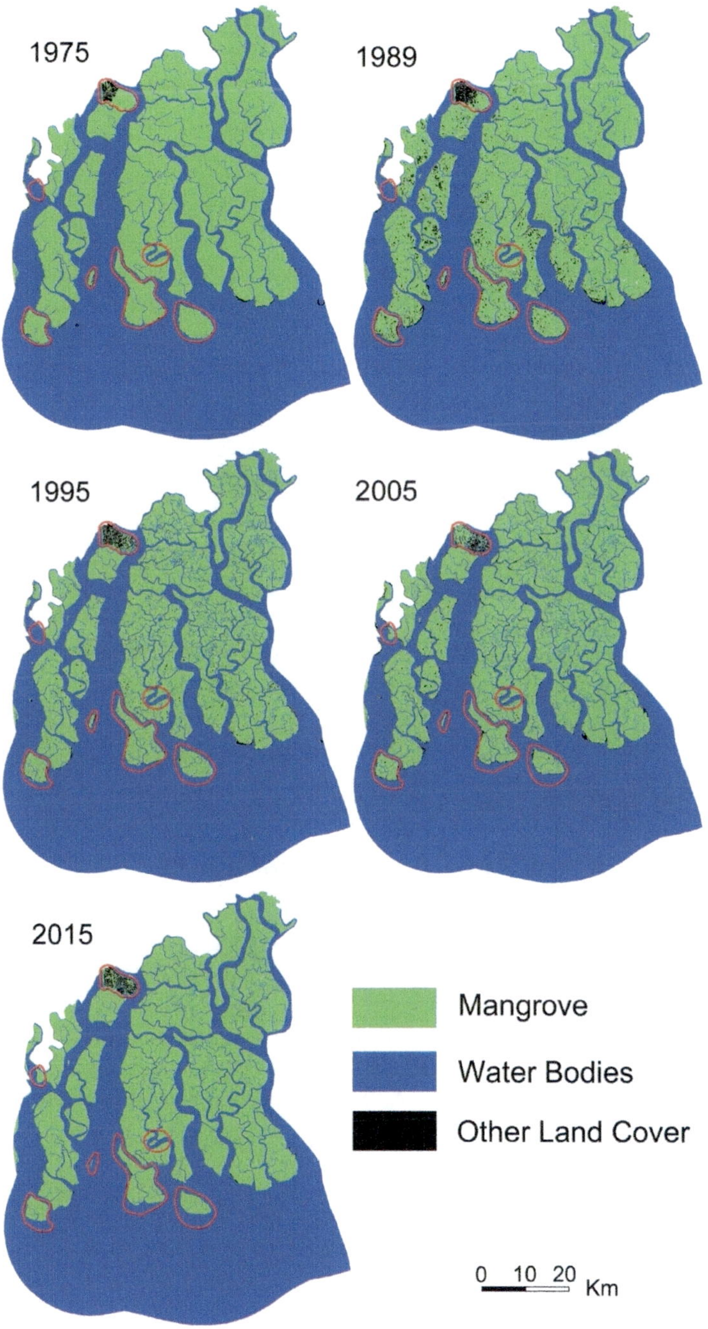

Fig. 3 Land use and land cover maps (1975–2015) generated through OBIA. Major changes are highlighted with red lines

other two land cover classes, they are grouped together and marked as 'Other land cover'. Finally, accuracy assessment of all the images were performed with the help of a separate set of 5 objects per class based on field GPS data. The overall accuracy of all the classified maps were more than 90%.

2.3.3 Change Detection

Post classification change detection analysis was performed considering land use and land cover map of each year. The changes were computed for each land cover in hectare and presented as in percentage (Tables 2 and 3). A change detection matrix was also computed using 1975 and 2015 maps for better understanding the class wise land cover change scenario (Table 4).

2.3.4 Analysis of Sea Level Data

The recorded annual mean sea level data for all the five selected nearby stations from India and Bangladesh (www.psmsl.org) were plotted against time (1975–2015) and best-fit lines were drawn. Similarly, trend line was also drawn to show the change in total land area. The trend lines were visually compared to understand the relationship between land cover loss and sea level rise.

3 Result and Discussion

3.1 Distribution of Land Use and Land Cover Classes

Land use and land cover maps generated for the years 1975, 1989, 1995, 2005 and 2015 of Sundarban World Heritage Site, shows that the mangroves have dominated among the land covers other than water bodies (Fig. 3). In 1975, there were about 41.3% of mangrove area and with time the mangrove area has gradually reduced to about 37.8% (Table 2). The total area under water bodies has increased throughout the analysis period (Table 2). The total area under sand, barren land and agriculture is less than 1% and therefore it is insignificant to determine the trend.

3.2 Pattern of Land Use and Land Cover Change

This section highlights the trend of areal coverage of different land use and land covers over the time period of 40 years (1975–2015). Mangroves show a sharp declining trend whereas water has an increasing trend (Table 2). Although, the

Table 2 Area-wise distribution (in %) of different land use and land cover in the study area (1975–2015)

LULC classes	1975	1989	1995	2005	2015
Mangroves	41.3	39.5	36.6	36.3	37.8
Water	58.4	59.4	62.6	63.1	62.0
Other land cover (Sand, Barren land, Agriculture)	0.3	1.1	0.8	0.7	0.3

Table 3 Land use land cover change detection matrix (1975–2015)

2015 (in %)	1975 (in %)						
		Mangrove	Water	Sand	Barren land	Agriculture	Total
	Mangrove	37.01	0.65	0.02	0.03	0.05	37.75
	Water	4.10	57.73	0.08	0.02	0.01	61.95
	Sand	0.01	0.02	0.01	0.00	0.00	0.04
	Barren land	0.09	0.00	0.00	0.02	0.02	0.13
	Agriculture	0.09	0.00	0.00	0.02	0.02	0.14
	Total	41.30	58.40	0.11	0.09	0.10	

Table 4 Land area change (1975–2015) in some disappearing islands

Year	Bulchery Island		Dalhousie Island		Bhangaduni Island		Halliday Island	
	Area (km^2)	% Area change	Area (km^2)	% Area change	Area (km^2)	% Area change	Area (km^2)	% Area change
1975	30.8		77.7		43.0		3.6	
1989	27.9	−9.4	72.7	−6.5	38.5	−10.4	3.0	−18.0
1995	26.7	−13.4	69.7	−10.3	35.3	−18.0	2.6	−29.2
2005	23.7	−22.9	64.7	−16.7	28.2	−34.5	1.2	−66.6
2015	21.0	−31.7	59.5	−23.4	23.3	−45.7	0.3	−91.5

values are insignificant compare to the main land covers (mangroves and water), agricultural activities in the periphery region shows increasing trend (Table 2). Visual interpretation of Landsat images of 1975 and 2015 also shows how a small patch of agricultural land has expanded to remove mangroves completely from Jharkhali Island located in the north-western boundary (Figs. 4 and 5). Agriculture, aquaculture activities and human settlements are on rise in this area for the last few decades (Samanta and Hazra 2012). It is quite unfortunate that, the region being under Sundarban World Heritage Site, such activities are increasing day by day.

The main objectives of change detection in remote sensing include identifying geographical location and type of change as well as quantifying the changes (Im and Jensen 2005). A change detection matrix (Table 3) has been prepared through cross tabulation of classified images (from 1975 to 2015) to understand inter class change of land use and land cover over the time period of 40 years. Around 58% of water bodies and around 37% of mangroves are remained unchanged over the time

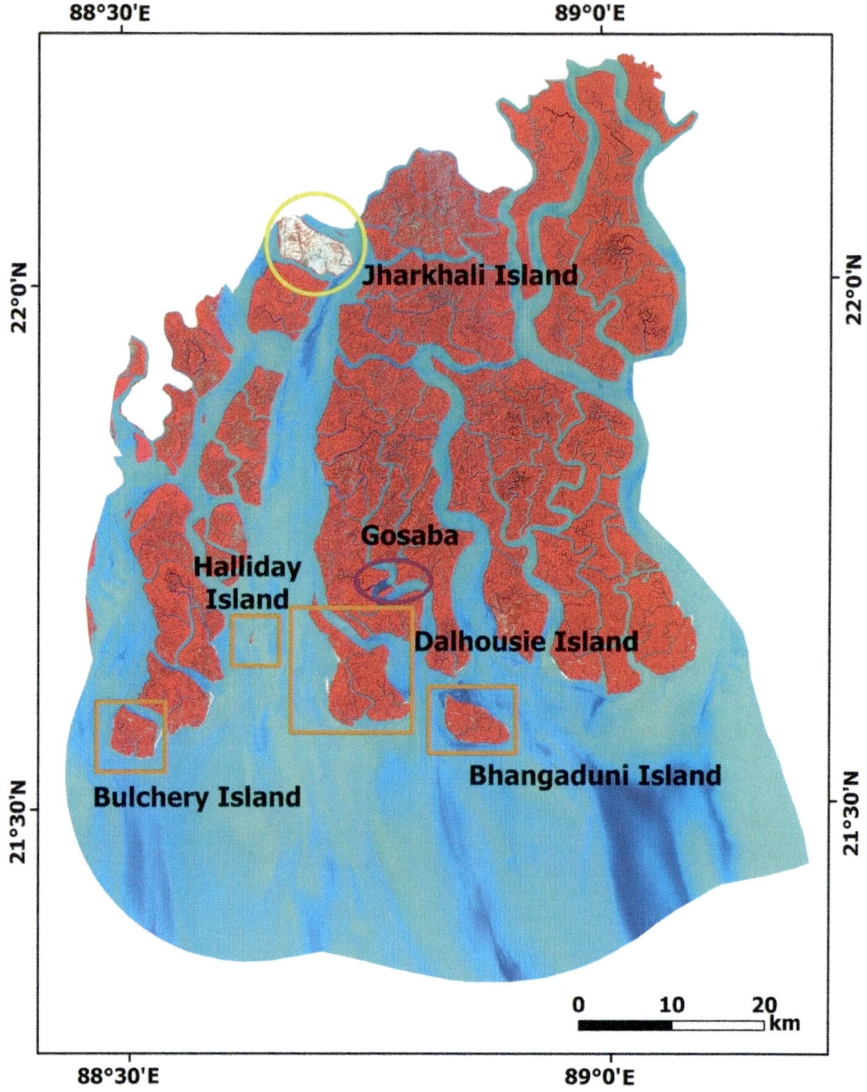

Fig. 4 Map showing location of the vulnerable islands selected for detailed analysis

period. Although, there is a small addition of 0.65% of mangroves from water bodies (Table 3), there are around 4.1% area of mangrove has been converted to water body (from 1975 to 2015). It is a major indication of mangrove loss may be due to rising sea water level (Islam et al. 2016; Snankar et al. 1996). The change detection matrix shows that about 0.09% area (~525 ha) of mangrove has been converted into agriculture. Conversion of mangrove to agricultural activity in the periphery region as shown in Fig. 5 may also pose further threat to the ecosystem.

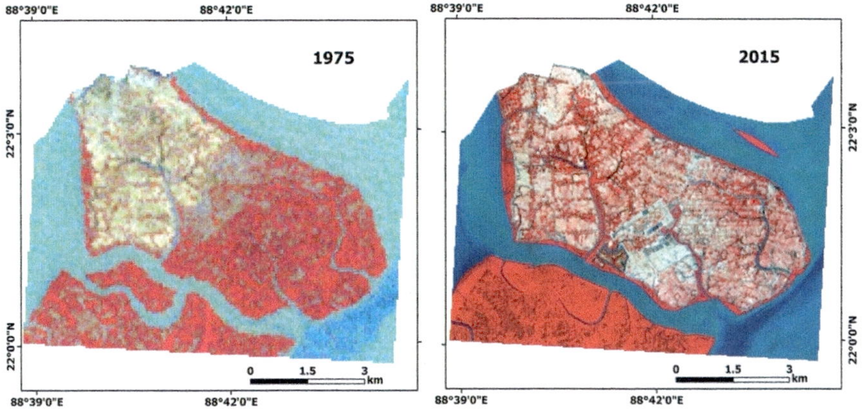

Fig. 5 Conversion of mangroves to agricultural lands in Jharkhali Island

The results also indicate that from during 1975 to 2015, total land area (except water) has been reduced by around 3.51% (Table 2).

3.3 *Identification of Disappearing Islands*

To identify the land area submerged under water as indicated in change detection matrix (Table 3), land area polygon maps generated through Object-based Image Analysis of time series Landsat images, were overlayed in Arc GIS software. For better visual interpretation, the land area polygons for each year were colour coded (blue in 1975 to red in 2015). Through this GIS based visualization four islands, viz. Bulchery Island, Dalhousie Island, Bhangaduni Island and Halliday Island have been found to be more vulnerable (Fig. 4).

It is observed that these islands (Fig. 6) are degrading day by day due to erosion which is more dominant than the accretion (Pramanik 2014). Seaward side is more vulnerable in all those islands mostly due to wave action. Considering, 1975 as the base year, it can be calculated that the land area has reduced by 32%, 23% and 46% in case of Bulchery, Dalhousie, and Bhangaduni Islands respectively (Table 4).

The most alertly situation is recorded in case of Halliday Island, where almost 91% of the land area has lost over the last 40 years. If the rate of erosion remains same, Halliday Island will lose its existence in the near future and same situation may happen in Bhangaduni, Dalhousie, Bulchery Islands as well.

Through visual interpretation, other important changes in the region may be highlighted. Figures 7 shows in Gosaba Reserve Forest how a chunk of land area has been completely eroded due to changing river morphology.

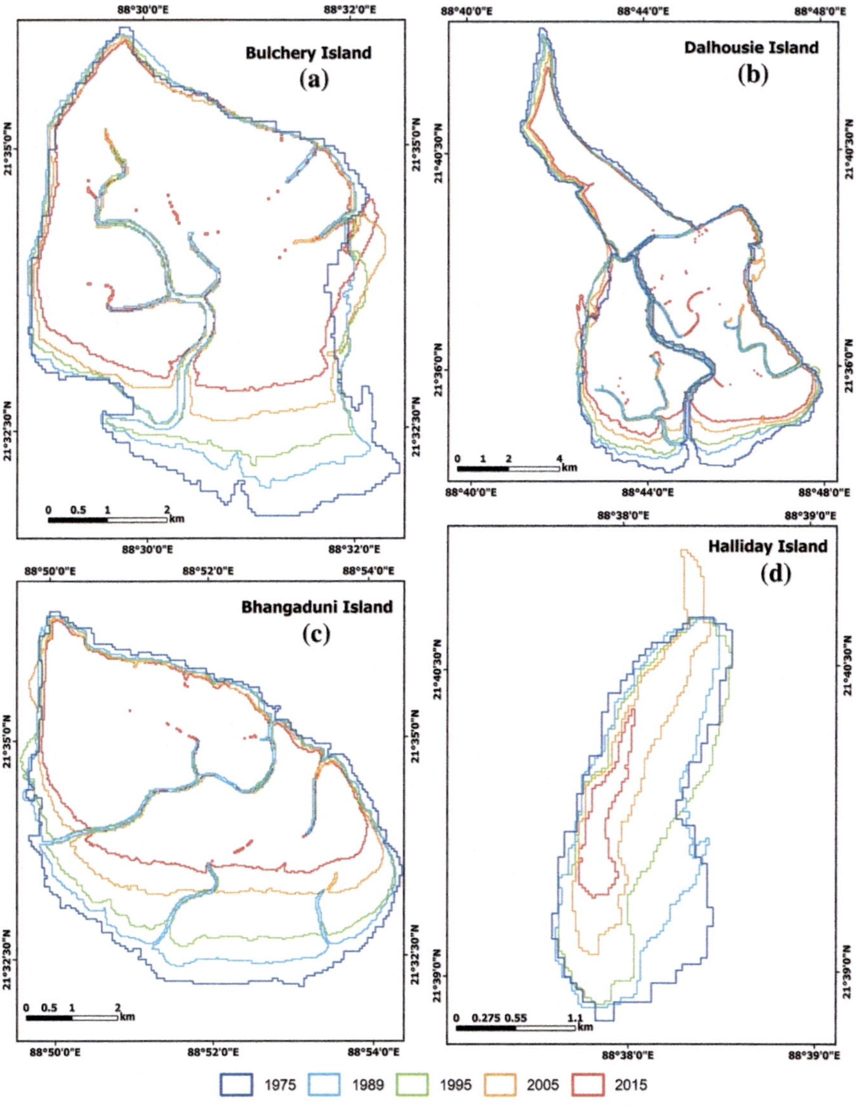

Fig. 6 Spatio-temporal pattern of land cover loss in **a** Bulchery Island, **b** Dalhousie Island, **c** Bhangaduni Island **d** Halliday Island

3.4 Possible Causes of Land Area Loss

From the discussion so far, it is clear that the loss of mangroves in the region is mainly attributed to the loss of land area. The sea facing mangrove margins migrate landward as mangroves die due to stress (weakened root structures, falling of trees,

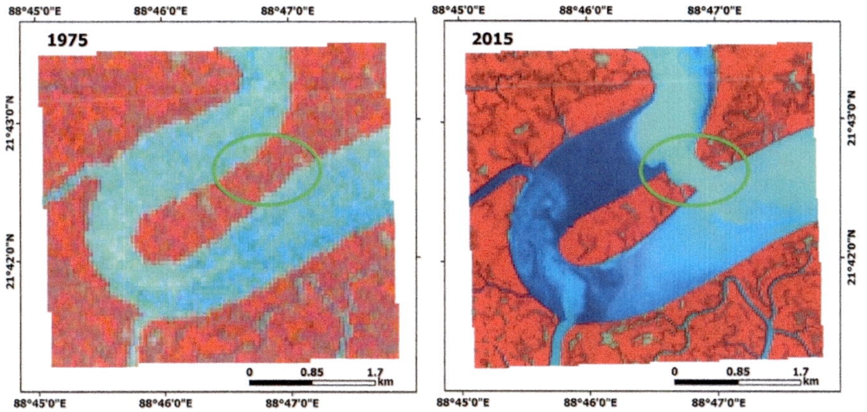

Fig. 7 Loss of mangrove cover due to changing river morphology identified in a part of Gosaba Reserved Forest

increased salinity for long duration, frequency and depth of inundation) caused by rising sea level (Ellison 1993; Lewis 2005). The annual mean sea level change data for all the selected stations show a positive trend, indicating the sea level is rising in the whole region, whereas, the total land area is showing a negative trend, i.e., the land area is gradually reducing (Fig. 8). Therefore, this inverse relationship strongly supports that land area reduction and the disappearance of these islands are related to sea level rise.

However, some studies have highlighted that not only sea level rise but also subsidence, wave action, sediment composition contribute to the land area loss in the region. According to Rahman et al. (2011), Sundarban delta front has undergone a net erosion of ~170 km^2 of coastal land in between 1973 and 2010, though the direction and extent of erosion and accretion rates very throughout. Other studies have been highlighted local factors such as wave action (Snankar et al. 1996) and sediment composition and compaction (Allison 1998) may be two other reasons of land loss. The most significant anthropogenic disturbance occurred in 1975 when India completed a dam on the river Ganges in Farakka, West Bengal. This dam has caused a significant reduction of water flow and sediment to the Sundarban coast of the Bengal delta (Mirza 1998). Coastal areas of Bengal delta are also undergoing a mean annual land subsidence of 15–50 mm (Stanley and Hait 2000) which may be a big factor of land area loss in near future. Another study on Grande Glorieuse Island (Indian Ocean) by Testut et al. (2016), highlighted that sea level is not necessarily the controlling factor, but also sediment supply and wave action can be key factors for shoreline changes. Moreover, role of frequent cyclone storms and other hazards cannot be ruled out (Islam et al. 2016).

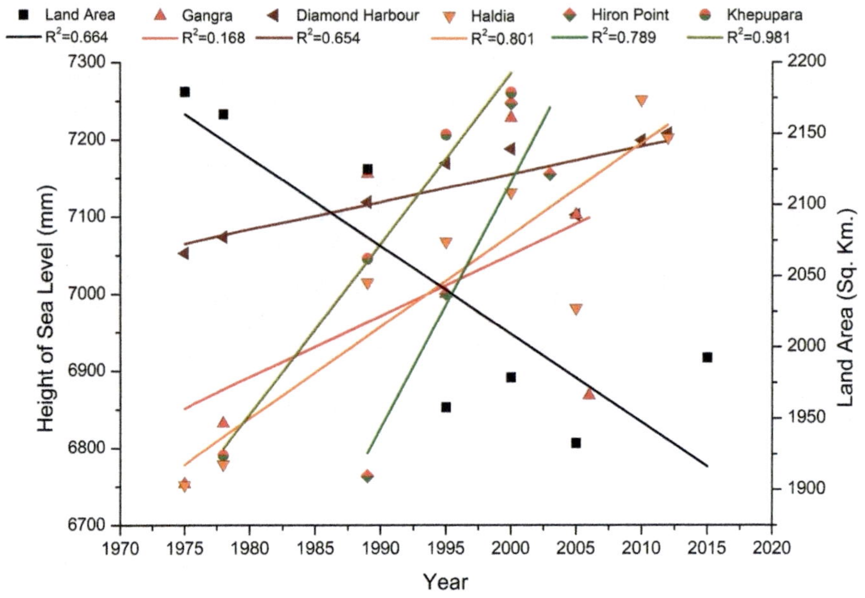

Fig. 8 Relationship between trend of sea level rise in some selected stations and land area change (1975–2015)

4 Conclusion

In the present study, an effort has been made to analyse spatio-temporal pattern of mangrove and other land cover change using time series Landsat images (1975–2015) using OBIA classification approach. The result shows that significant portion of land area covered with mangroves has been demolished over the study period and converted into water body. The inverse relationship between total land area change and the annual mean sea level trends for different stations strongly supports that with increasing height of sea level, land area as well as mangroves are reducing. The present finding also reiterates the similar claim by many researches (Rahman et al. 2011; Islam et al. 2016; Gilman et al. 2008). Both natural and human activities have an increasing trend in this coastal mangrove habitat. The agriculture and aquaculture activities are increasing in the periphery region. Although, it is beyond the scope of the present research, cyclone storms are also very common phenomena in the Sundarban region may further accelerate land degradation and mangrove loss.

The methodology adopted here has a huge potential for delineating and monitoring changing pattern of mangroves with better precision. Such results may be useful for mangrove conservation in the present context of global warming, sea level rise, natural disaster and ever increasing influence of human population in this fragile ecosystem. Finally, if the land degradation continues in the same manner, it will impose tremendous pressure on the ecosystem and many small islands may disappear in near future.

Acknowledgements Research grant provided by University of Delhi, Research Council is duly acknowledged. Biswajit Mondal is thankful to University Grants Commission for research fellowship.

References

Allison MA (1998) Historical changes in the Ganges—Brahmaputra delta front. J Coast Res 14 (4):1269–1275

Alongi DM (2008) Mangrove forests: resilience, protection from tsunamis, and responses to global climate change. Estuar Coast Shelf Sci 76(1):1–13

Blasco F, Gauquelin T, Rasolofoharinoro M, Denis J, Aizpuru M, Caldairou V (1988) Recent advances in mangrove studies using remote sensing data. Mar Freshw Res 49(4):287–296

Blasco F, Aizpuru M, Gers C (2001) Depletion of the mangroves of Continental Asia. Wetl Ecol Manag 9(3):245–256

Census (2011) www.censusindia.gov.in/2011census/dchb/DCHB.html. Accessed 01 Dec 2017

Chavez PS (1988) An improved dark object subtraction technique for atmospheric scattering correction of multispectral data. Remote Sens Environ 24(3):459–479

Chen G, Hay GJ, Carvalho LMT, Wulder MA (2012) Object based change detection. Int J Remote Sens 33(14):4434–4457

Church JA, Clark PU, Cazenave A, Gregory JM, Jevrejeva S, Levermann A, Merrifield MA, Milne GA, Nerem RS, Nunn PD, Payne AJ, Pfeffer WT, Stammer D, Unnikrishnan AS (2013) Sea level change. Climate change 2013: the physical science basis. Contribution of working group I to the fifth assessment report of the intergovernmental panel on climate change. In: Stocker TF, Qin D, Plattner G-K, Tignor M, Allen SK, Boschung J, Nauels A, Xia Y, Bex V, Midgley PM (eds) Cambridge University Press, Cambridge, United Kingdom and New York, NY, USA. https://doi.org/10.1017/CBO9781107415324.024

Conchedda G, Durieux L, Mayaux P (2008) An object based method for mapping and change analysis in mangrove ecosystem. ISPRS J Photogramm Remote Sens 63(5):578–589

Dass R, Priyanka, Devi S (2012) Image segmentation techniques. Int J Electron Commun Technol 3(1):66–70

Davis BA, Jensen JR (1998) Remote sensing of mangrove biophysical characteristics. Geocarto Int 13(4):55–64

District Statistical Handbook (2009) South 24-Parganas, Bureau of Applied Economics and Statistics, Government of West Bengal, India

Dornik A, Dragut L, Urdea P (2017) Classification of soil types using geographic object-based image analysis and random forest. Pedosphere. https://doi.org/10.1016/S1002-0160(17)60377-1

Ellison JC (1993) Mangrove retreat with rising sea level, Bermuda. Estuar Coast Shelf Sci 37 (1):75–87

Emch M, Peterson M (2006) Mangrove forest cover change in the Bangladesh Sundarbans from 1989–2000: a remote sensing approach. Geocarto Int 21(1):5–12

Field CD (1999) Mangrove rehabilitation: choice and necessity. Hydrobiologia 413:47–52

Ghosh A, Schmidt S, Fickert T, Nusser M (2015) The Indian Sundarban mangrove forests: history, utilization, conservation strategies and local perception. Diversity 7(2):149–169

Gilman EL, Ellison J, Duke NC, Field C (2008) Threats to mangroves from climate change and adaptation options, a review. Aquat Bot 89(2):237–250
Ha TP, Dijk HV, Visser L (2014) Impacts of changes in mangrove forest management practices on forest accessibility and livelihood: a case study in mangrove-shrimp farming system in Ca Mau Province, Mekong Delta, Vietnam. Land Use Policy 36:89–101
Hazra S, Mukhopadhyay A, Mukherjee S, Akhand A, Chanda A, Mitra D, Ghosh T (2016) Disappearance of the New Moore Island from the Southernmost Coastal Fringe of the Sundarban delta-a case study. J Indian Soc Remote Sens 44(3):479–484
Huang H, Jia X (2012) Integrating remotely sensed data, GIS and expert knowledge to update object-based land use/land cover information. Int J Remote Sens 33(4):905–921
Human Development Report (2009) District Human Development Report: South 24 Parganas. Development and Planning Department, Government of West Bengal, India
Im J, Jensen JR (2005) A change detection model based model based on neighbourhood correlation image analysis and decision tree classification. Remote Sens Environ 99(3): 326–340
Islam MA, Mitra D, Dewan A, Akhter SH (2016) Coastal multi-hazards vulnerability assessment along the Ganges deltaic coast of Bangladesh-a geospatial approach. Ocean Coast Manag 127:1–15
Jusoff K (2006) Individual mangrove species identification and mapping in port Klang using airborne hyperspectral imaging. J Sustain Sci Manag 1(2):27–36
Kamal M, Phinn S, Johansen K (2015) Object—based approach for multi-scale mangrove composition mapping using multi-resolution image data sets. Remote Sens 7(4):4753–4783
Lewis RR III (2005) Ecological engineering for successful management and restoration of mangrove forests. Ecol Eng 24(4):403–418
Lugo AE, Snedaker SC (1974) The ecology of mangroves. Annu Rev Ecol Syst 5:39–64
Mahiny AS, Turner BJ (2007) A comparison of four common atmospheric correction methods. Photogramm Eng Remote Sens 73(4):361–368
Maltus TJ, Mumby PJ (2003) Remote sensing of the coastal zone: an overview and priorities for future research. Int J Remote Sens 24(13):2805–2815
Mirza MMQ (1998) Diversion of the Ganges water at Farakka and its effects on salinity in Bangladesh. Environ Manag 22(5):711–722
Mishra M (2009) Integrated coastal zone management: a case study of selected coastal districts of Orissa. PhD thesis, Jawaharlal Nehru University. http://hdl.handle.net/10603/18140. Accessed 1 Dec 2017
Nabahungu NL, Visser SM (2011) Contribution of wetland agriculture to farmers livelihood in Rwanda. Ecol Econ 71:4–12
Navulur K (2007) Multispectral image analysis using the object-oriented paradigm. CRC Press Taylor and Francis Group, USA
Pramanik MK (2014) Assessment the impact of sea level rise on mangrove dynamics of Ganges delta in India using remote sensing and GIS. J Environ Earth Sci 4(1):117–127
Rahman AF, Dragoni D, El-Masari B (2011) Response of the sundarbans coastline to sea level rise and decreased sediment flow: a remote sensing assessment. Remote Sens Environ 115(12):3121–3128
Samanta K, Hazra S (2012) Landuse/landcover change study of Jharkhali Island Sundarbans, West Bengal using remote sensing and GIS. Int J Geomat Geosci 3(2):299–306
Snankar D, McCreary JP, Han W, Shetye SR (1996) Dynamics of the East India Coastal Current 1. Analytic solutions forced by interior Ekman pumping and local alongshore winds. J Geophys Res 101(C6):13975–13991
Stanley DJ, Hait AK (2000) Holocene depositional patterns, neotectonics and Sundarban mangroves in the western Ganges-Brahmaputra delta. J Coast Res 16(1):26–39

Testut L, Duvat V, Ballu V, Fernandes RMS, Pouget F, Salmon C, Dyment J (2016) Shoreline changes in a rising sea level context: the example of Grande Glorieuse Scattered Islands, Western Indian Ocean. Acta Oecologica 72:110–119

Thom BG (1984) Coastal landforms and geomorphic processes. In: Snedaker SC, Snedaker JG (eds) Mangrove ecosystem: research methods. UNESCO, Paris

Wang L, Sousa WP, Gong P (2004) Integration of object based and pixel based classification for mapping mangroves with IKONOS imagery. Int J Remote Sens 25(24):5655–5668

Wood AL, Butler JRA, Sheaves M, Wani J (2013) Sport fisheries: opportunities and challenges for diversifying coastal livelihoods in the Pacific. Marine Policy 42:305–314

Part II
Volunteered Geographic Information and Participatory GIS

3D Georeferencing of Historical Photos by Volunteers

Timothée Produit and Jens Ingensand

Abstract Historical photographs are a very rich source of information that can be useful in a variety of different contexts such as environmental analyses, land planning and illustration of landscape evolution. However to reach this goal the images must be accurately georeferenced. In this paper we propose to use the crowd to perform the 3D georeferencing of collections of historical images. To this goal we implemented a web 3D georeferencer that offers volunteers the possibility to semi-automatically identify 1. the location of the point from where a picture has been taken, 2. the three angles: tilt, roll and yaw and 3. the field of view. A virtual web-based globe is the central element in this tool that allows both for the georeferencing in three dimensions by volunteers and for the visualization of georeferenced images to assess the landscape variation through time. In this paper we evaluate the method and the georeferencer and give suggestions for further developments and exploitation of the database.

1 Introduction

Historical photographs contain information about both natural and man-made processes that have occurred in the past and that have an influence on todays world and future development. These photographs are often stored in archives that are difficult to access for researchers. Moreover, historical photographs are rarely scanned and stored in digital files. Another problem is the precise georeferencing of these photographs. Indeed if the exact location and orientation of a picture is known, the picture can be visualized as a 3D object for instance in a virtual globe. Moreover the georeferencing of a picture in three dimensions has an important value since it enables an exact identification and indexing of places that are visible in these images. Furthermore the georeferencing also has a value for scientists who can extract geographic

T. Produit (✉) · J. Ingensand
Territorial Engineering Institute (insit) University of Applied Sciences and Arts Western Switzerland (HES-SO), Route de Cheseaux 1, CH-1401, Yverdon-les-Bains, Switzerland
e-mail: timothee.produit@heig-vd.ch

J. Ingensand
e-mail: jens.ingensand@heig-vd.ch

A. Mansourian et al. (eds.), *Geospatial Technologies for All*,
Lecture Notes in Geoinformation and Cartography,
https://doi.org/10.1007/978-3-319-78208-9_6

data from the pictures using photogrammetric methods and thereby for instance recreate three-dimensional models of destroyed buildings or calculate the volume of a glacier at different times.

The rise of web-based 3D solutions opens opportunities for volunteers to generate and interact with three-dimensional data. Today, the public is increasingly familiar with virtual globes and 3D software such as video games. In the project that we present in this paper we focus on the georeferencing and visualization of historical images using a 3D virtual web-based globe (Produit and Ingensand 2016). In order to compute the location and orientation of a picture, a user needs to manually provide 3D coordinates of several visible locations in the picture. This task can for example be achieved by measuring 3D points in the real world using surveying instruments, by extracting the 3D locations in a GIS or by digitizing 3D points in a virtual globe. Hence, in the method that we present in this paper we want to use the familiarity with 3D software and to offer a tool for volunteers for the georeferencing of historical images. In this tool the goal is to identify and to digitize similar locations that are visible both in the picture and in a virtual globe. These correspondences allows us to compute the exact location and orientation of the picture and to visualize the picture as a 3D object in a virtual globe.

Our first case study was a set of 1500 postcards shared with us by the *Archive de la Construction Moderne (ACM)—EPFL* representing the swiss Alps shot in 1960.

This paper is structured as follows. First we will identify different ways of collecting 3D information by volunteers and georeferencing images. Thereafter we will explain our idea and the tools that we have implemented. Then, we will describe the first case study. In the following paragraphs we will analyze and evaluate how real-world users use the tools in order to georeference historical photos. Finally ideas for future work and for an exploitation of the collected information are presented.

2 State of the Art

2.1 Approaches to Collecting 3D Data Using Volunteers

The digitization of 3D objects can be regarded as a challenging task: most existing software that can be used for this task is generally dedicated to professionals. Several researchers have addressed the challenge to involve volunteers for the digitization of 3D objects and to design this task as easy as possible.

A first approach for volunteers to provide 3D data is to add attributes to 2D objects such as a buildings height or the type of a roof. These attributes can then be used for the rendering of a 3D object (Goetz and Zipf 2013). A second approach is to offer tools for volunteers to perform measurements in 3D. For instance a user can use the GPS of a smartphone to record a position in 3D (Brovelli et al. 2013). Pictures are a specific kind of data. Associated with a GPS tag, pictures can be used to compute 3D models. Some authors (Snavely et al. 2006; Strecha et al. 2010; Hartmann et al. 2016) show that pictures downloaded from photosharing platforms can be used to generate 3D point clouds. A similar technology is used in the project Mapillary. In this web platform volunteers can provide smartphone pictures that have been collected along a path. Using computer vision algorithms Mapillary generates several outputs based on the added pictures such as a dense 3D point cloud, however it is not possible yet to derive 3D GIS objects such as 3D buildings from the point cloud.

By providing attributes or measurements, a volunteer passively provides 3D data and does not have to navigate a 3D interface to digitize 3D data. Indeed, the digitization of 3D data is a challenging task for a volunteer. In the project *Building Maker* that has been developed until 2013, Google asked volunteers to draw buildings in oblique aerial images. The users were required to align and to reshape a 3D building model using 2D pictures. In this project, the georeferencing task remained accessible to volunteers. Uden and Zipf (2012) suggest to involve volunteers with 3D digitization skills in the project OpenStreetMap. To reach this goal, the authors provided a repository of 3D buildings models which can be associated with OpenStreetMap objects.

2.2 3D Georeferencing of (Historical) Photos

The 3D georeferencing of a picture requires that the 3D location (X, Y, Z) and 3D orientation (three angles) as well as some camera parameters such as the field of view are computed. For a similar task Google Earth requires the volunteer to manually provide the picture parameters *location, orientation* and *field of view*. The volunteer navigates the virtual globe and refines both the orientation and location parameters with a slider. This method however has the drawback that it has the potential to be time-consuming and less accurate.

The regular approach applied in the computer vision and photogrammetry communities is 2D-3D image orientation (camera orientation or space resection). The picture parameters are calculated using Ground Control Points (GCP) which are 3D points that have been measured in the field or in a map and that have corresponding 2D points in the image (see Fig. 1). In photogrammetry this concept is generally applied to compute the location of a collection of overlapping images in a bundle adjustment. 2D-3D image orientation is also applied in several projects dedicated to the extraction of geographic data from single pictures (Bozzini et al. 2012; Produit and Tuia 2012) or defined in the project Viewfinder (University of Southern California 2008) to insert an image in a 3D model. In this way, similarly to the project *Building Maker*, photogrammetric relations between the reality and a picture simplify the task of finding the image georeferencing. By using these relations the users does not need 3D modelling skills as the task of clicking points in a map or a virtual globe remains less complicated.

Fig. 1 Ground control points: 2D locations in the image and corresponding 3D locations in the virtual globe

3 Methodology

3.1 Georeferencing

Our concept, similarly to the project Viewfinder (University of Southern California 2008) relies on the idea to establish GCP of a single image using a virtual representation of the reality rather than measurements in the field. This implies that a virtual globe needs to be positioned roughly at the same location as a corresponding photo. Thereafter a user can digitize GCP both in the virtual globe and in the 2D photo using locations that are both visible in the photography and in the virtual globe. The key in this concept is to use the workforce as well as the place knowledge of volunteers in order to accomplish this task.

A photo often has a location name associated with it. This location name can be used for a rough georeferencing and positioning within a virtual globe. However, the most difficult task of the georeferencing process is to navigate the virtual globe in order to reach the location represented in the picture. To facilitate this process, we have identified four steps. All steps imply that the user is familiar with the scene represented in the picture and also approximately with the topography of the area.

- First, the volunteer navigates a 2D map. The images are marked in the map according to their a priori location which has been geocoded from a place name. The volunteer chooses a picture.
- Second, the volunteer provides an approximate location of the picture by indicating the 2D location of the camera position on a 2D map. If the original photo has been associated with a location name, this location name is used to automatically position the 2D map.
- Third, the user provides the orientation of the picture.
- Fourth, the provided location and direction are used to approximately position a virtual globe. At this point, if the two previous steps have been completed successfully, the volunteer should be able to see a similar scene in the 3D globe and in the 2D photograph. Now the task of the volunteer is to identify points that are

visible both in the virtual globe and in the photograph. These points must also be stable from the time when the picture had been taken until today. Examples for such points that are easy to identify are mountain peaks, churches, crossroads and shapes of water-bodies. Once that at least four GCP have been provided, the picture location can be computed.

- Finally, the user has to provide two more GCP to improve the accuracy. Functionalities to move and delete inaccurate GCP are also provided.

3.2 Validation of the Georeferencing

In the presented project, the alignment of the picture with the virtual landscape is a good indicator of the picture location and orientation accuracy. Moreover, the visualization of 3D images is central since it allows the visitors to compare the current virtual state of the landscape with the historical state shown in the historical picture. However, in order to have an optimal experience during this comparison, the 3D image must have a good alignment with the virtual globe. To ensure this alignment accuracy we have implemented two validations steps.

We implemented automatic checks to avoid that the volunteers provide incorrect locations. First, at least six GCP must be provided for each image. With six correspondences, more GCP are available than required to solve the equations. Hence, an error can be computed for each GCP. Namely, we compare the 2D coordinates digitized by the volunteer with the corresponding 2D coordinates computed by our algorithm. If these errors are above a certain threshold, the picture location cannot be saved.

A second check is based on the computed picture parameters. If the computed field of view is impossible (too small or too large) the picture location is not accepted.

Finally, a manual visual validation is performed by our team of advanced users. They visualize each picture and check its alignment with the virtual landscape. If the alignment is not satisfactory, the validator can improve the correspondences or reject the georeferencing.

4 Case Study

4.1 Implementation

The web-based prototype has been created using the open source virtual globe API (www.cesium.com). The virtual globe renders data (aerial images and a digital terrain model (DTM)) provided as web services by the Swiss Federal Office of Topography Swisstopo. The aerial images have a resolution ranging from 25 to 50 cm and the DTM has a resolution of 1 m. The virtual globe is used for the georeferencing

Fig. 2 Screenshot of the 3D interface. A picture of the Swiss alps shot in 1910 by E. Spelterini from an hot-air balloon. The picture is visualized as an overlay on top of the virtual landscape

but also for the 3D visualization of historical images. Historical images thereby can be compared with the current virtual landscape (Fig. 2).

The 2D-3D image orientation algorithm uses the GCP provided by the volunteer to compute the camera position from which the 3D locations of the image corners can be derived. The camera parameters are stored in a PostgreSQL/PostGIS database. The image is stored as a 3D entity in the *gltf* 3D format which is the format used by Cesium.

The website can be accessed at: https://smapshot.heig-vd.ch.

4.2 Case Study Setting

The prototype was officially made public on February 1st 2017. An exhibition was organized by the *Archives de la Construction Moderne (ACM), EPFL*. The ACM also provided the first photo collection; 1200 postcards shot between 1960 and 1970 showing the swiss Alps. These photos were scanned and delivered with the corresponding metadata which typically contained the photographer's name, a precise or approximate date and a place name. The geocoded place name allowed us to provide an a priori image location.

A key issue was the acquisition of visitors and volunteers. In order to attract volunteers, we provided a set of 50 already georeferenced photos that could be visualized in the virtual globe. Hence, the first visitors were able to see the results and to understand the goal of the project. Second, to reach the volunteers a media campaign was organized by both universities (EPFL and HES-SO) communication teams. Hence, the project's aim was published in newspapers, radio, TV and web media. This first 1200 photos were georefenced within 10 weeks.

After this first collection had been published, several other collections were retrieved from open data repositories. The collections contain both terrestrial and aerial pictures taken between 1850 and 1970. Currently, the prototype stores 2500 images among which 2300 have been georeferenced by 90 registered volunteers.

4.3 Hypotheses

In the presented project volunteers are asked to provide the exact position and orientation of a historical picture. To reach this goal, the volunteers have to navigate a 3D virtual globe, click on points which are both visible on the surface of the DTM and in the picture. In this context we have established two hypotheses which are about user participation and spatial data quality:

1. Volunteers are interested in the georeferencing task: they are attracted by the problematic of historical images metadata acquisition and visualization. Moreover the volunteers like spending time executing these tasks.
2. Volunteers are able to perform the task: the proposed tasks are sufficiently easy for volunteers and the volunteers provide a correct georeferencing.

These two hypotheses are critical—if either the users do not show any interest or the collected data shows a low quality, the outcome would be less valuable. Another important point in the analysis of these hypotheses is the understanding of user habits in order to improve the platform at subsequent stages.

4.4 Data Collection

The platform records information about the visits and the participation of volunteers. For instance, for each georeferenced image the time it takes for a specified user to accomplish the georeferencing task is recorded. We also used Google Analytics in order to collect and analyze statistics about the users.

5 Results

5.1 Volunteers' Activity

The first major result from the analyzed data is that some platform visitors acquired during the media campaign were converted into volunteers who were able to use the 3D georeferencer to provide the location and orientation of images.

The number of visits and the number of new volunteers is strongly related to the publications in the media. Each publication was followed by a peak of georeferenced pictures (up to 100 pictures during one single day, see Fig. 3).

As in other crowdsourcing projects (e.g. Neis et al. (2013); Sauermann and Franzoni (2015)) there are few volunteers who provide most of the georeferencing and there are many volunteers who provide the georeferencing of only a few images (see Fig. 4). The most active users were using the platform every day. Some users stopped after a time, either because there were no more images in their area of interest or the images became too challenging (e.g. presumably in regions they are less familiar with or images with few prominent landmarks) or because they get tired.

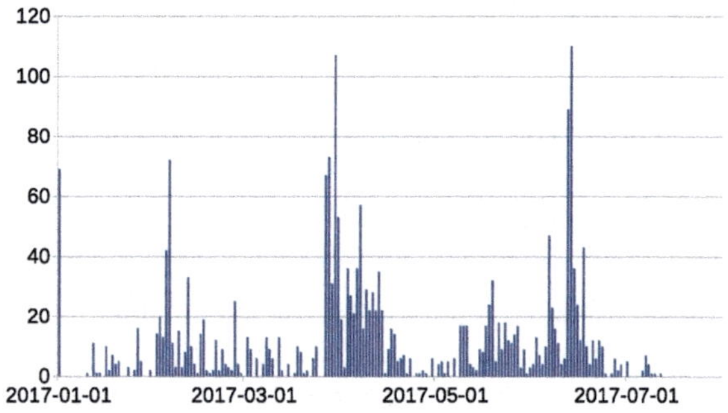

Fig. 3 Number of georeferenced images per day. Peaks occur after a publication in the media

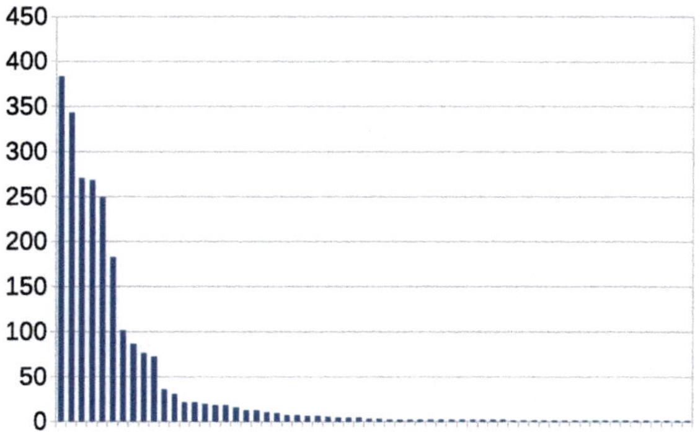

Fig. 4 Ranking of the volunteers. The y-axis shows the number of images per volunteer. Few volunteers provide the georeferencing of the largest portion of images. Many volunteers tested the platform and provided the georeferencing of a small number of images

5.2 Georeferencing Assessment

Every image's georeferencing has been checked by our team. About 50% of the images were correctly georeferenced, the remaining images were slightly improved. This validation has the goal to ensure the quality of the alignment with the virtual globe but also to understand the main sources for errors:

- **Incorrect correspondences**: The indicated control point is not the same in the picture and in the virtual globe. The landscape variation over time, the similarity of the land-cover and the quality of the rendering may have caused this problem. A typical error is a confusion of buildings which might be caused by the fact that the virtual globe does not provide 3D buildings, but only an aerial image draped over a DTM. Moreover the buildings in the Alps were often more scattered in the past while the density has drastically risen during the last 50 years. Another example of this type of error is a lake shore which moved a hundred meters due to a change of its level.
- **Roof tops**: As aforementioned, 3D buildings are not available yet in the virtual globe. The volunteers who are less accustomed to geographic data often click on a roof top in the picture and on a point at ground level in the virtual globe. This error generates a GCP height error of several meters which can have a high impact for pictures that have been taken close to the buildings (Fig. 5).
- **Correspondences scattering**: To improve the accuracy and the alignment, the correspondences should be scattered across the entire picture. Sometimes the volunteers provide correspondences only in a portion of the picture where the correspondences are easy to find.
- **Connected pointer**: Once that a camera position is computed with the four initial GCP, the mouse cursor in the virtual globe is connected with a pointer in the

Fig. 5 Incorrect GCP provided on a building. The roof top is clicked in the image, while the same location is provided at ground level in the virtual globe

Fig. 6 Due to some inaccurate initial GCP, the mouse pointer on the left indicates a location which is slightly shifted

picture. Namely, if the user moves the mouse in the virtual globe, the mouse pointer position in the image is automatically calculated. We implemented this functionality to ease the task of the volunteer to find additional correspondences. However, we also noticed that this feature can mislead the volunteer: if the initial position is not exact due to the fact that some initial GCP are inaccurate, the picture pointer indicates an incorrect location. It appears that some volunteers trust the picture pointer too much and provide incorrect correspondences (Fig. 6). This situation occurs mainly when the correspondences are difficult to define; for instance in areas with an undiscriminating land cover.

In general, few pictures were rejected. Indeed, the automatic checking discussed in Sect. 3.2 prevents the submission of completely incorrect locations.

Some pictures are difficult or even impossible to locate. The most difficult images are the ones having a very rough or incorrect a priori location (due to different places having similar names or metadata errors). In this scenario, only a person who recognizes the landscape in the picture can perform the georeferencing. Images that are impossible to georeference are the ones that do not match the settings of our georeferencing algorithm. Currently our implementation of the collinearity equations is not adapted to:

- Vertical aerial images
- Cropped images
- Photomontages

In these situations, the algorithm fails or computes an incorrect location. In the future, our goal is to improve the georeferencing algorithm in order to cope with the first scenario. The cropped images issue can also be solved if the volunteer provides more GCP. However, photomontages will remain an issue due to the fact that

Fig. 7 Photomontage: the foreground and the background pictures were cut from two different pictures and pasted together. This image was georefenced with GCP marked only in the background. (*Source* Library of Congress, Views of Switzerland)

they look like real photos and the volunteer can spend a considerable amount of time trying to identify their location. For instance in the published collections some collages of two different pictures had been identified (Fig. 7).

5.3 Volunteers Habits

5.3.1 Area of Interest

Our assumption is that volunteers provide the georeferencing of images in their regions of interest. For each volunteer we extracted the exact location of *their* pictures and the date of the georeferencing. Hence we were able to visualize the evolution of the volunteer's area of interest.

In Fig. 8 we compared for each volunteer the dispersion (standard deviation) of his five initial images with the dispersion of all images that the same user had georefenced. This result concerns only the most active volunteers who have provided more than 10 images. We computed the dispersion histogram for every volunteer. The first images (green) generally have a dispersion smaller than 20 km and few dispersion values are larger than 20 km. It means that generally the five first images provided were located in a area of 20 km in diameter. The dispersion of all images is more linear, but small dispersions still occur more often. This implies that the volunteers who continue to provide picture georeferencing move to other regions either

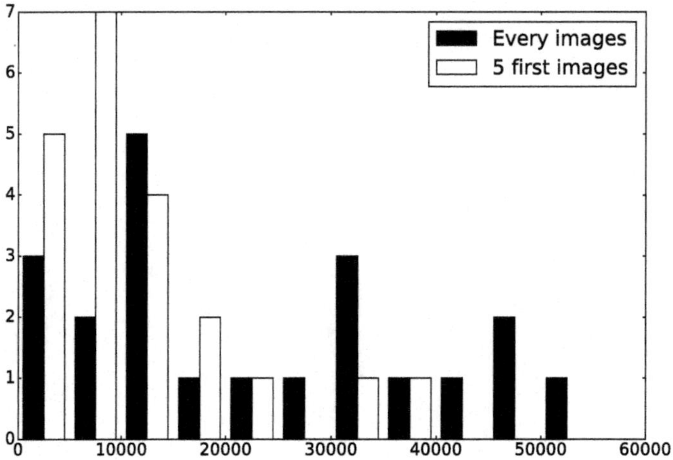

Fig. 8 The dispersion in (m) is represented on the x-axis. The histogram of the dispersion of the five first images is compared to the dispersion of every image. The dispersion of the first five images is generally smaller which indicates that the volunteers start to work on a specific region and tend to extend their region of interest afterwards

because there are no more images available in the initial region of interest or because the users want to explore other areas. Interestingly we also noted that the most active and talented volunteers were able to provide the georeferencing of pictures everywhere in Switzerland as long as the a priori location provided by the place name was accurate.

5.3.2 Browser and Device

The web-based platform was optimized for the Chrome web browser which provides the best performance for the rendering of the virtual globe. The georeferencing task is only comfortable with a desktop or laptop computer (smartphone and tablet screens are too small to perform the georeferencing task). Indeed 75% of the visitors used a desktop or laptop computer and 44% of the users used the Chrome web-browser. The sessions of the users who used the Chrome web browser also lasted longer. This result suggests that the rendering speed might refrain visitors. In general, the information about the browser and used device will help us to specify priorities regarding the improvement of the platform's user experience.

5.3.3 Target Audience

Another interesting point is the audience of the prototype. This analysis allows for an optimized targeting of potential volunteers. 75% of the visitors are male and the most represented age group is between 35 and 44 years old which represents 30% of the visitors. A typical visitor is thus mainly in a group which can be described as "technology-friendly". This finding is in contradiction to our initial guess that the platform would mainly attract retired citizen (who could be more interested in visualizing historical landscapes and would have more time to perform the georeferencing task).

6 Discussion

6.1 Volunteers' Interest for Georeferencing

The launch of this project can be considered as successful. This success was helped by the coverage in regular media such as radio and TV. The media's interest in the project was risen by the media campaign created by two universities at the same time. Furthermore, this project can be considered as relatively visual and easy to explain. Finally, in the context of climate change and urbanization assessment, the project demonstrates the value of historical images as a witness for slow change.

At this stage, we do not exactly know the motivation of our volunteers. Their motivation could be related to the will to help archive managers to gain information about the images. A second possible motivation can be the usage and sharing of their geographic knowledge. Finally, among the volunteers who give us feedbacks, some of them told us that the georeferencing is a way to spend time in a meaningful way.

6.2 Ability of the Volunteers to Perform the Georeferencing Task

The first sets of image collections show that the volunteers are able to provide accurate location of pictures. The visual validation by our trained team is the best way to ensure the quality of the provided georeferencing. Moreover, the validation also helped us to understand the volunteers' common mistakes. In the next version of the platform, the validators will be able to provide the reason of the correction and rejection to the volunteers. In this way, one important goal will be to train the volunteers in order to improve their skills and to reduce the validation workload. We will also provide a tutorial explaining common mistakes. This will help the volunteers to understand how to improve the georeferencing accuracy.

The validation remains both time consuming and expensive and it can be considered as the main limitation of the project. Hence, in the future we aim at implementing a validation by the crowd itself (e.g. similarly to the Wikipedia project) and thereby offering tools to expert volunteers to improve or reject entries.

We also aim at recording the time spent by users to perform the georefencing. This indicator will help us computing the value of the volunteers' activity (sum of the time spent), but also to understand if the volunteers improve their skills and if modifications of the georeferencing screen ease the georeferencing task.

6.3 Volunteers Habits

The logging of the volunteers' activity is a way of understanding the volunteers' interests and habits. A deep comprehension of the volunteers is a key to keeping them active, and to collect information about the most efficient acquisition channels. In the future we will also analyze other parameters such as time habits, which can indicate if the volunteers mostly work at home or at their work place.

The sum of all information about the volunteers can be used to target specific audiences for instance in a social media campaign.

7 Conclusions and Perspectives

Our project is one of the first crowdsourcing project which actively involves volunteers in generating 3D data. Due to the fact that the georeferencing task has been separated into smaller and easier tasks, we can demonstrate that volunteers are able to generate quality 3D data. A major issue is the final validation of the 3D data which is currently manually performed by our team. In other projects such as Wikipedia, the validation by the crowd itself proves to be a working concept. This feature would be a major improvement of our platform which already includes an efficient automated validation system that minimizes the submission of completely incorrect data. The georeferencer is also at an early stage. Our objective is to make it more robust and available for every place on earth. The semi-automatic 3D georeferencing also has a considerable potential. Indeed, the similarity between pictures can be computed with state of the art computer vision algorithms. Thereby an a priori 3D location can be provided that is more accurate than the geocoding of place names that is currently used.

Another current challenge is to keep the platform active. With the documented results, we draw the interest of archivists who can: (1) improve the visibility of their collections and (2) improve the metadata of their photos with the georeferencing. Moreover archivists can be provided with the exact 3D location of the picture, the exact extent of the area visible in the image and the place names that are visible in each image. The crowd can also be asked to correct or improve metadata such as the image title or caption. During the georeferencing task, a volunteer has a focused look at the entire image. He may therefore notice interesting elements that he could share. Hence, in the next stage of the project, we will improve the utility of the platform for archivists and continuously add new collections to keep the crowd active.

In the future, we aim at improving the use of the georeferenced pictures. Indeed, single historical images are already used by geoscientists and land planners to understand the evolution of the landscape. Our goal will be to draw their attention to the project and demonstrate how they can benefit from it. Regarding citizen science projects we have proved that the presented project already solves two problems regarding Volunteered Geographic Information for scientific projects (Ingensand et al. 2016): we have found volunteers and they are able to use the platform. The remaining problem is the data quality in terms of the accuracy of the generated geographic data. The visual alignment of the image with the virtual globe is a good starting point but it is not sufficient for a scientific purpose. We will have to work on two different aspects. First the improvement of the georeferencing quality provided by the users and second, the development of advanced photogrammetric functionalities for scientists who can improve the georeferencing accuracy themselves.

Our project has the goal to create a time machine that visualizes the past derived from historical images. Regarding this goal we imagine several improvements. (1) We can add other layers showing historical data such as maps or aerial and satellite images as base data for the virtual globe. (2) In some famous areas, we notice that historical pictures have a very high time resolution: many pictures with similar viewpoints are shot every year. Such densely photographed regions open opportunities regarding the creation of time lapses. Finally another objective is to provide a comparison of a historical picture with the real world rather than the virtual globe. Current advances in Augmented Reality (AR) technologies open the possibility to use our database in order to compare the reality with the past for instance during a city tour or a hike.

References

Bozzini C, Conedera M, Krebs P (2012) A new monoplotting tool to extract georeferenced vector data and orthorectified raster data from oblique non-metric photographs. Int J Herit Digital Era

Brovelli M, Minghini M, Zamboni G (2013) Participatory GIS: experimentations for a 3d social virtual globe. Int Arch Photogram Remote Sens Spat Inf Sci

Goetz M, Zipf A (2013) The evolution of geo-crowdsourcing: bringing volunteered geographic information to the third dimension. In: Crowdsourcing geographic knowledge. Springer

Hartmann W, Havlena M, Schindler K (2016) Towards complete, geo-referenced 3d models from crowd-sourced amateur images. ISPRS Annals Photogram Remote Sens Spat Inf Sci

Ingensand J, Composto S, Nappez M, Produit T, Ertz O, Oberson M, Rappo D (2016) Challenges in VGI for scientific projects. PeerJ Preprints

Neis P, Zielstra D, Zipf A (2013) Comparison of volunteered geographic information data contributions and community development for selected world regions. In: Future internet

Produit T, Ingensand J (2016) A 3d georeferencer and viewer to relate landscape pictures with VGI. In: AGILE international conference on geographic information, LINK-VGI workshop

Produit T, Tuia D (2012) An open tool to register landscape oblique images and generate their synthetic model. In: Open source geospatial research and education symposium (OGRS)

Sauermann H, Franzoni C (2015) Crowd science user contribution patterns and their implications. In: Proceedings of the national academy of sciences

Snavely N, Seitz SM, Szeliski R (2006) Photo tourism: exploring photo collections in 3d. ACM Trans Graph

Strecha C, Pylvninen T, Fua P (2010) Dynamic and scalable large scale image reconstruction. In: Conference on computer vision and pattern recognition

Uden M, Zipf A (2012) Openbuildingmodels—towards a platform for crowdsourcing virtual 3d cities. In: 7th 3D GeoInfo conference

University of Southern California (2008) Viewfinder: how to seamlessly flickrize google earth. http://interactive.usc.edu/projects/viewfinder/

Patterns of Consumption and Connectedness in GIS Web Sources

Andrea Ballatore, Simon Scheider and Rob Lemmens

Abstract Every day, practitioners, researchers, and students consult the Web to meet their information needs about GIS concepts and tools. How do we improve GIS in terms of conceptual organisation, findability, interoperability and relevance for user needs? So far, efforts have been mainly top-down, overlooking the actual usage of software and tools. In this article, we critically explore the potential of Web science to gain knowledge about tool usage and public interest in GIScience concepts. First, we analyse behavioural data from Google Trends, showing clear patterns in searches for GIS software. Second, we analyse the visits to GIScience-related websites, highlighting the continued dominance of ESRI, but also the rapid emergence of Web-based new tools and services. We then study the views of Wikipedia articles to enable the quantification of methods and tools' popularity. Fourth, we deploy web crawling and network analysis on the ArcGIS documentation to observe the relevance and conceptual associations among tools. Finally, in order to facilitate the study of GIS usage across the Web, we propose a linked-data inventory to identify Web resources related to GI concepts, methods, and tools. This inventory will also enable researchers, practitioners, and students to find what methods are available across software packages, and where to get information about them.

Keywords GIS · Geographic information science · GIScience · GIS operations Web science · Google Trends · Wikipedia · ArcGIS · Linked data

A. Ballatore (✉)
Department of Geography, Birkbeck, University of London, London, UK
e-mail: a.ballatore@bbk.ac.uk

S. Scheider
Human Geography and Planning, Universiteit Utrecht, Utrecht, The Netherlands
e-mail: s.scheider@uu.nl

R. Lemmens
Faculty of Geo-Information Science and Earth Observation (ITC), University of Twente, Enschede, The Netherlands
e-mail: r.l.g.lemmens@utwente.nl

A. Mansourian et al. (eds.), *Geospatial Technologies for All*,
Lecture Notes in Geoinformation and Cartography,
https://doi.org/10.1007/978-3-319-78208-9_7

1 Introduction

The Web offers invaluable resources for researchers, practitioners, and students of geographic information science (GIScience) and spatial data science. To meet their information needs, users search and consume online information originating from technical manuals, software documentation, academic websites, blogs, forums, discussion boards, as well as social media. The same set of GIScience ideas, ranging from core concepts (Kuhn 2012) to methods such as buffer and interpolation, are found in a vast number of heterogeneous, incompatible software suits, such as ArcGIS, QGIS, R, and Carto. Hundreds of GIS tools exist, and it is currently not known how much and when they are actually used. Analysing usage patterns would be immensely useful to improve the conceptual organisation, usability, and findability of these tools, as well as the methods and concepts that underpin them. Knowledge about to what extent GIS software, tools, and methods attract the attention of users would be valuable to ground research in this direction: Researchers, developers, and practitioners could relate their work to information needs in a data-driven way.

GIS users would benefit from a mapping between tools, concepts, and Web pages that describe them. For example, spatial analysts could grasp the workings of methods at an abstract level across software, identifying suitable tools more effectively. Teachers could indicate to students the variety of ways in which similar concepts and methods are implemented in real software packages. Software developers could better integrate their products to existing software, making their tools more findable and better linked to the GIScience concepts that they use. Several initiatives aimed at structuring GIS concepts, methods, and tools with a rather top-down approach, only observing the tools and their formal definitions, without considering behavioural data (Lemmens 2006; Gao and Goodchild 2013; Kuhn and Ballatore 2015; Scheider et al. 2017).

In this article, we take the Web as a resource to study the patterns of consumption of GIS-related information, focussing on tools, software packages, organisations, as well as more abstract GIScience concepts. By adopting a Web science approach (Hendler et al. 2008), this study focusses on the following research questions:

- To what extent are Web sources useful to study GIS usage?
- What is the relative popularity of GI tools and organisations?
- How are tools associated with each other?
- What is the popularity in GIS methods and concepts?
- How can we connect Web resources to GIScience concepts and methods using linked data?

After reviewing existing efforts in understanding and mapping GIScience usage (Sect. 2), we report on this study in five parts, organized as follows. First, Google Trends data about GIS is explored critically in Sect. 3. A pool of highly visible GIS-related websites is studied in Sect. 4, relying on data from Web analytics firms Alexa

Internet and SimilarWeb. Section 5 then focusses on the popularity of Wikipedia articles related to GIScience, charting topics that attract high, medium, and low interest. Subsequently, Sect. 6 performs a network analysis of the most visited GIS website, i.e., the documentation of ArcGIS. As a way to improve the organisation and findability of these resources, we then outline the proof-of-concept of a linked-data inventory, which highlight commonalities and relationships across these GIScience resources (Sect. 7). As part of this study, we also tested NLP methods, such as topic models (Blei 2012), on a corpus of GIScience websites, but as these did not seem to yield interesting results, we left them out of this article. Finally, Sect. 8 draws conclusions and directions for future work.

As part of our efforts to make GIS more semantically structured, all the resources created as part of this study are available in an online repository as open knowledge.[1]

2 Related Work

GIScience principles are in use in a plethora of tools. Currently, we face a lack of up-to-date knowledge on which GI tools exist, how they link to each other and to underlying core concepts (Kuhn and Ballatore 2015). This is necessary to know how tools should best be used in a given context (Hofer et al. 2017), and how we can translate between GIS workflows (Bernard et al. 2014; Ludäscher et al. 2006), abstracting from particular software packages (Hinsen 2014; Scheider and Ballatore 2018). Currently, all we have is a vague idea about different GIS software products and their associated (and often closed) worlds of terminology (Steiniger and Hunter 2013). Better linkage would have positive effects in both GIScience practice and education.

The World Wide Web constitutes a network of resources that can be exploited for Web science (Hendler et al. 2008) and, more generally, for data-driven science (Hey et al. 2009). Its wealth of inter-connected, distributed, user-generated content makes it an obvious candidate for studying usage patterns of informational resources and tools (Castellano et al. 2013), on a scale which is unprecedented and may be impossible to reach with traditional usability or empirical user studies (Kveladze et al. 2013).

Empirical studies in GIScience that make use of the Web and social media to explore human behaviour abound. They include estimating the location of tweeting users (Hecht et al. 2011), or harvesting geospatial information about places from social media feeds (Stefanidis et al. 2013; McKenzie et al. 2015) and from text corpora (Hollenstein and Purves 2010), and are based on mature, well-established methods (Ferrara et al. 2014). New approaches for extracting semantic information from unstructured texts (Blei 2012; Ramage et al. 2009) have been used to describe and link information resources about GI tools and methods (Hu et al. 2015; Gao and Goodchild 2013). Web statistics derived from search engines like Google can inform researchers across disciplinary boundaries (Stephens-Davidowitz 2013).

[1]https://github.com/simonscheider/GISTrends.

Yet, it is debatable to what extent the "unstructured" Web can be a reliable empirical resource for estimating GIScience content consumption patterns. Foundational critique about the big data hype was raised in recent years (Boyd and Crawford 2011), addressing the representational bias in human language texts on the Web (Caliskan et al. 2017), which can lead to severe estimation errors in a data-driven science (Lazer et al. 2014). Furthermore, the missing structure of Web information, i.e., the lack of "semantic rails for the data train", were recently criticised (Janowicz et al. 2014), making it hard to pre-select data and tools in a way that accounts for their inherent biases, and thus to separate signal from noise (Scheider et al. 2017). The linked data paradigm may offer a strategy to counter this weakness of pure bottom-up methods, in so far as it provides an infrastructure for sharing unstructured as well as structured and semantically precise information about tools and data (Brauner 2015; Hofer et al. 2017; Scheider and Ballatore 2018), including the classification of GIS functions. Besides the informal classifications in several GIS text books, a few efforts have presented approaches for formally classifying GIS functions (Albrecht 1998; Lemmens 2006; Brauner 2015).

A strategy for integrating bottom-up and top-down approaches to research on GI usage is still lacking (Scheider et al. 2017), and a critical exploration of GIScience online resources is overdue. Hence, in this study, we deploy a Web science approach to inspect what online information about GIScience and GI tools is consumed. What follows is a first mapping of GIScience online, based on behavioural data from a number of complementary sources, assessing their usefulness and reliability.

3 GIS Software Tools on Google Trends

Google Trends[2] offers aggregate search statistics generated in the Google ecosystem, and is a valuable source for studying the behaviour of users on the Web, for example to predict economic patterns (Choi and Varian 2012), analyse consumer behaviour (Goel et al. 2010), and explore cultural changes (Stephens-Davidowitz 2013). The service provides relative search frequencies for arbitrary terms at a weekly resolution since 2004. Results are aggregated per country, and are given as an index where 100 denotes the highest frequency measured for the given terms over time. A maximum of five terms can be compared against one another.[3] Since the volume is given only as a relative index from 100, and Google rounds off volumes that are below a certain resolution threshold, term frequencies can easily drop to zero. For this reason, the selection of comparable terms is essential for this method to provide interpretable results.

Since GIS users commonly rely on the Web as an information resource to find out about software, tools, methods and their intended usage based on the Google

[2]https://trends.google.com.

[3]https://medium.com/@pewresearch/using-google-trends-data-for-research-here-are-6-questions-to-ask-a7097f5fb526.

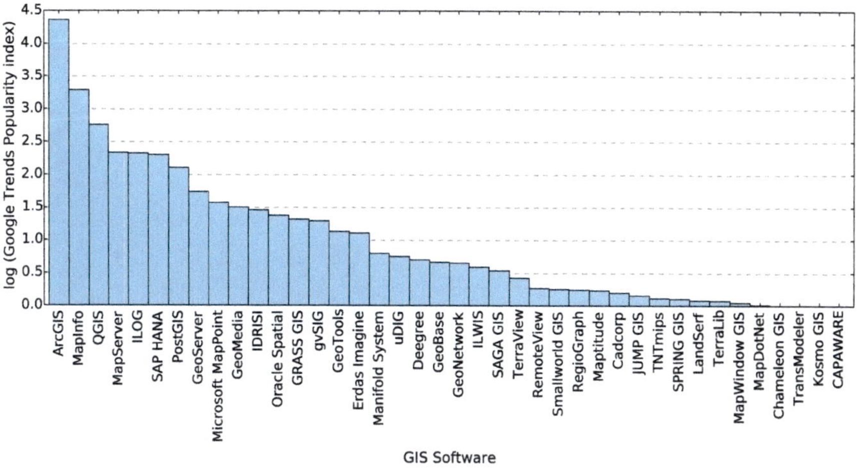

Fig. 1 Relative popularity of GIS software names on Google Trends, compared with the search term "ArcGIS" and averaged over the entire Google Trend history (2004–2017). On the y axis, we show the natural logarithm of the popularity index compared among five terms including "ArcGIS" during the entire period. Data collected in November 2017

search engine, relative volume of searches for GIScience-related keywords and topics provide an indicator for the prominence of topics and tools. In this section, we focus on searches for GIS tools, selecting their official names as keywords for software packages and tools which we gathered from the Web as described in Sect. 7. To collect Google Trends data, we devised a method that selects four keywords at a time against a reference keyword with comparably high volume, averaging relative trends over the entire recording period (from 2004 to 2017). This way, it becomes possible to compare a larger set of keywords, circumventing the problem that search volumes are not provided as absolute numbers. To ensure the interpretability of the results, we only search for individual keywords, and not for topics, i.e., aggregates of keywords identified by Google.

Figure 1 displays an averaged relative search volume index over all GIS software tools, measured against the reference term "ArcGIS", since this term was used most often. We used a logarithmic scale because search volume differs a lot between terms. Note that we had to exclude the term "AutoCAD", because its search volume is several magnitudes higher than that of any GIS tool, making the comparison difficult. Furthermore, in the case of polysemic names that coincide with frequent search terms like "Grass", we added the string "GIS" to restrict the search to the desired semantic field. Results appear meaningful, suggesting that ArcGIS is most the popular GIS tool, followed by MapInfo, and QGIS. PostGIS, Intergraph's GeoMedia, and GeoServer have a considerably lower but still comparable search volume, while tools like the deegree (sic) map server and ILWIS obtain much lower online attention. Similarly, exactly 4 tools are in fact too infrequently searched to be comparable with the reference term.

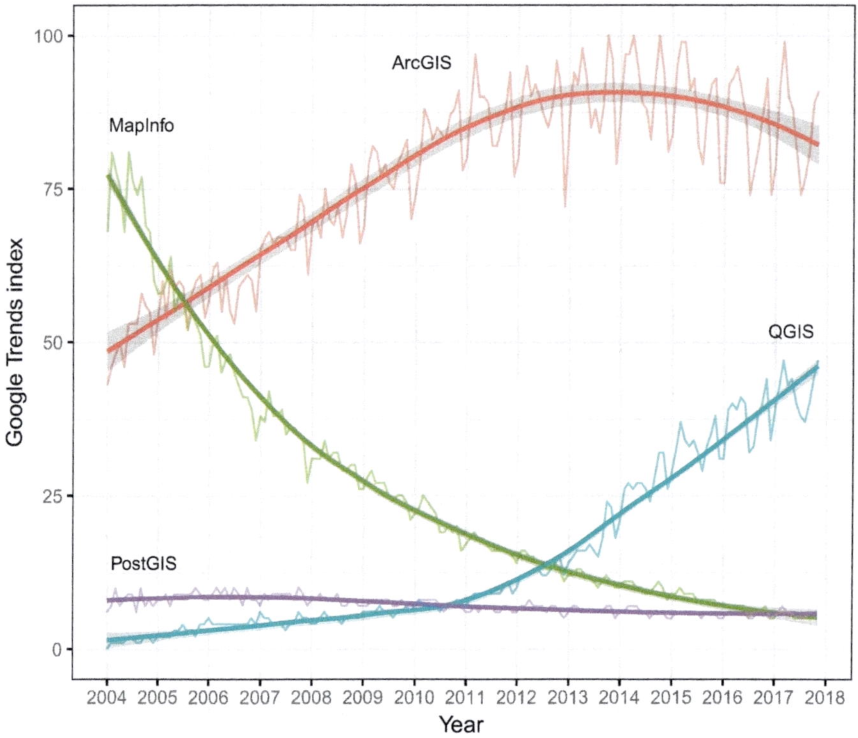

Fig. 2 Relative popularity of selected GIS software product names over the entire Google Trends history (2004–2017). The trend lines are produced with a LOESS regression. Data collected in November 2017

As illustrated in Fig. 2, the temporal trends for these tool names clearly show that, while QGIS started to grow rapidly in 2011, searches for MapInfo have been continuously decreasing since 2004. More surprisingly, interest in ArcGIS enjoyed robust growth until 2014, and then levelled off and started to decline.

While trends for software products yield meaningful results, this is unfortunately not the case for the GIS tools. We carried out the same kind of comparison on all ArcGIS tools contained in the popular toolboxes "Spatial Analyst", "Conversion Tools", and "Analysis Tools", compared with the reference term "ArcGIS". On the surface, it seems that some tool names are very frequently searched. At closer inspection, however, these tool names are highly polysemic. The most searched toolnames are "Aggregate", "Corridor", "Watershed" and "Visibility" with an index greater than or equal to 50. However, it is apparent that these terms have meanings beyond GIScience, and therefore the results bear large amounts of noise. Searches for "Table To Excel" might be popular for reasons entirely unrelated to GIS, and therefore cannot say anything about the usage of the ArcGIS tool of this same name. Adding software names to these tool names (e.g., "ArcGIS Aggregate") delimits the semantic field correctly, but the low volume of searches makes all scores fall to zero. A similar problem arises when searching for more general GIScience topics, such as the term "Kriging".

In summary, Google Trends analysis works fairly well with unambiguous, distinctive, and relatively popular terms (e.g., "ArcGIS", "QGIS", "Kriging"), but is utterly unusable for more polysemic terms used in many semantic contexts, such as "join", "buffer", and "interpolation". Apart from mainstream tools, other searches appear to be too infrequent to identify discernible signals.

4 GIScience Top Websites

As the Web is a prominent locus of information production and consumption, in this section we investigate which websites offer GIScience-related information and quantitatively observe their popularity. In this analysis, the data is collected from two sources: Alexa Internet is a US-based online marketing company that collects detailed statistics on online resources.[4] SimilarWeb is a London-based company that offers analogous web analytics resources.[5] These companies gather a variety of indicators of online behaviour to estimate the traffic to websites along different facets, including spatial, temporal, and demographic variables. Taking website *wikipedia.org* as an example, Alexa Internet states that it is the fifth most visited website worldwide.[6] Along the same lines, SimilarWeb estimates that it is the eleventh most visited website, with about 6.6B visits per month.[7] In most instances, Alexa Internet produces rankings that are significantly higher than those by SimilarWeb. This data can be used to quantify the engagement of audiences with websites, and observe trends in web-based consumer behaviour.

To draw a picture of GIScience content online, we selected a pool of websites based on the tools discussed in Sect. 3, starting from a Wikipedia-based list of GIS tools. To broaden the scope beyond tools, we included a range of specialist magazines (GIS Geography and GIM International), and a set of notable organisations that produce online content related to GIScience (e.g., the Open Geospatial Consortium and the Ordnance Survey). All these websites contain GIScience-related content, including product descriptions, software documentation, tutorials, examples, and discussions. While this pool cannot be exhaustive in its current form, we believe it captures a significant portion of top online content that most GIS practitioners and students consult.

From a methodological perspective, the data provided by Alexa Internet and SimilarWeb present limitations. The websites operate as black boxes, and it is hard to ascertain the accuracy of the estimates. Moreover, the data does not provide

[4]https://www.alexa.com.

[5]https://www.similarweb.com.

[6]https://www.alexa.com/siteinfo/wikipedia.org.

[7]https://www.similarweb.com/website/wikipedia.org.

statistics about subsets of websites, limiting the analysis to websites that are thematically focussed. For example, it is possible to obtain data about *stackexchange.com*, but not about *gis.stackexchange.com*, which would be more relevant to this study. Similarly, several software tools do not have dedicated domains, but are hosted at large repositories. For example, the software package PySAL is hosted on *readthedocs.io* (*pysal.readthedocs.io*), and it is hard to obtain traffic statistics. For this reason, many potentially relevant sub-domains had to be excluded. That said, we consider this data to be sufficient as an indication of the magnitude of online popularity of these resources.

We collected engagement information for a pool of 55 GIScience-related websites, of which 18 were discarded for lack of data. Table 1 summarises the results of this analysis: For each website, the table indicates the average worldwide rank calculated from the ranks from the two sources, thus reducing bias. In the interest of brevity, the ranks and visit counts were heavily rounded to the thousands or millions. To the best of our knowledge, other websites that we initially considered ranked more than six millionth on either platform, without enough data to produce estimates, and were therefore removed. The table also includes the SimilarWeb estimate of monthly visits, and not unique visitors, i.e. the same web user can generate more than one visit.

The top countries indicated by Alexa Internet are selected based on the absolute volume of visits, hence countries with large populations tend to dominate. The US, China, and India are top countries for most websites, with some notable exceptions, e.g., Italy, Algeria, and other countries for specific websites. A set of important, but non-thematically specific, websites about technologies like Oracle, Python, and R is included at the end of table, also providing a reference point for the GIScience websites. The pool of websites is available on the GitHub repository, and can be re-used for similar analyses.

Unsurprisingly, the websites of ArcGIS and ESRI emerge as the most popular in the pool, with about 19M monthly visits, and ranking between 5,000th and 19,000th in the world. Another traditional GIS, MapInfo by Pitney Bowes, also maintains a popular position, but it is hard to estimate visits specific to the tool, and not to other branches of the company. More interestingly, emergent competitors to ESRI are visible, including aggressive Web start-ups Mapbox and Carto, which attract respectively 2.9M and 724,000 monthly visits. Free and open source GI tools (Steiniger and Hunter 2013) reach high visibility, spearheaded by desktop-based QGIS (1.4M monthly visits). Web mapping JavaScript libraries Leaflet and OpenLayers have become extremely popular since the late 2000s. Mature tools, such as GDAL, GeoTools, PostGIS, MapServer, and GeoServer, obtain between 230,000 and 50,000 monthly visits, suggesting persistent engagement by their communities of users. The other websites obtained lower ranks and visits, and are therefore not discussed in detail.

Table 1 Popular GIScience websites, according to Alexa Internet and SimilarWeb data, as of 15 November 2017. The ranks are a measure of global popularity of the websites. The visits are a monthly estimate from SimilarWeb for October 2017

Product/Organisation	Website	Average rank	*Alexa rank*	*SimWeb rank*	Monthly visits	Alexa top countries
ArcGIS	arcgis.com	5K	3K	6K	14.2M	US, Canada
ESRI	esri.com	13K	8K	19K	4.9M	US, China
Mapbox	mapbox.com	16K	16K	16K	2.9M	US, China
Ordnance Survey	ordnancesurvey.co.uk	44K	55K	34K	1.3M	UK, US
QGIS	qgis.org	45K	32K	58K	1.4M	US, China
Leaflet	leafletjs.com	57K	50K	65K	1.5M	US, China
Carto	carto.com	60K	42K	77K	724K	US, Spain
OS GEO, GRASS GIS	osgeo.org	99K	68K	131K	640K	US, China
OpenLayers	openlayers.org	110K	76K	141K	371K	China
GIS Geography	gisgeography.com	162K	85K	240K	372K	US, India
Intergraph	intergraph.com	203K	136K	270K	164K	US, India
PostGIS	postgis.net	214K	147K	281K	231K	US, Belgium
GDAL	gdal.org	237K	135K	340K	180K	China, Japan
GeoServer	geoserver.org	270K	172K	367K	130K	Brazil, US
Erdas Imagine	hexagongeospatial.com	286K	203K	370K	175K	US, South Africa
OGC	opengeospatial.org	392K	268K	516K	132K	US, Spain
MapServer	mapserver.org	516K	353K	679K	79K	Italy, India
GIM International	gim-international.com	596K	329K	864K	67K	India, US
Directions Magazine	directionsmag.com	608K	268K	947K	57K	US, India

(continued)

Table 1 (continued)

Product/Organisation	Website	Average rank	*Alexa rank*	*SimWeb rank*	Monthly visits	Alexa top countries
GeoTools	geotools.org	658K	499K	817K	57K	Algeria, Russia
Geography UK	geography.org.uk	806K	546K	1.1M	52K	UK
uDIG	udig.refractions.net	960K	960K	–	–	US, India
TurfJS	turfjs.org	992K	616K	1.3M	–	US, India
gvSIG	gvsig.com	1M	635K	1.4M	–	Spain, Russia
Spatial.ly	spatial.ly	1.1M	1M	1.3M	–	–
GeoNetwork	geonetwork-opensource.org	1.2M	943K	1.4M	–	Germany
TerrSet, formerly IDRISI	clarklabs.org	1.5M	520K	2.5M	–	India
52 North	52north.org	1.7M	750K	2.6M	–	–
Cadcorp	cadcorp.com	2.2M	1.2M	3.2M	–	India
Manifold System	manifold.net	2.3M	1.3M	3.3M	–	US
R Spatial	r-spatial.org	3.5M	2.3M	4.7M	–	–
OpenJump	openjump.org	3.6M	2M	5.2M	–	–
Deegree	deegree.org	4.1M	2.8M	5.3M	–	–
Oracle[a]	oracle.com	685	369	1K	63.4M	US, India
Python[a]	python.org	2K	879	3K	35.1M	US, China
R[a]	r-project.org	11K	7K	15K	7.2M	US, China
Pitney Bowes, MapInfo[a]	pitneybowes.com	43K	28K	59K	1.9M	US, UK

[a]These entries include non-spatial content, and therefore tend to rank higher. For example, *oracle.com* covers all Oracle products, not just Oracle Spatial and Graph

5 GIScience Content in Wikipedia

In our mapping of GIScience Web resources, we dedicate particular attention to Wikipedia, which represents without doubt a prominent entry point to much Web content. Wikipedia articles are highly heterogeneous, and cover from very general (e.g., geography) to very specific technical topics, such as Moran's *I*. For this analysis, we selected a pool of Wikipedia articles in English that are related to GIScience. As GIScience is by its nature a multi-disciplinary, porous domain, we selected a very broad range of topics by crawling the website from a set of highly central seed pages,[8] and collecting the links to other articles for two edges in the network. This procedure generated a list of 1,073 articles, which we manually scanned and classified as either GIScience-related or not. For example, we included *location intelligence* and *contour line*, while *personal computer* and *animal cognition* were discarded as only marginally relevant to this analysis. When in doubt, we included the article, recognising the degree of subjectivity in this classification.

As a result of this process, we obtained a list of 349 relevant pages. In this analysis we focus exclusively on page views, and not on other indicators, such as number of edits and article length. Because of its constrained structure, Wikipedia articles are thematically delimited, and page views provide an indicator of interest in a given topic. However, the data has indeed known limitations that should not be ignored. The page view counts are sensitive to current events that can generate short-lived bursts of views, as well as to polysemy, when pages with unrelated topics with some of the same keywords are opened by mistake. Links on the main page of Wikipedia can also boost views without other explanatory factors.[9] In sum, we consider these problems acceptable in our set of GIScience-related articles, which—alas—do not seem to obtain mainstream visibility on the Web. For each page, we retrieved usage statistics from the Wikimedia API, focussing on monthly views from October 2016 to October 2017.[10] The average monthly views were then calculated as a proxy of interest in the article topics.

In the set of the 349 pages, the number of monthly views ranges from 15 to about 117,000, with a median of 1,055. To provide context to this data, the most popular 50 pages in Wikipedia currently obtain between 6.9M and 1M monthly views.[11] As expected in hypertext-based data, the distribution is heavily skewed towards a small set of pages that attract most of the views, with a tail of low-traffic pages. The top 10% of the pages generate about 65% of the total views in the set. Table 2 shows a summary of this analysis, ordering the Wikipedia articles by monthly views. Based

[8]Seed pages include *Geographic information science*, *Category: Geographic information systems*, *List of geographic information systems software*, and *Geoinformatics*.

[9]https://en.wikipedia.org/wiki/Wikipedia:Pageview_statistics#Accuracy_of_the_tools.

[10]https://wikitech.wikimedia.org/wiki/Analytics/AQS/Pageviews.

[11]https://tools.wmflabs.org/topviews/.

on Jenks natural breaks, we grouped the pages into five classes, ranging from very high volume of views to very low. Some articles in the last group were omitted for the sake of brevity. The complete table can be found in the GitHub repository.

The most visited articles, having more than 42,000 views per month, include geographic coordinate systems, GPS, GIS, latitude, and longitude. The difference in interest between GIS and GIScience is staggering, with respectively 71,000 and 2,200 views, suggesting that, while GI *systems* keep attracting a broad audience, GI *science* remains a small academic discipline, particularly when compared with its cognate disciplines of geography (69,000 views) and data science (52,000 views). Similarly, crowdsourcing remains a highly consulted article (30,000 views), while the more specific volunteered geographic information (VGI) is a niche topic, with only 1,000 monthly views.

Unlike the GIScience websites covered in Sect. 4, the articles in this analysis show how Wikipedia tend to have good coverage of topics at a high level of abstraction (e.g., thematic map) and software packages (e.g., QGIS), but minimal inclusion of GI methods, such as a buffer and weighted overlay. This helps explaining why the ESRI and ArcGIS websites still take the lion's share of GIScience online traffic. We hope that the data reported in this analysis can help GIScience practitioners and students guide efforts to make the discipline more visible online, increasing the coverage, connectedness, and quality of GIScience-related articles.

6 The Structure of the ArcGIS Documentation

The online documentation of ArcGIS is the most visited GIS-related website (see Sect. 4), and therefore offers the opportunity of studying a software tool in detail. First, we scraped the website *arcgis.com*, collecting 928 documentation pages about the popular software suite. These pages include tool documentation, tutorials, and various forms of technical explanations, mixing applied and scientific content. As visible in the ArcGIS graphical interface, the tools are grouped in arbitrary toolboxes, such as the Spatial Analyst. The documentation describes different versions of the tools, and therefore, to avoid duplication, we restricted the analysis to a popular major version (10.x), for a total of 285 pages about tools. For example, the popular buffer tool is documented in a Web page.[12]

To observe the semantic associations between the tools, we run a network analysis on the tool-related pages, aiming at identifying which tools tend to be used together. A manual inspection of the links shows a rather sparse network, without clearly interpretable, non-trivial patterns. Hence, we perform a *graph selection* (Stell and Worboys 1999) which connects pages through at most one intermediate page. That is,

[12]http://www.desktop.arcgis.com/en/arcmap/10.3/tools/analysis-toolbox/buffer.htm.

Table 2 Wikipedia articles about GIScience-related topics, grouped by number of monthly views. In each group, the articles are sorted in descending order by monthly visits. Different colours are used to denote a concept, a software, a tool and an organisation. Please note that some articles are referred to with more than one title, obtaining different views (e.g., *Global Positioning System* and *GPS*). The prefix for the pages is https://en.wikipedia.org/wiki/

Monthly views (thousands)	Wikipedia articles [size of group]
Very high [42, 120]	Geographic coordinate system, Global Positioning System, **R (programming language),** Geographic information system, Geography, Latitude, Map, Cluster analysis, Data science, Longitude, Topology [11]
High [20, 42)	Surveying, Census, Map projection, Remote sensing, Crowdsourcing, Cartography, **SAP HANA,** Tessellation, Universal Transverse Mercator coordinate system, Ontology (information science), Raster graphics, Human geography, Data visualization, Contour line, **OpenStreetMap,** Data model [16]
Medium [6, 20)	Satellite navigation, **Garmin,** Geotechnical engineering, National Geospatial- Intelligence Agency, Aerial photography, Geomorphology, Geotagging, Satellite imagery, Geodesy, **ArcGIS,** Heat map, Scale (map), Spatial analysis, GPS, Geoid, Geophysics, Kriging, Digital elevation model, **TomTom,** Geodetic datum, R-tree, Quadtree, Political geography, List of geographic information systems software, Choropleth map, Geolocation, Location-based service, **Esri, Ordnance Survey, Public Land Survey System,** History of geography, Well-known text, Thematic map, **Bing Maps, QGIS,** Cadastre, **Geohash,** Citizen science, Gazetteer, **Wikimapia,** Geomatics, Spatial database, GeoJSON, Web Mercator, Cultural geography, **Landsat program,** Geospatial analysis [47]
Low [2, 6)	Outline of geography, Geospatial intelligence, Geoinformatics, GIS file formats, Spatial reference system, Lambert conformal conic projection, Geographical distance, **Google Sky,** Inverse distance weighting, Moran's I, ISO 19115, Maps, Maidenhead Locator System, LIDAR, Geography Markup Language, ISO 10005, **Ingres (database),** Development geography, Geostatistics, **Google Moon,** Georeferencing, List of GIS data sources, **What3words,** Geolocation software, Scientific visualization, GIS, SVG, GeoTIFF, Regional geography, Population geography, Jenks natural breaks optimization, **MapInfo Professional,** Virtual globe, Crime mapping, Image rectification, Triangulated irregular network, WGS84, Web Feature Service, **USGS,** List of programs for point cloud processing, **PostGIS,** Datum (geodesy), Big Data, Philosophy of geography, **CartoDB, Erdas Imagine, ArcMap, GDAL, GRASS GIS,** Meridian arc, Geographic information science, Global Map, Geodynamics, Cartographer, Behavioral geography, Orthogonal projection, **GeoServer** [57]

(continued)

Table 2 (continued)

Monthly views (thousands)	Wikipedia articles [size of group]
Very low [.01, 2)	Health geography, DE-9IM, Global navigation satellite system, Geodemographic segmentation, WikiMapia, Minimum bounding rectangle, Geographic profiling, Geovisualization, Modifiable areal unit problem, Urban informatics, ArcGIS Server, Spatial index, Web Coverage Service, Data model (GIS), GPS receiver, Cartographic generalization, British national grid reference system, Geomarketing, Spatiotemporal database, Simple Features, Location intelligence, Grid (spatial index), Environmental geography, Vector Map, Polygons, Treemap, Satellite geodesy, MrSID, Land administration, ArcInfo, Georeference, Geoportal, SpatiaLite, Volunteered geographic information, Spatial query, USGS DEM, Data Mining, Geocode, Vector tiles, CityEngine, Counter-mapping, NAD83, Indicators of spatial association, Buffer (GIS), Mapnik, Oracle Spatial and Graph, GeoMedia, Geographic information systems, MapInfo Corporation, GIS and public health, Viewshed, Digital Earth, GvSIG, GeoSPARQL, SAGA GIS, Cartographic relief depiction, ... [218]

paths between pages $p_0 \rightarrow p_1 \rightarrow p_2$ correspond to a second-order edge $p_0 \rightarrow p_2$ in the resulting graph. The second-order edges between ArcGIS tool Web pages are summarised in Fig. 3. Meaningful patterns start emerging when the second-order graph is further cleaned from obvious hubs, such as toolbox and tutorial pages that are highly inter-linked. A link from one tool to another tool means here that there is either a direct Web link between corresponding tool Web pages, or over one intermediate page, where the latter can also be a non-tool page (e.g., a page describing general principles of the software). Node and label sizes are scaled relative to the node degree in the network.

It is possible to see in this network that there are several tools acting as central nodes. The node with the highest degree is *Reclassify* from the Spatial Analyst toolbox (SAT), with an in-degree of 142, followed by *Save-to-layer-file* (18), *Make-feature-layer* (17), *Copy-features* (15) from the Data Management toolbox (DMT) (see Table 3). The node centrality and connectivity pattern reveals an insight: In raster analysis, the *Reclassify* tool is actually a central means to transform a raster layer based on its cell values. It therefore acts as an interface between all kinds of raster tools, such as map algebra operations. This tool has other tools pointing to it, but does not point itself to other pages (see out-degree in Table 3).

Furthermore, layer operations from the Data Management toolbox are central for all kinds of GIS workflows to deal with layers as inputs and outputs. Lastly, one can see a meaningful cluster containing the spatial analyst tools *Kriging*, *Trend*, *Spline*,

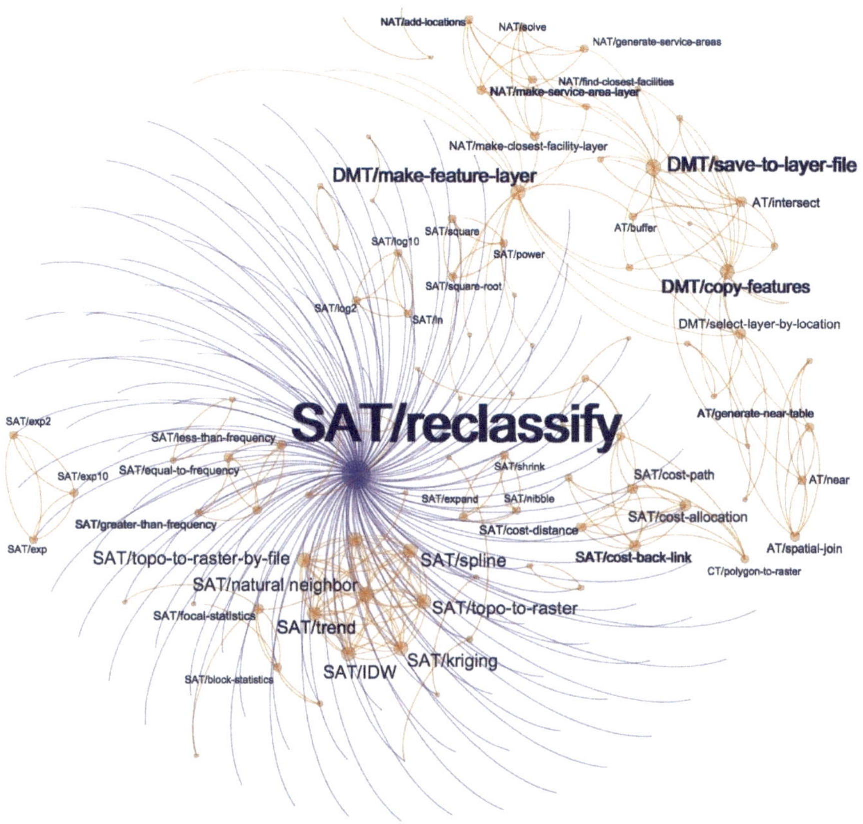

Fig. 3 Second-order links between ArcGIS tool Web pages, showing their degree in terms of node size and colour from orange (low) to blue (high), taken from the toolboxes Spatial Analyst (SAT), Data Management (DMT), Network Analyst (NAT), Analysis (AT), Conversion (CT), Geocoding (GT). The network layout was obtained with the Fruchterman-Reingold algorithm

IDW and *Topo-to-raster* and *Natural Neighbor*. These are tools that can be used to interpolate surfaces in Digital Elevation Models (DEM). Even smaller subclusters are nicely interpretable, such as the cluster of *Cost-Distance*, *Cost-Back-Link*, *Cost-Allocation*, which together form a set of highly interdependent tools for least cost path analysis on cost surface raster layers. Note also that clusters partially overlap with and link different toolboxes. This method can be used to analyse connections between tools, making implicit knowledge emerge from the website network.

Table 3 Node degrees in the second-order graph of ArcGIS tool Web pages

Tool	Degree	Out-degree
SAT/reclassify	142	0
DMT/save-to-layer-file	18	2
DMT/make-feature-layer	17	2
DMT/copy-features	15	0
SAT/idw	15	8
SAT/spline	15	8
SAT/topo-to-raster	15	8
SAT/spline-with-barriers	15	8
SAT/trend	15	8
SAT/topo-to-raster-by-file	15	8
SAT/kriging	15	8
SAT/natural-neighbor	15	8
SAT/cost-allocation	9	6
SAT/cost-back-link	9	6

7 Linked Inventory of GIS Tools

To systematize studies of these Web resources and to share our results about usage patterns of GIS software, tools and concepts, we suggest a way to unambiguously describe and identify the involved resources with linked data. For this purpose, we designed a comprehensive linked inventory that describes GIS tools and their implementations across different packages (e.g., ArcGIS, GRASS, and R).[13] This dataset was used as a basis for all Web analyses performed in previous chapters, and contains resources derived as a result of this study. To generate the inventory, an initial set of GIS software packages was identified from Wikipedia articles,[14] and then enriched with links from DBPedia.[15] For example, in Listing 1, ArcGIS is described with standard RDF vocabularies.

```
dbp:ArcGIS a dbo:Software;
  dbo:developer dbp:Esri;
  foaf:homepage <http://www.esri.com/software/arcgis>;
  foaf:isPrimaryTopicOf <https://en.wikipedia.org/wiki/ArcGIS>;
  foaf:name "ArcGIS".
```

Listing 1 Describing GIS software products using linked data

To obtain information about the tools contained in each software, we additionally scraped manuals on the Web, for example that of GRASS GIS.[16] When pos-

[13]http://geographicknowledge.de/vocab/GISTools.ttl, [.rdf]

[14]https://en.wikipedia.org/wiki/Comparison_of_geographic_information_systems_software.

[15]See for example http://dbpedia.org/page/ArcGIS.

[16]https://grass.osgeo.org/grass72/manuals/keywords.html.

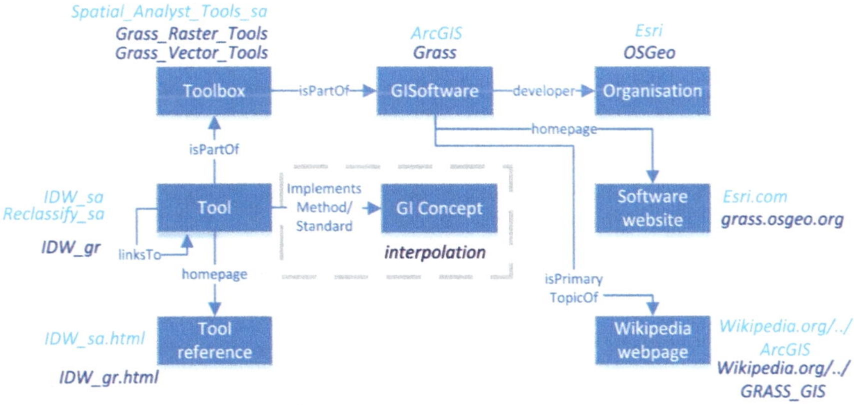

Fig. 4 Overview of the linked data inventory. The boxes represent the tool vocabulary, while examples of GIS tools are in italic. The dashed area represents future work

sible we used scripts within a given package to generate tool inventories, linking them to a preliminary subset of software packages based on our Web study, e.g., *arcpy* in ArcGIS. For example, Listing 2 shows how we used `dct:isPartOf` from Dublin Core terms to nest tools within toolboxes and packages such as ArcGIS. Finally, we enriched this dataset with tool network information scraped from web texts and their hyperlinks (see Sect. 6). This enabled us to link tools to webpages (using `foaf:homepage`) and to encode their network structure (with the SIOC term `sioc:links_to`) into linked data. Figure 4 shows a schematic representation of the linked data inventory as also described in Listing 1 and 2. The linked data approach facilitates the interconnection of tools and their descriptions and can form the basis for further connections with GI concept definitions in text books, tutorials, curricula, etc.

```
@prefix tools:<http://geographicknowledge.de/vocab/GISTools.rdf#>.
@prefix sioc:<http://rdfs.org/sioc/ns#>.
@prefix dct:<http://purl.org/dc/terms/>.
@prefix foaf:<http://xmlns.com/foaf/0.1/>.

tools:Spatial_Analyst_Tools_sa a gis:Toolbox;
  dct:isPartOf dbp:ArcGIS;
  rdfs:label "Spatial Analyst Tools(sa)".
tools:IDW_sa a gis:Tool;
  dct:isPartOf tools:Spatial_Analyst_Tools_sa;
  foaf:homepage <http://desktop.arcgis.com/.../idw.htm>;
  sioc:links_to tools:Reclassify_sa.
```

Listing 2 Capturing GIS tools, toolboxes, websites and Web links as linked data

Once extended beyond this proof-of-concept, we hope that this resource will support education and research purposes, becoming a basis for further research on GIS tools usage patterns.

8 Conclusions

In this article, we explored the Web science approach to gather new knowledge about the consumption of online information about GIS tools, software, and concepts. As part of our efforts to improve the conceptual organisation of GIS, we critically examined Google Trends data about the popularity of tools, and the top websites that host GIScience content, based on publicly available Web-analytics data. Subsequently, we studied two notable websites, including behavioural data about Wikipedia articles and the network structure of the ArcGIS online documentation. Based on this study, we designed the structure of a linked-data inventory, which connects these Web resources across GIS software, tools, and concepts, and presented examples of its use.

In sum, the Web scientific approach allowed us to discover patterns buried in behavioural and structural aspects of websites, producing some interesting findings. Google Trends allows granular tracking of software popularity, confirming the dominance of ESRI products, but also the emergence of new tools and companies. Alexa Internet and SimilarWeb enable the estimation of visits to GIS websites. In Wikipedia, we can observe the popularity over time of a plethora of topics, ranging from software to scientific concepts and methods. Our analysis also suggests much higher popularity for term "GIS" as opposed to "GIScience", potentially directing efforts to better represent the discipline online. Finally, the network analysis of online documentation allowed us to capture meaningful functional relationships between tools that are not immediately apparent, and which may be used as a basis to recommend tools.

However, this study also highlighted several limitations of Web science. Noise caused by semantic ambiguity of keywords limits the interpretability of some analyses, particularly in the case of Google Trends. Moreover, this approach focused on large-scale online information consumption, which is at best a proxy to user behaviours, such as GIS usage and adoption. The latter can only be measured in a direct way based on traditional research methods, such as local log files, surveys and interviews, which are restricted to a small scale. In this sense, access to corporate data would be immensely beneficial to understand tool usage (but unlikely to happen). Finally, we realized that the Web science approach is heavily dependent on what software organisations and the majority of users deem relevant, and this may just not what an analyst needs in a particular situation.

For future research, we envisage several worthwhile directions. It is paramount to produce more structured information about the relevance of GIS tools, methods, and concepts, boosting the precision and recall of user searches (Ballatore et al. 2016), instead of relying on unstructured data such as texts. For this reason, the inventory we outlined in this article should be incrementally extended to reach broader coverage of existing tools, embedding them into a coherent conceptual framework. Furthermore, to support data scientists and students, we must increase the semantic depth of our inventory, capturing the functionality of tools and related concepts (Scheider and Ballatore 2018), which is only partially possible with the Web scientific method.

This would result in a better linkage between methods (e.g., buffer and interpolation) and their software implementations, for example in R and ArcGIS.

Finally, in order to map GIS software, tools, and related websites, more comprehensive analyses are needed, increasing the completeness of our mapping with input from the GIScience community. For this purpose, crowdsourcing would facilitate information gathering and error-correction, supporting the iterative revision of our assumptions. A near-complete, maintainable set of tools, software, and websites will allow researchers and practitioners to find suitable resources, monitoring the evolution of this broad technical landscape.

References

Albrecht J (1998) Universal analytical GIS operations: a task-oriented systematization of data structure-independent GIS functionality. In: Onsrud H, Craglia M (eds) Geographic information research: transatlantic perspectives. Taylor and Francis, New York, pp 577–591

Ballatore A, Kuhn W, Hegarty M, Parsons E (2016) Spatial approaches to information search. Spat Cogn Comput 16(4):245–254

Bernard L, Mäs S, Müller M, Henzen C, Brauner J (2014) Scientific geodata infrastructures: challenges, approaches and directions. Int J Digit Earth 7(7):613–633

Blei DM (2012) Probabilistic topic models. Commun ACM 55(4):77–84

Boyd D, Crawford K (2011) Six provocations for Big Data. In: Decade in internet time: symposium on the dynamics of the internet and society. Oxford Internet Institute, Oxford, UK

Brauner J (2015) Formalizations for geooperators–geoprocessing in spatial data infrastructures. PhD thesis, TU Dresden, Dresden, Germany

Caliskan A, Bryson JJ, Narayanan A (2017) Semantics derived automatically from language corpora contain human-like biases. Science 356(6334):183–186

Castellano G, Fanelli A, Torsello M (2013) Web usage mining: discovering usage patterns for web applications. In: Velsquez J, Palade V, Jain L (eds) Advanced techniques in web intelligence—2. Studies in computational intelligence, vol 452. Springer, Berlin, pp 75–104

Choi H, Varian H (2012) Predicting the present with Google Trends. Econ Rec 88(1):2–9

Ferrara E, De Meo P, Fiumara G, Baumgartner R (2014) Web data extraction, applications and techniques: a survey. Knowl-Based Syst 70:301–323

Gao S, Goodchild MF (2013) Asking spatial questions to identify GIS functionality. In: Fourth international conference on computing for geospatial research and application (COM. Geo). IEEE, pp 106–110

Goel S, Hofman JM, Lahaie S, Pennock DM, Watts DJ (2010) Predicting consumer behavior with web search. Proc Natl Acad Sci USA 107(41):17486–17490

Hecht B, Hong L, Suh B, Chi EH (2011) Tweets from Justin Bieber's Heart: the dynamics of the location field in user profiles. In: Proceedings of the SIGCHI conference on human factors in computing systems, CHI '11, New York. ACM, pp 237–246

Hendler J, Shadbolt N, Hall W, Berners-Lee T, Weitzner D (2008) Web science: an interdisciplinary approach to understanding the web. Commun ACM 51(7):60–69

Hey T, Tansley S, Tolle KM et al (2009) The fourth paradigm: data-intensive scientific discovery, vol 1. Microsoft Research, Redmond, WA

Hinsen K (2014) Computational science: shifting the focus from tools to models. F1000Research 3(101)

Hofer B, Mäs S, Brauner J, Bernard L (2017) Towards a knowledge base to support geoprocessing workflow development. Int J Geogr Inf Sci 31(4):694–716

Hollenstein L, Purves R (2010) Exploring place through user-generated content: using Flickr tags to describe city cores. J Spat Inf Sci 1:21–48
Hu Y, Janowicz K, Prasad S, Gao S (2015) Enabling semantic search and knowledge discovery for ArcGIS Online: a linked-data-driven approach. AGILE 2015. Springer, Berlin, pp 107–124
Janowicz K, Van Harmelen F, Hendler JA, Hitzler P (2014) Why the data train needs semantic rails. AI Mag 36(1):
Kuhn W (2012) Core concepts of spatial information for transdisciplinary research. Int J Geogr Inf Sci 26(12):2267–2276
Kuhn W, Ballatore A (2015) Designing a language for spatial computing. In: Bacao F, Santos MY, Painho M (eds) AGILE 2015: geographic information science as an enabler of smarter cities and communities. Springer, Berlin, pp 309–326
Kveladze I, Kraak M-J, van Elzakker CP (2013) A methodological framework for researching the usability of the space-time cube. Cartogr J 50(3):201–210
Lazer D, Kennedy R, King G, Vespignani A (2014) The parable of Google Flu: traps in big data analysis. Science 343(6176):1203–1205
Lemmens R (2006) Semantic interoperability of distributed geo-services. PhD thesis, TU Delft, Delft, The Netherlands
Ludäscher B, Lin K, Bowers S, Jaeger-Frank E, Brodaric B, Baru C (2006) Managing scientific data: from data integration to scientific workflows. Geol Soc Am Spec Papers 397:109–129
McKenzie G, Janowicz K, Gao S, Yang J-A, Hu Y (2015) POI pulse: a multi-granular, semantic signature-based information observatory for the interactive visualization of big geosocial data. Cartogr: Int J Geogr Inf Geovis 50(2):71–85
Ramage D, Hall D, Nallapati R, Manning CD (2009) Labeled LDA: a supervised topic model for credit attribution in multi-labeled corpora. In: Proceedings of the 2009 conference on empirical methods in natural language processing, vol 1. Association for Computational Linguistics, pp 248–256
Scheider S, Ballatore A (2018) Semantic typing of linked geoprocessing workflows. Int J Digit Earth 11(1):113–138
Scheider S, Ostermann FO, Adams B (2017) Why good data analysts need to be critical synthesists. Determining the role of semantics in data analysis. Future Gener Comput Syst 72:11–22
Stefanidis A, Crooks A, Radzikowski J (2013) Harvesting ambient geospatial information from social media feeds. GeoJournal 78(2):319–338
Steiniger S, Hunter AJ (2013) The 2012 free and open source GIS software map—A guide to facilitate research, development, and adoption. Computers, Environment and Urban Systems 39:136–150
Stell JG, Worboys MF (1999) Generalizing graphs using amalgamation and selection. In: Güting RH, Papadias D, Lochovsky F (eds) Advances in spatial databases: 6th international symposium, SSD'99 Hong Kong, China, July 20–23, 1999 proceedings. Springer, Berlin, pp 19–32
Stephens-Davidowitz SI (2013) Essays using Google Data. PhD thesis, Harvard University, Cambridge, MA

Charting the Geographies of Crowdsourced Information in Greater London

Andrea Ballatore and Stefano De Sabbata

Abstract Crowdsourcing platforms and social media produce distinctive geographies of informational content. The production process is enabled and influenced by a variety of socio-economic and demographic factors, shaping the place representation, i.e., the amount and type of information available in an area. In this study, we explore and explain the geographies of Twitter and Wikipedia in Greater London, highlighting the relationships between the crowdsourced data and the local geodemographic characteristics of the areas where they are located. Through a set of robust regression models on a sample of 1.6M tweets and about 22,000 Wikipedia articles, we identify level of education, presence of people aged 30–44, and property prices as the most important explanatory factors for place representation at the urban scale. To some extent, this confirms the received knowledge of such data being created primarily by relatively wealthy, young, and educated users. However, about half of the variability is left unexplained, suggesting that a broader inclusion of potential factors is necessary.

Keywords Information geography · Crowdsourcing · Volunteered geographic information · Geo-demographics · Twitter · Wikipedia

1 Introduction

Over the past decade, the diffusion of crowdsourcing platforms and GPS-enabled smartphones has enabled the large-scale production of spatial information. This phenomenon has been variously characterised as spatial crowdsourcing, volunteered geographic information (VGI), spatial social media, and user-generated content

A. Ballatore (✉)
Department of Geography, Birkbeck, University of London, London, UK
e-mail: a.ballatore@bbk.ac.uk

S. De Sabbata
School of Geography, Geology, and the Environment, University of Leicester, Leicester, UK
e-mail: s.desabbata@le.ac.uk

A. Mansourian et al. (eds.), *Geospatial Technologies for All*,
Lecture Notes in Geoinformation and Cartography,
https://doi.org/10.1007/978-3-319-78208-9_8

(UGC) (Sui et al. 2012; See et al. 2016). Among others, Wikipedia articles, OpenStreetMap vector data, and geo-located tweets are popular data sources for countless studies in geography, demography, sociology, and even seismology (e.g., Earle et al. 2010; Zagheni et al. 2014).

Although the potential of these data sources is apparent, much research analysed very few platforms, such as OpenStreetMap (Mashhadi et al. 2015), often paying limited attention to the geo-demographic context in which the data was produced. Studies of information geographies have so far focused on large spatial units, such as countries (Graham et al. 2015a), with few works focusing on the urban or regional scale. The latter is however of particular importance, as VGI tends to be produced in urban areas (Hecht and Stephens 2014). When using VGI, it is necessary to consider the socio-spatio-temporal processes that supported its generation, thinking about what data is missing, and not only about what is visibly present.

As part of our ongoing efforts to chart geographies of digital information (Ballatore et al. 2017; Graham et al. 2015a), this article investigates the spatial structure of two popular VGI sources at the urban scale. In particular, we consider geo-located Twitter posts and Wikipedia articles in Greater London, comparing and contrasting their spatial distribution. After providing descriptive statistics, the data from both sources is then studied in relation to a set of socio-demographic variables that characterise Greater London. Through a number of regression analyses, we explore the factors exhibit a similar presence or absence of information, including day-work population, ethnic composition, education level, and property prices, which might indicate a relationship to underlying geographies.

The term VGI encompasses diverse sources of spatial information that vary dramatically in terms of demographic, thematic, spatial, and temporal coverage. One of the objectives of this study is precisely to highlight the commonalities and differences of two very different data sources observed in the same geographic area. Moreover, this study analyses the *place representation*, i.e., the digital information available to characterise areas in heterogeneous data sources, regardless of the demographic characteristics of producers. For this reason, apart from a handful of prolific bots, we include all data available for each spatial unit of analysis, to see to what extent an area is either data-rich or data-poor.

This study contributes to the knowledge of VGI sources, providing findings about what areas are over- and under-represented in these two sources. These insights are relevant to VGI users, providing evidence about datasets' geographical structure, representativeness, and therefore fitness-for-purpose for studies and applications. Producers can also benefit by updating their platforms for a more equal place representation, along similar lines of studies of gender inequality in Wikipedia, which prompted a number of initiatives to increase participation of women.[1] From a more social-scientific perspective, this study can contribute to the study of digital divides and "informational ghettos" (Shaw and Graham 2017, p. 4), data-poor areas that persist even in wealthy, digitally over-represented global hubs like London. Our initial hypothesis–only partially confirmed–is that content in both Twitter and Wikipedia

[1] https://en.wikipedia.org/wiki/Gender_bias_on_Wikipedia.

tends to be more representative of wealthier urban areas, inhabited by a younger, and more educated, and less ethnically diverse population than average.

The remainder of this paper is organised as follows. Section 2 summarises critically related work in this area. The socio-economic datasets used to characterise the geography of London are described in Sect. 3. Subsequently, Sects. 4 describes our study of Twitter and Wikipedia located in Greater London, starting with a visual analysis and then continuing with a set of regression models. Section 5 summarises our findings. Finally, conclusions and future work directions are drawn in Sect. 6.

2 Related Work

After a decade of research, much is known about VGI sources. Some sources are spatially-explicit, aiming at a spatial coverage, while others are spatially-implicit, embedding locational information as simply one of the attributes being expressed (Antoniou et al. 2010). For example, Wikipedia and GeoNames aim at comprehensive coverage of cities, while geo-located tweets and Instagram photos are the by-product of a mediated communication process between users located in cities. Unsurprisingly, urban areas tend to be better covered than rural ones (Hecht and Stephens 2014), with the exclusion of sources limited to highly specific themes (e.g., hiking).

Different crowdsourcing platforms attract different demographic groups in terms of age, gender, income, education, area of residence, interests, and motivations, shaping the properties of the resulting dataset, each displaying its own idiosyncrasies (Acheson et al. 2017). Despite early claims of radical democratization and inclusion, user communities tend to be skewed towards Western, wealthy, educated, white, and male users (Crampton et al. 2013), although exceptions exist—countering the general trend, Wikimapia enjoys more uptake among Indian and Middle-Eastern users (Bittner 2017). Beyond the VGI niche, the characteristics of social media users are widely studied, particularly in relation to their similarities and differences to the general population.

The 330M monthly active users of Twitter tend to be wealthier and more educated than the average population, and they generate 500M tweets a day. From a statistical perspective, Twitter users are actually not representative of *any* particular population (Blank and Lutz 2017), but their sheer number and ease of access to large samples of the data attract researchers and marketeers. As Sloan and Morgan (2015) point out, 0.85% of the Twitter feed output is geotagged with coordinates, which amounts to roughly 4M tweets a day, produced by a population only marginally different to the overall platform population. Geo-located tweets have been used for a variety of purposes, sensing for example urban activities (Lansley and Longley 2016), emotions (Quercia et al. 2012), and beer-related behaviour (Zook and Poorthuis 2014).

Using geo-located tweets, Longley and Adnan (2016) conducted a geo-demographic analysis of tweets in London, identifying sub-groups in the user population and measuring the heterogeneity and the connectedness of places. Hahmann et al. (2014) investigated the spatial relationship between geo-located tweets and points of interest, showing correlations at the local scale for certain topics (e.g., "train station", "airport") and not for others ("pub", "bakery"). Using an approach similar to the one we adopt in our study, Li et al. (2013) explore tweets and Flickr photos' spatio-spatial distribution in California, showing that photos are denser in natural parks and that tweets tend to originate from areas with educated, high-income people.

Our second source for the study of place representation is Wikipedia. Despite a prolonged decline in the number of contributors (Halfaker et al. 2013), the crowd-sourced encyclopaedia is one of the top ten most visited websites worldwide, reaching more than 270M views per day,[2] and hosts 5.5M articles in English, edited by about 130,000 monthly active editors. Wikipedia shows an extreme gender imbalance, with about 84% of male editors, and less so in the readership, which is about 40% female (Hill and Shaw 2013). From a geographical perspective, in 2013, about 730,000 articles in English were associated with a geo-location. The bulk of the editing of these articles occurs in the Global North, also for articles about places in the Global South, exhibiting a staggering bias towards Western European and North American contributors (Graham et al. 2015b). For this reason, the location of Wikipedia editing activity can be used as a proxy to knowledge capital of countries (Stephany and Braesemann 2017). Editing of spatial features such as cities tends to be performed by local editors, as observed for OpenStreetMap (Johnson et al. 2016).

Much of this research aims at understanding the population of data producers, while the theme of place representation has been studied only marginally. The informational geographies of crowdsourced, VGI datasets have been charted at the global level (Graham et al. 2015a), drawing attention to the common bias towards relatively young, educated, wealthy users located in the Global North. To the best of our knowledge, no study has comparatively explored the properties of diverse VGI datasets at the urban scale, and their relationship with socio-economic texture of the places that the data describe. Moreover, the relationship between Twitter and Wikipedia from a spatial perspective has not been directly studied before.

3 Datasets

The area of our study is Greater London, which has a population of 8.87M, extended over 1,569 km^2. Three groups of datasets were collected and harmonised: socio-economic data from the UK Census, geo-located tweets, and Wikipedia articles. The summary statistics for the three datasets are shown in Table 1, showing minimum, median, maximum, mean, and standard deviation for all relevant variables. All variables are captured at the level of spatial unit selected for the analyses, detailed in the remainder of this section.

3.1 London Demographic Data

The UK Census, the latest of which occurred in March 2011, provides detailed socio-economic information about London. Census data is structured in Output Areas

[2]https://www.alexa.com/siteinfo/wikipedia.org.

Table 1 Descriptive statistics for the socio-economic data for Greater London for 983 MSOAs, 1.6M geo-referenced tweets and 22,411 Wikipedia articles, both grouped in the MSOAs. (*) Black, Asian, and minority ethnic. (†) Post-high school qualifications

Statistic (983 MSOAs)	Min	Median	Max	Mean	St. Dev.
Socio-economic variables (UK Census 2011)					
Area (ha)	29.40	114.50	2,243.00	159.94	186.10
Population	5,184	8,156	14,719	8,315.30	1,448
Workday population	3,444	6,789	360,075	8,826	14,500
Age 0–15 (%)	6.05	19.77	35.90	19.83	4.14
Age 16–29 (%)	10.61	21.29	52.73	22.36	5.73
Age 30–44 (%)	13.44	24.86	38.26	25.25	4.36
Age 45–64 (%)	10.06	20.98	31.96	21.34	4.01
Age 65+ (%)	2.40	10.37	27.23	11.22	4.12
BAME* population (%)	3.80	37.30	93.90	39.42	19.31
Household one person (%)	12.60	30.50	56.40	30.82	7.22
Household couple (%)	8.70	19.00	30.30	18.89	4.41
Hh. couple with dep. child (%)	4.60	18.80	32.20	18.28	5.12
Qualification 4 or above† (%)	8.19	28.05	62.62	30.32	11.83
House price (2012, £)	130,000	284,000	2,930,000	333,675	186,641
Twitter					
Number of tweets	18	413	161,050	1,617	8,063
Number of Twitter users	9	128	58,286	678	3,03
Twitter entropy	0.70	3.93	10.14	4.13	1.51
Twitter Gini coefficient	0.08	0.54	0.90	0.55	0.14
Wikipedia					
Number of Wikipedia articles	0	9	1,857	22.80	87.50
Wikipedia cumulative length (bytes)	0	47,606	17,042,103	155,839	735,909
Wikipedia cumulative edits	0	450	68,739	1,088	3,512
Wikipedia cumulative minor edits	0	4	2,499	94.95	192.92

(OA), each covering between 40 and 149 households, corresponding to an average of 300 people.[3] For small area statistics, OAs are grouped into Lower Layer Super Output Areas (LSOA), which contain from four to six OAs, with a mean population of 1,500. Given the spatially uneven distribution of Twitter and Wikipedia data, the OAs and even LSOAs are too granular, leaving many areas without data. For this reason, we use Middle Layer Super Output Areas (MSOA), which further aggregates LSOAs into contiguous groups, with a minimum population of 5,000 and a national mean of 7,200.

[3]https://www.ons.gov.uk/methodology/geography/ukgeographies/censusgeography.

Greater London contains 983 MSOAs, whose boundaries were collected from the UK Data Service,[4] while the Census 2011[5] aggregated variables used in our study were collected from the London Datastore[6] and Nomis.[7] The variables include the MSOA's area, total and workday population, age composition, household size, education, and house prices (see Table 1). While some variables tend to have parametric distributions (e.g., percentage of residents aged from 0 to 15), others are heavily skewed towards large outliers, which we will take into account in our analyses. Notably, some areas of London have extremely high workday population: notably, during the day, the City of London hosts about 360,000 workers, while having just 9,400 residents. Similarly, house prices are skewed by multi-million-pound properties that are common in Central London.

Because of the high dimensionality and complexity of the Census data, geodemographic classifications have been produced as a way to summarise the population into a set of discrete classes. Notably, the London Output Area Classification (LOAC) categorises each OA in Greater London into eight super-groups, such as "Urban Elites" and "Settled Asians", further classified into groups (Singleton and Longley 2015). This classification is useful to detect the demographic structure of the urban space, and can be related to the place representation observed in the informational geographies.

3.2 Twitter Data

All geo-referenced tweets produced in Greater London were collected from the Twitter API from October 2015 to May 2016, for a total of 2,076,588 tweets, produced by 222,719 users, excluding re-tweets. As we are interested in place representation and not in specific user behaviours, we retain low-activity users. The only category of users that we exclude from the analysis is high-activity bots, whose tweets do not capture the manual information production we intend to observe.

To identify bots, we combine two heuristics, measuring for each user (1) the number of tweets per day, and (2) the percentage of repeated tweets, assuming that bots, for advertising purposes, generate a high number of tweets, and tend to repeat the same content more than human users. Hence, we selected users that generated more than 10 tweets per day, and whose 10% of tweets were repeated at least once. These thresholds were identified by trial and error, and then observing a sample of excluded users to make sure they were all bots (e.g., *trendinaliaGB*). This process filtered out 1.4% of users, corresponding to 22.7% of tweets.

The remaining dataset of 1,589,819 tweets was generated by 219,604 users. As expected, most users produced very few tweets: The number of tweets per user range from 0 to 1,950, with a median of just 2. From a linguistic viewpoint, 91.6% of tweets are in English, with the other larger groups being undefined (3.4%), Spanish (1.6%),

---

[4]https://borders.ukdataservice.ac.uk/.

[5]https://www.ons.gov.uk/census/2011census.

[6]https://data.london.gov.uk/.

[7]https://www.nomisweb.co.uk/.

and French (0.8%). The tweets were then grouped in the MSOAs of Greater London. The number of tweets per spatial unit ranges from 9 to about 160,000 in Westminster, with a median of about 400 tweets (see Table 1). The number of users active in each unit follows a similarly skewed distribution. We also calculated Shannon entropy and the Gini coefficient as measures respectively of contribution diversity and inequality.

3.3 *Wikipedia Data*

The second VGI source in this study consists of geo-referenced Wikipedia articles located in Greater London. At the time of writing, Wikipedia only allows for geo-tags in the form of points, and even large geographical entities are geo-tagged to a point. For example, the article about the Palace of Westminster is associated with a latitude/longitude point.[8] The decision about where to locate entities is a combination of the platform guidelines and the editors' arbitrary choices. As a result, the same entity can be pin-pointed in different locations in different language editions. Such inconsistencies are common in collaborative editing (Ballatore and Mooney 2015). For instance, at the time of writing, "England" in the English Wikipedia[9] is geo-tagged on the River Thames near the Palace of Westminster, whereas the Italian version geo-tags "Inghilterra" somewhere in the Borough of Bromley, near Biggin Hill. By contrast, the German Wikipedia selects a geometric centroid located east of Birmingham.

Using the databases available on Wikimedia Toolforge,[10] we extracted 22,411 articles, including features such as monuments, notable buildings, parks, and headquarters of organisations. About 41% of articles are in English, followed by 6% in French, 6% in German, and the remainder 47% in other languages. After grouping them in the MSOAs, it is possible to note that the articles are sparser than the tweets, with 32 spatial units (3%) without any article (see Table 1). Furthermore, 34% of units contain fewer than 6 articles. The densest parts of the distribution are found in an MSOA in Westminster (1,857 articles), and in the City of London (1,466).

4 Explaining Crowdsourced Geographies

To understand the factors that shape the geography of the Twitter and Wikipedia data, we start by observing the properties of their spatial distribution. Figures 1 and 2 show the number of tweets and Wikipedia articles in Greater London, scaled by workday population. While both distributions show, as largely expected, high density in Central London, the maps also suggest differences, for example in Southern parts of the city, which deserve more investigation.

To relate Twitter and Wikipedia data to the demographic geography of London, we display in Fig. 3 the London Output Area Classification (LOAC) by Singleton and Longley (2015), as a summary of the demographic characteristics of each area.

[8]https://en.wikipedia.org/wiki/Palace_of_Westminster.

[9]https://en.wikipedia.org/wiki/England.

[10]https://tools.wmflabs.org.

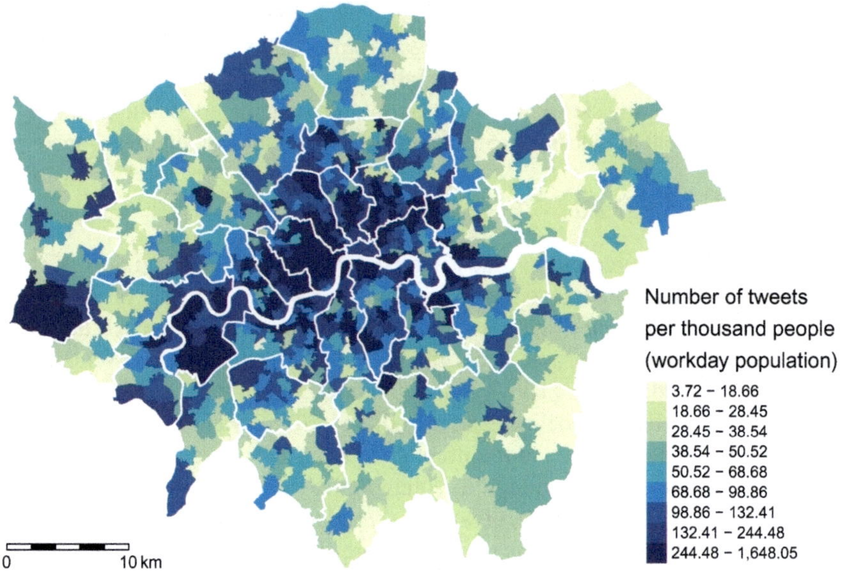

Fig. 1 Distribution of 1.6M geo-located tweets in Greater London, scaled by workday population, for 983 MSOAs. The data is grouped into 9 quantiles. The boundaries of the boroughs are outlined in white

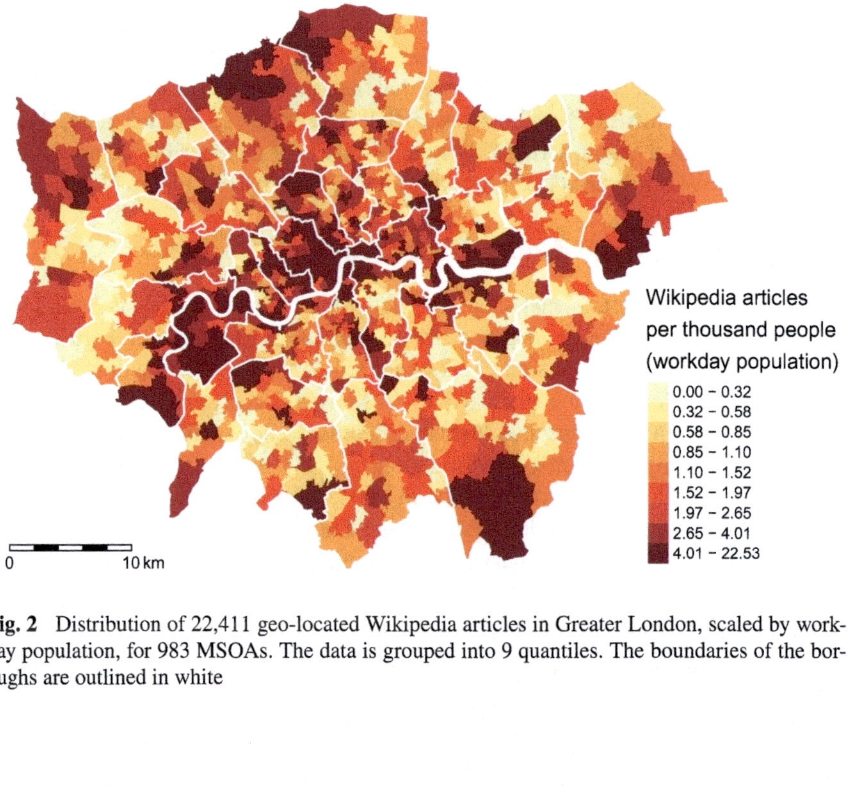

Fig. 2 Distribution of 22,411 geo-located Wikipedia articles in Greater London, scaled by workday population, for 983 MSOAs. The data is grouped into 9 quantiles. The boundaries of the boroughs are outlined in white

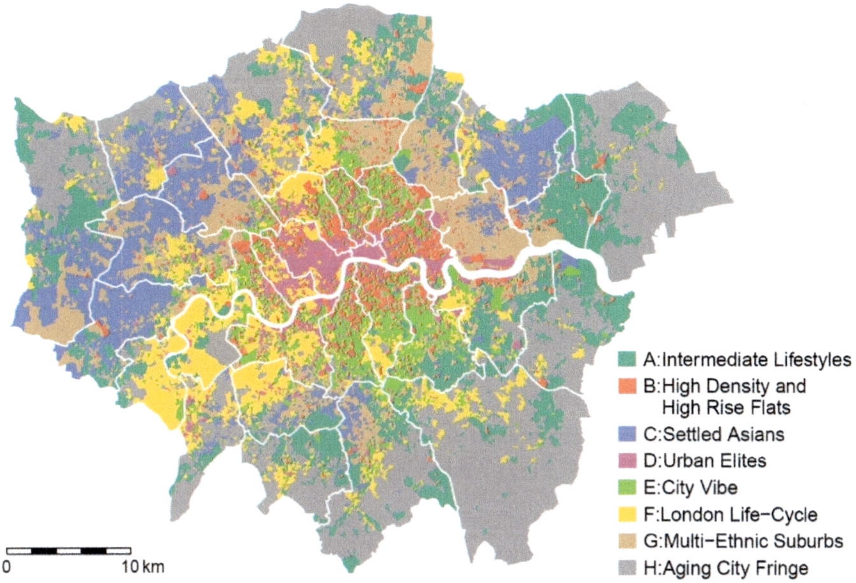

Fig. 3 London Output Area Classification (LOAC), showing super-groups. Source: Singleton and Longley (2015)

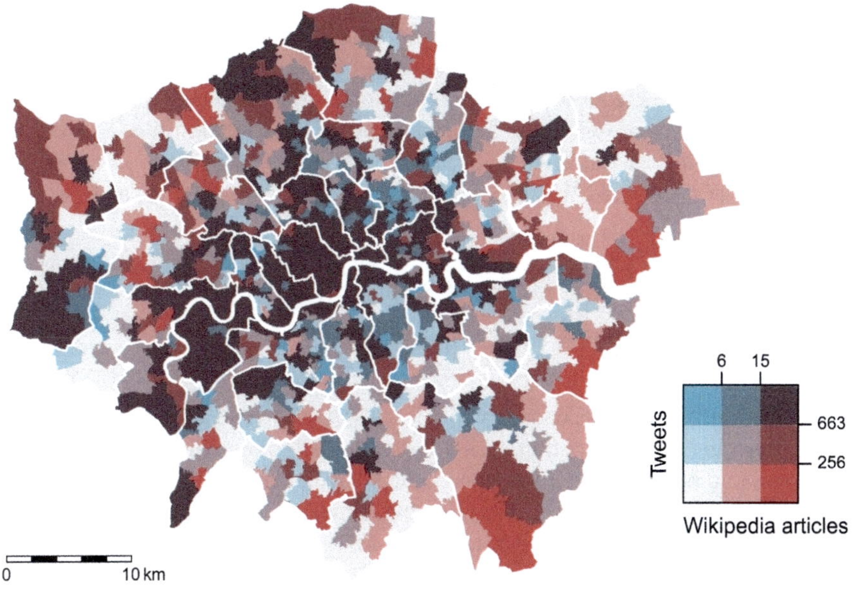

Fig. 4 Comparison of the distribution of 1.6M tweets and 22,411 Wikipedia articles in Greater London in 983 MSOAs. The boundaries of the boroughs are outlined in white

In order to allow for a visual comparison of tweets and Wikipedia articles, Fig. 4 displays a bi-variate choropleth map generated from the intersections of the three quantiles of each of the two variables. The darkest areas in this map represent the highest quantiles of both Twitter and Wikipedia content, indicating the data-richest areas. These areas tend to correspond with OAs classified as “Urban Elites” (i.e., young professionals in science, technology, and finance) and “London Life-Cycle” (i.e., relatively low numbers of students and households with dependent children, highly qualified professionals, predominantly white) in the City and Westminster, as well as Richmond, Merton, and the south part of Newham.

Heathrow Airport, located at the Western edge of Greater London, shows extremely high content density, as already noted by Longley and Adnan (2016). Many areas in Southwark and Hackney, classified as “City Vibe”, that is, single professionals and students in communal establishments, display a high density of Twitter content, but relatively low Wikipedia content. This might be due to not only to socio-demographic characteristics, but to the relatively low density of notable urban features in those districts. More generally, low-tweet areas seem to align with OAs classified as “Intermediate Lifestyles” and “Aging City Fringe”, both associated with households in later stages in life-cycle.

This visual examination indicates that within Greater London there are substantial differences in the amount of content representing the different areas of the metropolis, broadly following the spatial distribution of the demographic characteristics that have been linked to VGI content production in the literature (e.g., Crampton et al. 2013). Our initial hypothesis is that content in both Twitter and Wikipedia tends to be representative of wealthier areas, inhabited by younger sections of the population, with access to higher levels of qualifications. In the remainder of this section, we perform a series of regression analyses to explain the relationships between the socio-demographic characteristics discussed above, as independent variables, and the number of tweets and Wikipedia articles as dependent variables, aggregated at the level of MSOA.

4.1 Variable Selection and Normalisation

The selection of the independent variables was based on a correlation analysis: Fig. 5 illustrates the distribution of and correlations between the variables used in the regression models. The normalised values have been created using the natural logarithm (when [*log*] is added to the variable name) and the inverse hyperbolic sine (when [*ihs*] is added to the variable name) (Burbidge et al. 1988; Pence 2006).

The percentage of population between 16 and 29 and between 30 and 44 were initially considered, but only the latter was included in the models, as it shows higher correlation with both dependent variables. This was then combined with the house prices in the first models (Tw1, Tw2, and Wk1 below), which is the variable that show lower correlation among the other independent variables here considered. The percentage of households with dependent children and the percentage of population

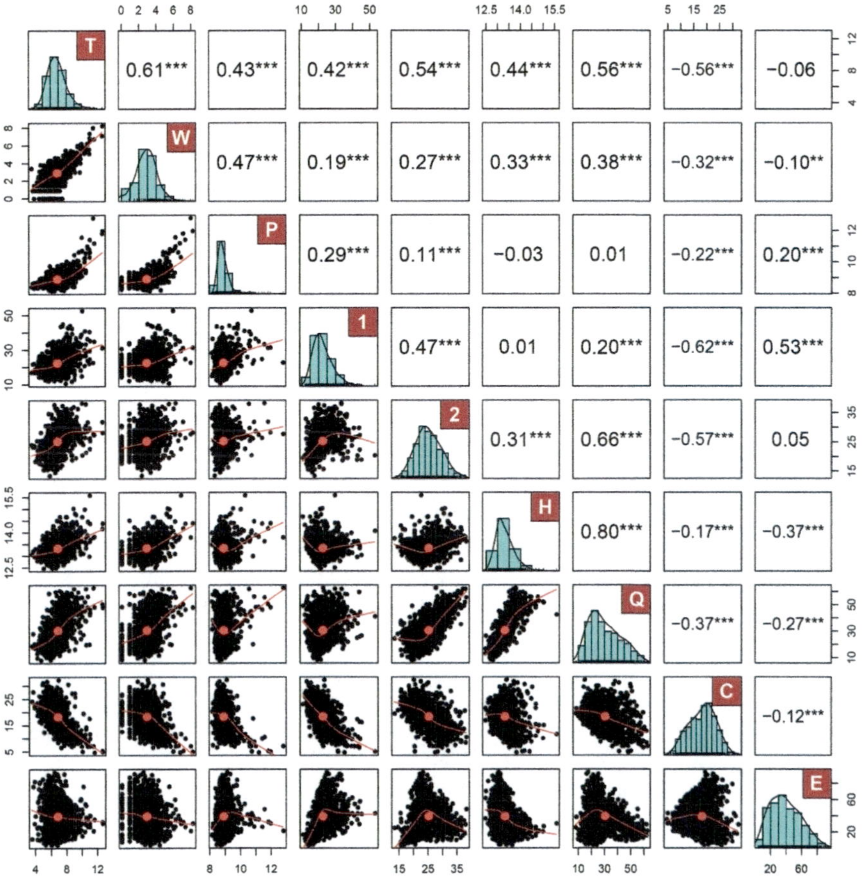

Fig. 5 Distribution and correlations between tweets *[ihs]* (T), Wikipedia articles *[ihs]* (W), workday population *[log]* (P), percentage of population aged between 16 and 29 (1), and between 30 and 44 (2), level four qualifications or above (Q), couples with dependent children (C), minority groups (E), and house prices *[ihs]* (H), in 983 MSOAs in Greater London. Significance level: $^{*}p < 0.1$; $^{**}p < 0.05$; $^{***}p < 0.01$

with level 4 qualifications or above are then combined in subsequent models (Tw3 and Wk2 below), as they show a lower correlation between each other.

All models also include workday population independent variable, to account for the varying presence of people in each MSOA. Workday population was preferred to resident population due to its higher correlation with the dependent variable. Finally, we compare the amount of content present in the two VGI platforms, which is a novel approach to studying these information geographies (model TwWk below). It is important to note that these models are understood as explanatory of relationships to common underlying geographies, without claims to causality.

4.2 Twitter Models

The first model in Table 2 (Tw1) includes the percentage of population between 30 and 44 and house prices as independent variables. However, the residuals of the model are not normally distributed (Shapiro-Wilk test, $W = 0.99, p < 0.001$), due to a handful of MSOA overrepresented in the Twitter dataset. To overcome this issue, we devised Model Tw2, which replicates Model Tw1, but excluding all MSOAs having an average of 10 or more tweets per day. In this second model, the independent variables account for 48% of the variation in the number of tweets. Model Tw2 is fit and robust, as the residuals are normally distributed ($W = 1, p = 0.808$), and satisfy the homoscedasticity assumption (Breusch-Pagan test, $BP = 1.56, p = 0.668$). The errors are slightly positively correlated, but this does not raise concerns (Durbin-Watson test, $DW = 1.65, p < 0.001$), and no multicollinearity has been identified in this model (average VIF is 1.07).

Model Tw3 includes the percentage of households with dependent children and the percentage of population with level 4 qualifications or above as independent variables (see Table 2). As above, the residuals are not normally distributed ($W = 0.99$, $p < 0.001$), thus we create Model Tw4 by excluding all MSOAs having an average of 10 or more tweets per day. In this model, the independent variables account for 55%

Table 2 Linear regression models to explain the spatial variation in the 1.6M geo-located tweets over 983 MSOAs in Greater London. Four models were devised (standard errors between parentheses, significance levels: $^{*}p < 0.1$; $^{**}p < 0.05$; $^{***}p < 0.01$)

	Dependent variable			
	Number of tweets [ihs]			
	(Tw1)	(Tw2)	(Tw3)	(Tw4)
Workday pop. [log]	1.451*** (0.057)	1.296*** (0.078)	1.287*** (0.056)	1.225*** (0.074)
Age 30–44 (%)	0.092*** (0.006)	0.089*** (0.006)		
House price [ihs]	0.977*** (0.066)	0.902*** (0.067)		
Qual. 4+ (%)			0.040*** (0.002)	0.040*** (0.002)
Couple w/child. (%)			−0.071*** (0.005)	−0.064*** (0.005)
Constant	−21.397*** (0.927)	−19.039*** (1.160)	−4.530*** (0.547)	−4.144*** (0.689)
Observations	983	922	983	922
R^2	0.612	0.483	0.660	0.548
Adjusted R^2	0.611	0.481	0.659	0.546
Res. Std. Error	0.783 (df = 979)	0.712 (df = 918)	0.733 (df = 979)	0.666 (df = 918)
F Statistic	515.135*** (df = 3; 979)	285.995*** (df = 3; 918)	632.371*** (df = 3; 979)	370.803*** (df = 3; 918)

of the variation in the number of tweets. Model Tw4 is fit and robust, as the residuals are normally distributed ($W = 1, p = 0.471$), and satisfy the homoscedasticity assumption ($BP = 9.27, p = 0.026$). The errors are slightly positively correlated, but this is not a cause for concern ($DW = 1.74, p < 0.001$), and no multicollinearity has been identified (average VIF is 1.12). By observing the spatial distribution of residuals of models Tw2 and Tw4, we did not find evidence of spatial clustering through local Moran's I.

Based on Model Tw2, a one percent increase in the population aged 30–44 is linked to an increase the number of tweets produced in the MSOA of about 9%. Similarly, a ten percent increase in house prices is linked to an increase the number of tweets in the MSOA of about 9%. Based on Model Tw4, other things being equal, a one percent increase in the population with level 4 qualifications or above is linked to an increase the number of tweets of about 4%, and a one percent increase in households composed by couples with dependent children is linked to a decrease the number of tweets of about 6%.

4.3 Wikipedia Models

After having assessed the distribution of tweets, we proceed to observe possible explanatory factors of the presence of Wikipedia articles in a given MSOA in London. For the statistical modelling, we excluded the 3% of MSOAs that do not contain any Wikipedia article. We generated two regression models, Model Wk1 and Model Wk2, which are robust and fit (see Table 3). The residuals of Model Wk1 are normally distributed (Shapiro-Wilk test, $W = 1, p = 0.219$), and satisfy the homoscedasticity assumption (Breusch-Pagan test, $BP = 5.05, p = 0.168$). The errors are independent (Durbin-Watson test, $DW = 1.89, p = 0.088$), and no multicollinearity has been identified (average VIF is 1.08). The independent variables account for 44% of the variation in the number of Wikipedia articles, when aggregated by MSOA.

Similarly, the residuals of Model Wk2 are normally-distributed residuals ($W = 1$, $p = 0.158$), and satisfy the homoscedasticity assumption ($BP = 3.52, p = 0.318$), In this model too, the errors appear to be independent ($DW = 1.86, p = 0.04$), and no multicollinearity has been identified (average VIF is 1.21). The independent variables account for 45% of the variation in the number of tweets, when aggregated by MSOA. As for the models presented in the previous section, the spatial distribution of residuals shows no sign of spatial clustering.

Based on Model Wk1, other things being equal, a one percent increase in the population aged 30 to 44 is linked to an increase the number of Wikipedia articles in the MSOA of about 2%. Similarly, a ten percent increase in house prices is linked to an increase the number of Wikipedia articles in the MSOA of about 8%. Based on Model Wk2, other things being equal, a one percent increase in the population with level 4 qualifications or above is linked to an increase the number of Wikipedia articles in the MSOA of about 3%, and a one percent increase in households composed

Table 3 Linear regression models to explain the spatial variation in the 22,411 Wikipedia articles over 983 MSOAs in Greater London. Two models were devised (standard errors between parentheses, significance levels: $^{*}p < 0.1$; $^{**}p < 0.05$; $^{***}p < 0.01$)

	Dependent variable:	
	Number of Wikipedia articles [ihs]	
	(Wk1)	(Wk2)
Workday pop. [log]	1.344*** (0.062)	1.312*** (0.065)
Age 30–44 (%)	0.023*** (0.007)	
House price [ihs]	0.822*** (0.071)	
Qual. 4+ (%)		0.029*** (0.003)
Couple w/child. (%)		−0.018*** (0.006)
Constant	−20.522*** (1.012)	−9.265*** (0.633)
Observations	951	951
R^2	0.445	0.452
Adjusted R^2	0.443	0.450
Res. Std. Error (df = 947)	0.844	0.838
F Statistic	252.669*** (df = 3; 947)	260.448*** (df = 3; 947)

by couples with dependent child is linked to a decrease the number of Wikipedia articles in the MSOA of about 2%.

4.4 Comparison Between Twitter and Wikipedia

The geographies of place representation that we analysed above can be directly compared. For this purpose, we created a model using the number of Twitter posts as the dependent variable and the number of Wikipedia article, to observe to what extent they converge, and where they differ. As this is a simple regression, the choice of either variable as dependent or independent does not affect the outcome of the analysis, and thus it was made arbitrarily Model Tw-Wk is fit and mostly robust. The residuals normally distributed (Shapiro-Wilk test, $W = 1, p = 0.649$), and satisfy the homoscedasticity assumption (Breusch-Pagan test, $BP = 4.7, p = 0.03$). The errors appear to be slightly positively correlated, but this is not a cause for concern (Durbin-Watson test, $DW = 1.34, p < 0.001$). Overall, the variability in the num-

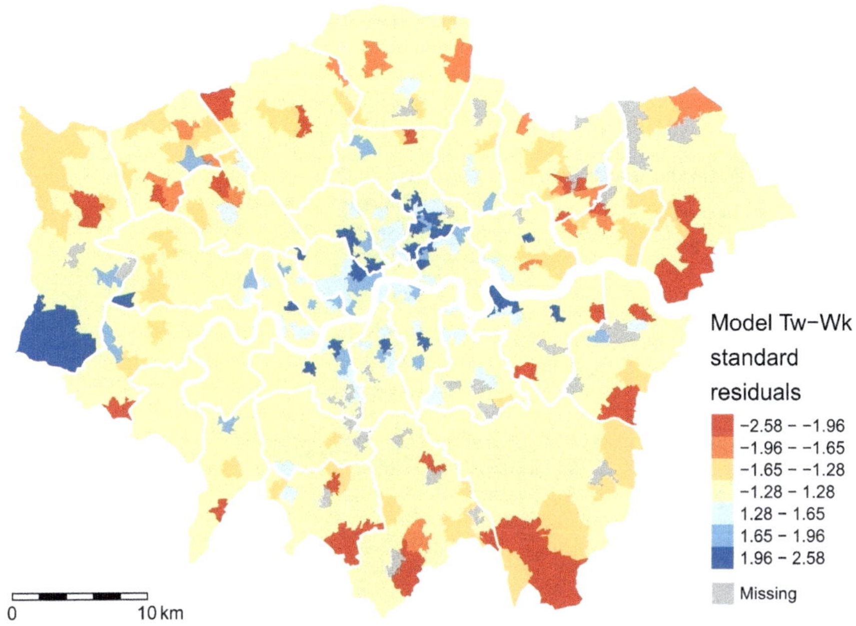

Fig. 6 Distribution of residuals for Model Tw-Wk, which relates Twitter and Wikipedia data. Blue areas have more tweets than expected based on the Wikipedia distribution, while red areas display lower Twitter density compared to Wikipedia

ber of Wikipedia articles accounts for 49% of the variation in the number of tweets, when aggregated at the MSOA level.

Based on Model Tw-Wk, other things being equal, a ten percent increase in the number of Wikipedia articles is linked to an increase in the number of tweets produced in the MSOA of about 8%. Indeed, these are not to be interpreted as causal relationships. For this model, the map in Fig. 6 shows the spatial distribution of residuals. Unlike the previous models, this map shows some clustering in Central London and Heathrow Airport, where Twitter activity is higher than expected based on Wikipedia content, whilst the outskirts exhibit lower tweet density (see Table 4).

Table 4 Linear regression model Tw-Wk, which relates Twitter and Wikipedia data, over MSOAs in Greater London (standard errors between parentheses, significance levels: $^{*}p < 0.1$; $^{**}p < 0.05$; $^{***}p < 0.01$)

	Dependent variable
	Number of tweets [ihs]
Number of Wikipedia articles [ihs]	0.778*** (0.026)
Constant	4.562*** (0.082)
Observations	951
R^2	0.490
Adjusted R^2	0.490
Res. Std. Error	0.896 (df = 949)
F Statistic	913.117*** (df = 1; 949)

5 Discussion

The exploratory and explanatory analyses discussed above highlight a number of aspects of the information geographies of Twitter and Wikipedia at the urban scale. Overall, the explanatory analyses in Sect. 4 confirm our general hypothesis based on the literature and the exploratory analysis. Crowdsourced information generated in Greater London exhibits a significant bias towards areas characterized by a wealthier, younger, and higher-qualified population (Crampton et al. 2013).

All models exhibit similar explanatory power, but the variability of content in both Twitter and Wikipedia seems to be more closely liked to level of education and average house prices, when aggregated at the MSOA level. At the same time, the standardized estimates presented in Table 5 suggest that the presence of population has a stronger influence on Wikipedia content than on Twitter, while both the percentage of people aged 30–44 and households composed by couples with

Table 5 Standardized beta estimates (β values) for the models Tw2, Tw4, Wk1, and Wk2, useful to evaluate the explanatory weight of independent variables

Dependent variable	Models			
	Tw2	Tw4	Wk1	Wk2
Workday pop. [log]	0.397	0.375	0.529	0.516
Age 30–44 (%)	0.391		0.087	
House price [ihs]	0.337		0.293	
Qual. 4+ (%)		0.455		0.301
Couple w/child.		−0.314		−0.083

dependent children have a stronger influence (positive and negative, respectively) on Twitter content. By contrast, house prices exert a strong influence on the presence of Wikipedia content. This might be linked to the tendency in Wikipedia content to document landmarks, heritage sites, and notable buildings, which tend to correspond to high property value.

The level of education seems to be a crucial factor for both datasets, suggesting that low-education level, indeed, constitute a barrier to accessing these platforms and take part in the digital production. Yet, our models highlight that Twitter and Wikipedia do not share the same geography. The fact of being different platforms used for different purposes is reflected in their informational geographies and in the way they over- and under-represent places. In particular, our comparison of geo-located tweets and Wikipedia articles indicate how the former is shaped more by the presence of population with given characteristics, while the latter by notable urban features.

However, these findings must be qualified: The proposed regression models are robust, but account for about 44–55% of the variability, suggesting that much of the variation is not captured by socio-demographic variables alone. More explanatory factors, such as tourism-related activities and amenities, must be included to better capture the place representation outcomes. Interestingly, while most explanatory variables we considered bear some relationships with both crowdsourced datasets, the ethnic composition of each MSOA does not exhibit a link to neither. This could be related to the comparatively low level of residential segregation in the UK (Johnston et al. 2007).

This study contributes to the deeper and paramount issue of representativeness in crowdsourced, "big data" sources. As Blank (2016) argues, Twitter users are generally younger and wealthier than the rest of the population, and do not strictly represent any demographic group. Therefore, Twitter users cannot be used to corroborate claims about society at large. For this reason, knowing the platform biases is important to inform researchers about what they can and, most cogently, cannot expect from such data. In a similar spirit, we argue that the information geographies we investigate in this study are unique and—strictly speaking—only representative of themselves. Hence, studying their socio-demographic biases is key to support their effective usage in research.

6 Conclusions

In this article, we investigated the information geographies of two well-known crowdsourced data sources, Twitter and Wikipedia, observing their place representation in Greater London. A set of 1.6M geo-referenced tweets and about 22,000 Wikipedia articles located in Greater London was studied with respect to socio-economic variables. MSOAs were selected as the spatial unit of analysis, allowing for a granular analysis of the spatial variation in both tweets and articles. Linear regressions revealed that about half of the variability can be explained through variabilities in the level of education of residents, share of population aged group 30–44, and property prices. These factors explain half of the variability, while the other half remains unknown, calling for further work.

This study focussed only on one city. A comparative approach with other cities, in the UK and elsewhere, is necessary to observe the geographical variation in the relationships that between place and its information. Moreover, our models need to include land use and tourism data in order to capture the role of urban function and mass travel, prominent in many data-rich areas in central London. In the early stages of this analysis, we used an Output Area Classification (OAC) as a compact description of the demographic structure of the city (Singleton and Longley 2015), and we plan to conduct further quantitative analysis, exploring patterns beyond what is visible. Furthermore, we intend to explore the spatial clustering of the residuals in Fig. 6. Geographically weighted regressions (GWR) might provide more detailed explanations of the relationship between Twitter and Wikipedia. Characteristics of the built environment, such as building density and presence of other urban features, are likely to provide good independent variables for the analysis.

As future work, more comparative analyses between urban information geographies are needed to reveal the socio-spatial structure of these data sources. To refine these models beyond simple Census variables, more interaction between GIScience and quantitative human geography is needed. Charting the factors that shape these geographies and their place representation can produce insights about the real value, limits, and uncertainty when using such informational assets for research and knowledge extraction. In a context where the problematic use of unrepresentative data is widespread (Bowker 2014), more research is needed to devise analytical techniques to reduce the spatial and social biases embedded in all emergent, online datasets.

Acknowledgements The demographic data used in this work have been provided by the Greater London Authority and Nomis under the Open Government Licence v2.0. The content analysed in this article was produced by Twitter users and Wikipedia contributors, and obtained through the web services by Twitter, Inc. and Wikimedia Foundation, Inc., under the respective licences. The maps contain data from CDRC LOAC Geodata Pack by the ESRC Consumer Data Research Centre; National Statistics data Crown copyright and database right 2015; Ordnance Survey data Crown copyright and database right 2015.

References

Acheson E, De Sabbata S, Purves RS (2017) A quantitative analysis of global gazetteers: patterns of coverage for common feature types. Comput Environ Urban Syst 64:309–320

Antoniou V, Morley J, Haklay M (2010) Web 2.0 Geotagged Photos: assessing the spatial dimension of the phenomenon. Geomatica 64(1):99–110

Ballatore A, Graham M, Sen S (2017) Digital hegemonies: the localness of search engine results. Ann Am Assoc Geogr 107(5):1194–1215

Ballatore A, Mooney P (2015) Conceptualising the geographic world: the dimensions of negotiation in crowdsourced cartography. Int J Geogr Inf Sci 29(12):2310–2327

Bittner C (2017) Diversity in volunteered geographic information: comparing OpenStreetMap and wikimapia in Jerusalem. Geo J 82(5):887–906

Blank G (2016) The digital divide among twitter users and its implications for social research. Soc Sci Comput Rev 679–697

Blank G, Lutz C (2017) Representativeness of social media in great britain: investigating Facebook, Linkedin, Twitter, Pinterest, Google+, and Instagram. Am Behav Sci 61:741–756

Bowker GC (2014) Emerging configurations of knowledge expression. In: Gillespie T, Boczkowski PJ, Foot KA (eds) Media technologies: essays on communication, materiality, and society. MIT Press, Boston, MA, pp 99–118

Burbidge JB, Magee L, Robb AL (1988) Alternative transformations to handle extreme values of the dependent variable. J Am Stat Assoc 83(401):123–127

Crampton JW, Graham M, Poorthuis A, Shelton T, Stephens M, Wilson MW, Zook M (2013) Beyond the geotag: situating 'big data' and leveraging the potential of the GeoWeb. Cartogr Geogr Inf Sci 40(2):130–139

Earle P, Guy M, Buckmaster R, Ostrum C, Horvath S, Vaughan A (2010) OMG earthquake! Can Twitter improve earthquake response? Seismol Res Lett 81(2):246–251

Graham M, De Sabbata S, Zook MA (2015) Towards a study of information geographies: (im)mutable augmentations and a mapping of the geographies of information. Geo Geogr Environ 2(1):88–105

Graham M, Straumann RK, Hogan B (2015b) Digital divisions of labor and informational magnetism: mapping participation in wikipedia. Ann Assoc Am Geogr 105(6):1158–1178

Hahmann S, Purves RS, Burghardt D (2014) Twitter location (sometimes) matters: exploring the relationship between georeferenced tweet content and nearby feature classes. J Spat Inf Sci 2014(9):1–36

Halfaker A, Geiger RS, Morgan JT, Riedl J (2013) The rise and decline of an open collaboration system: how wikipedia's reaction to popularity is causing its decline. Am Behav Sci 57(5):664–688

Hecht B, Stephens M (2014) A tale of cities: urban biases in volunteered geographic information. In: Proceedings of the eighth international AAAI conference on weblogs and social media, pp 197–205

Hill BM, Shaw A (2013) The Wikipedia gender gap revisited: characterizing survey response bias with propensity score estimation. PloS one 8(6):e65782

Johnson IL, Lin Y, Li TJ-J, Hall A, Halfaker A, Schöning J, Hecht B (2016) Not at home on the range: peer production and the urban/rural divide. In Proceedings of the 2016 CHI Conference on Human Factors in Computing Systems—CHI '16, pp 13–25

Johnston R, Poulsen M, Forrest J (2007) The geography of ethnic residential segregation: a comparative study of five countries. Ann Assoc of Am Geogr 97(4):713–738

Lansley G, Longley PA (2016) The geography of Twitter topics in London. Comput Environ Urban Syst 58:85–96

Li L, Goodchild MF, Xu B (2013) Spatial, temporal, and socioeconomic patterns in the use of Twitter and Flickr. Cartogr Geogr Inf Sci 40(2):61–77

Longley PA, Adnan M (2016) Geo-temporal Twitter demographics. Int J Geogr Inf Sci 30(2):369–389

Mashhadi A, Quattrone G, Capra L (2015) The impact of society on volunteered geographic information: the case of OpenStreetMap. In: Jokar Arsanjani J, Zipf A, Mooney P, Helbich M (eds) OpenStreetMap in GIScience. Springer, Berlin, pp 125–141

Pence KM (2006) The role of wealth transformations: an application to estimating the effect of tax incentives on saving. BE J Econ Anal Policy 5(1)

Quercia D, Capra L, Crowcroft J (2012) The social world of Twitter: topics, geography, and emotions. In International Conference on Web and Social Media, ICWSM. AAAI Press, Palo Alto, CA, pp 298–305

See L, Mooney P, Foody G, Bastin L, Comber A, Estima J, Fritz S, Kerle N, Jiang B, Laakso M et al (2016) Crowdsourcing, citizen science or volunteered geographic information? The current state of crowdsourced geographic information. ISPRS Int J Geo-Inf 5(5):55

Shaw J, Graham M (eds) (2017) Our digital rights to the city. Meatspace Press, Oxford, UK

Singleton AD, Longley P (2015) The internal structure of Greater London: a comparison of national and regional geodemographic models. Geo Geogr Environ 2(1):69–87

Sloan L, Morgan J (2015) Who tweets with their location? Understanding the relationship between demographic characteristics and the use of geoservices and geotagging on Twitter. PloS one 10(11):e0142209

Stephany F, Braesemann F (2017) An exploration of wikipedia data as a measure of regional knowledge distribution. In Social Informatics: 9th International Conference, SocInfo 2017, Oxford, UK. Springer, Berlin, pp 31–40

Sui DZ, Elwood S, Goodchild M (eds) (2012) Crowdsourcing geographic knowledge: volunteered geographic information (VGI) in theory and practice. Springer, Berlin

Zagheni E, Garimella V, Weber I (2014) Inferring international and internal migration patterns from Twitter data. In World Wide Web 2014 Companion. ACM, New York, pp 439–444

Zook M, Poorthuis A (2014) Offline brews and online views: exploring the geography of beer tweets. In: Patterson M, Hoalst-Pullen N (eds) The geography of beer. Springer, Berlin, pp 201–209

Supporting the Monitoring of Cheetahs in Kenya by Mobile Map Apps

Jan Jedersberger, Christian Stern, Mary Wykstra and Gertrud Schaab

Abstract The cheetah (*Acinonyx jubatus*) is Africa's most endangered large felid. In Kenya, cheetahs are extirpated from 25% of their historic range with 75% of the population residing outside protected areas. Action for Cheetahs in Kenya (ACK) endeavours to understand cheetah ecology and threats in human-dominated landscapes. The NGO is a suitable candidate for benefitting from the development and deployment of mobile map apps on smartphones to support the monitoring of species, also by so-called citizen scientists. The apps enable a digital workflow of data collection, transfer and analysis. This paper describes the process of developing custom mobile map apps from the conceptualizing of a system from data collection to data storage. We discuss the implementation as well as on the ground testing which included a usability study. The software environment from Esri's platform tools were used but aiming at a low-cost solution which supports both secure monitoring in the field and the management of sensitive data. Due to hardware constraints the implemented system cannot be considered a state-of-the art version, but for ACK it is a big step towards digital data collection by means of an app family and data management in a database. A Web frontend allows for input on cheetah sightings also from outside the organization and serves the purpose of visualizing observation efforts for potential donors. While software development took place mainly in Germany, the usability study following the installation of the mobile app monitoring system targeted twelve ACK staff members on the ground in Kenya. Before the testing the users expected faster and less work in particular in the field and an easier reporting to office staff and streamlined data collection. After testing the mobile apps or the data management routines, answers shifted to pointing out benefits of faster data transfer and in-time data access. The evaluation provided valuable insights in the needs for reaching a routine level and high quality data recording and management.

J. Jedersberger · C. Stern · G. Schaab (✉)
Faculty for Information Management and Media (IMM), Karlsruhe University of Applied Sciences (HsKA), Karlsruhe, Germany
e-mail: gertrud.schaab@hs-karlsruhe.de

M. Wykstra
Action for Cheetahs in Kenya (ACK), Nairobi, Kenya

A. Mansourian et al. (eds.), *Geospatial Technologies for All*,
Lecture Notes in Geoinformation and Cartography,
https://doi.org/10.1007/978-3-319-78208-9_9

Keywords Species monitoring · App development · Data management Usability study · Nature conservation

1 Introduction

The cheetah (*Acinonyx jubatus*) is not only Africa's most endangered large felid (Durant 2004), but one of the most wide-ranging carnivore species. There are currently an estimated 7,100 individuals found in 9% of its historical distributional range across Africa and southwestern Asia (Durant et al. 2015). The largest meta-populations occur in Eastern and Southern Africa, where Kenya with 1,200–1,400 individuals (Durant et al. 2017) is critical as the connection between two genetic subspecies. In Kenya, cheetahs are extirpated from 25% of their historic range with 75% of the population residing outside protected areas (Kenya Wildlife Service 2010). Threats are increasingly persistent (cp. Andresen et al. 2014; Durant et al. 2017) decreasing viable connectivity between protected and unprotected areas (Kenya Wildlife Service 2010). Durant et al. (2017) urge the International Union for Conservation of Nature (IUCN) to up-list the cheetah from Vulnerable to Endangered on the Red List which requires conservation researchers to collect more information on cheetah ecology and threats in human-dominated and highly fragmented landscapes.

Action for Cheetahs in Kenya (ACK) is a non-profit organization (NGO) focused on cheetah research expanding across the marginal areas of cheetah populations. ACK's community outreach activities mitigate human-wildlife conflicts and promote environmental care for maintaining connectivity between cheetah populations. In two focus areas ACK has been repeatedly collecting standardized monitoring data (cp. Bunce et al. 2008) since 2005 and 2010 (Wambua 2008), respectively. Between 2004 and 2007, ACK conducted the first Kenya-wide cheetah survey. Data collection during the rapid-survey was achieved by filling printed forms (Mutoro et al. 2016) and digitizing at a later point using spreadsheet tables. While ACK analyzed the impact of human factors on range-wide cheetah presence, Kuloba (2011) used environmental predictors for a Kenyan wide modelling, and Masseloux et al. (in press) tested occupancy modelling methods. The results on cheetah distribution in Kenya and critical areas augment regional and national strategic planning for cheetahs (Mutoro et al. 2016). Currently ACK is initiating a second national cheetah survey for Kenya that will use improved methods.

The higher GNSS (global navigation satellite system) signal accuracy due to deactivating the selective availability in combination with advancements in handheld computers allowed for a digital, spatially-referenced monitoring in the field since the beginning of the new millennium (cp. Haklay 2013). However, only the widespread use of tablets and smartphones triggered the development and deployment of mobile map apps supporting the monitoring of species (cp. Jepson and Ladle 2015). In this context, participatory sensing refers to data collection using the modern mobile phones as scientific instruments (Mousa et al. 2015). Benefits of Internet connectivity and the availability of sensors—including the GNSS receiver (commonly termed GPS)—allow for the acquisition and sharing of geospatial contents by anyone (Brovelli et al. 2016), the capability to send valuable supplementary information like high-quality photographs or descriptions (Saag et al. 2010), and the uploading of data to an online database in real-time or with a short time lag if offline, thus avoiding errors in data transcription or loss of data forms (Adriaens et al. 2015). A map interface can support orientation in the field as well as georeferenced data entry if objects are not accessible or the GNSS signal is limited (Brovelli et al. 2016). Small conservation organizations like ACK can benefit from this development enabling a digital workflow with accelerated and more efficient data collection, transfer and analysis (cp. Mihanyar et al. 2016).

Applications that have the capabilities to be understood and operated in a relatively simple way—referring here to monitoring data and coordinated participants—by using mobile applications, are suitable for what is called ‘citizen science’, where volunteers contribute to scientific projects in various ways (Newman et al. 2016). Citizen science supports nature conservation (e.g. Forrester et al. 2017; Ellwood et al. 2017) and has recently been used as a tool in cheetah conservation (van der Meer et al. 2017). The largest motivations for people to engage in citizen science projects are to learn more about topics of their personal interests and to help to conserve nature both locally and on a global scale (Ott 2015). There are also portals for communities with discussion forums (e.g. iNaturalists.org) which can act as a first step towards quality data contributions to citizen science projects. Established portals provide the chance to start a project for species monitoring where volunteers can provide basic information on the sighting and its location and upload a photo for verification. In citizen science, concerns exist about the data quality (see Burgess et al. 2017) but it can be argued that reliability depends on training opportunities for the volunteers (Ellwood et al. 2017; Forrester et al. 2017). Citizen science does not necessarily need to target the general public (cp. Forrester et al. 2017), as in the case where ACK field scouts cannot be considered professional scientists and are trained to collect specific data within the community study scope.

During the first national cheetah survey in Kenya, ACK faced several challenges with data collection and entry: the data forms had to be distributed and collected from the field, it took hours to enter (digitize) the data, it was cumbersome to identify errors, and the process was error prone and slow in creating final reports. During the second national cheetah survey, the aim is a state-of-the-art usage of GIS tools not only in all stages of planning and field data collection, but also when generating geospatial data, and in occupancy and gene flow modelling (Mutoro et al. 2016), to produce scientifically sound knowledge. “Going digital” in this case refers to paperless data collection with the ability to collect, record, manage and store the data efficiently. It includes harmonizing the myriad of the regularly used forms on patrols, transects, or in interviews for cheetah monitoring and national survey studies to gain consistent data, improved usability, and standardized workflows.

Informed conservation management for the cheetah requires reliable status assessments and inferences on their ability to survive in human-influenced landscapes. Some of the challenges of field work include the ability to record data efficiently. To further develop the data collection protocol of ACK this paper demonstrates the process of developing custom mobile map apps for transitioning from collecting data with paper forms to digital data collection via smartphones covering the conceptualizing of a system from data collection to data storage, its implementation as well as its testing on the ground which included an evaluation of how the developed tools can support and are received by the end users.

2 Methods

2.1 Study Area and Data Collection

Figure 1 shows Kenya with its cheetah distribution ranges and the location of the two study sites of ACK: 'Salama' and 'Samburu'. 'Salama' has experienced a severe decline in cheetahs, while 'Samburu' has one of the largest cheetah densities in Kenya (Durant et al. 2017). At the two sites, data is collected using sightings or tracks of cheetah and other wildlife during patrols (any 6 km, 2–5 days per week), or as predefined transects on a designated day/time (1 per week). In addition, scat of cheetahs is collected by field scouts and detection dog teams and subsequently analysed at a laboratory to determine individual DNA and relationships of the different cheetah populations within the country.

Field scouts also conduct interviews with local community members about their experiences with predators, especially focusing on human-wildlife conflicts. Outside of the regular study sites, national surveys combine the various methods of transect surveys and community interviews to collect range-wide data on cheetah presence, prey availability and threats to cheetah survival based on a 20 km × 20 km sampling grid. In addition, any data that is collected by field teams or citizen science participants on a national scale enhances geodatasets on cheetah presence or human development features.

2.2 Conceptualizing a State-of-the-Art Solution

The development environment for Esri's ArcGIS software platform was selected for this project (based on Jedersberger 2017). AppStudio for ArcGIS allows the user to create mobile apps with GIS functionalities by means of the Qt framework and its mark-up language QML. JavaScript codes can also be included to provide specialized functionality. AppStudio offers a do-it-yourself tool for developing cross-platform mapping apps, facilitated through out-of-the-box templates. The here

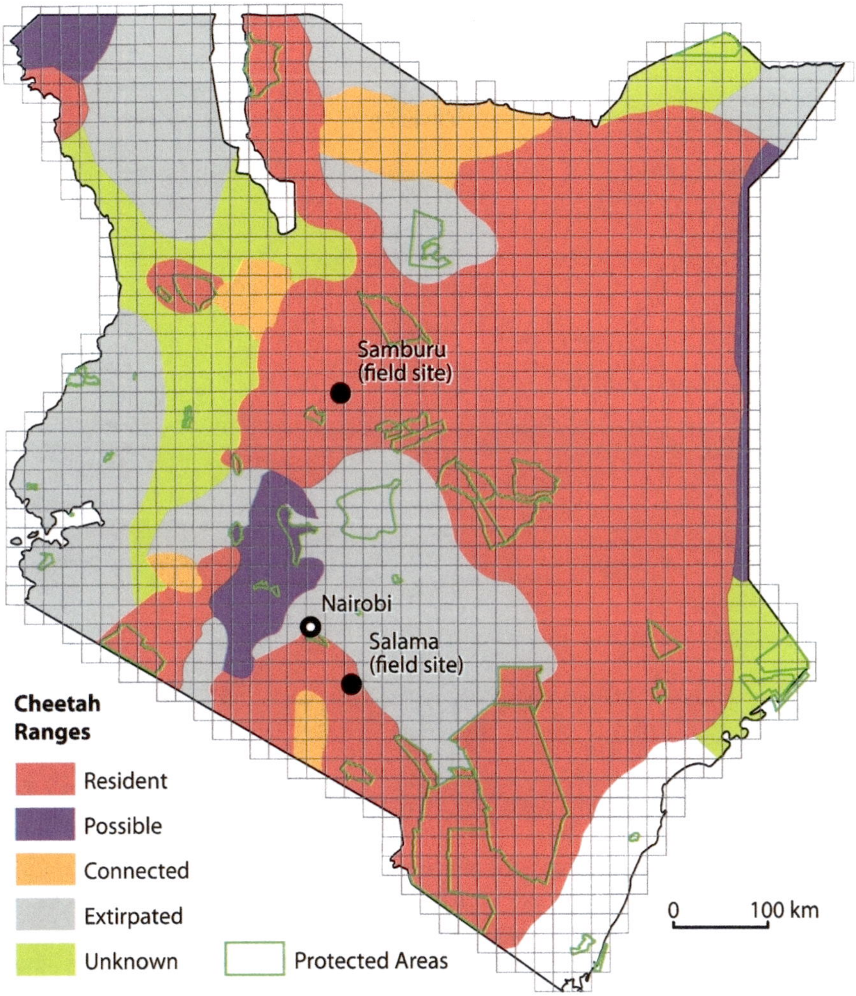

Fig. 1 Kenya with its cheetah distribution ranges overlaid by the sampling grid and the location of the two ACK study sites 'Salama' and 'Samburu'

described native apps are based on the Quick Report template. However, editing and modifications are required due to limitations of mere app configuration. AppStudio Player is used for testing the apps on various platforms (here different Android versions) and device models (smartphones brands). ArcGIS Online was selected for transferring the data recorded in the field to the database and compiling the actual apps for testing and operation in the field.

For conceptualizing a mobile mapping system for ACK (Fig. 2) three distinct user groups were envisaged: ACK field staff (i.e. scouts for monitoring wildlife and conducting interviews), ACK office staff (training and supervising the scouts,

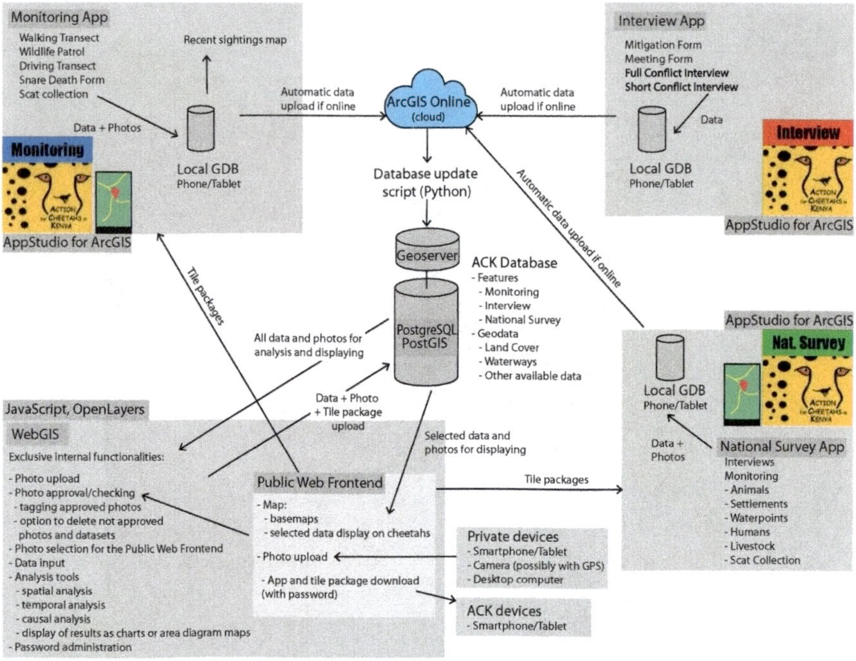

Fig. 2 Conceptualized mobile app monitoring system for supporting the cheetah monitoring work of ACK

responsible for data maintenance and analysis), and external users (contributing data on cheetah sightings).

For ACK's field work three apps are required: The app on 'Wildlife Monitoring' accommodates the monitoring of cheetahs and other predators (including tracks and scat) as well as prey animals during routine patrols. The app on 'Interviews' records interviews by ACK with the local population, i.e. when communicating about human-wildlife conflict and at meetings and events. The third 'National Survey' app contains functionalities of the other two apps plus additional details to support range-wide species distribution modelling. Each app consists of several forms: The monitoring app covers four forms for walking transects, wildlife patrols, driving transects and one related to mortalities caused by e.g. poachers. The interview app offers three forms on conflict, on mitigation and for meetings. The national survey app includes one interview and six monitoring forms to aid in mapping of biotic and abiotic influences of the cheetah ecosystem.

App development aims to harmonize workflow, which allows the reuse of coding scripts. A map for displaying the location of the collected data offers two basemaps, a map with topographic features and satellite imagery. It is also possible to work offline to accommodate scarce network connection in the field. All data requires an associated location which uses both GNSS signals for documenting location directly, and also pinpointing on the map where poor GNSS signal is an issue. The smart device's camera is used to document sightings, and, in case of the national survey app voice recordings are able to be stored for later analysis.

As the operation of ACK is depending on donors, a low-cost solution is needed to support both a secure monitoring in the field and the management of sensitive data. For the collected data a free object-relational database was designed to run on a server. The data collected by an app is uploaded to the cloud (ArcGIS Online) when the smartphone is online and subsequently deleted from the app's local database. A backend routine based on a Python script was designed for managing the data transfer and import from ArcGIS Online to the new database.

A public Web frontend will allow data entry on cheetah sightings by external users and enable the public to view selected monitoring efforts. By publishing selected photographs including annotations it can be used to inform and attract donors. The set-up of a customized internal WebGIS would serve ACK office staff by customizing additional functionality to support and work with the new monitoring suite of tools.

2.3 *Challenges Encountered During Implementation*

Software development followed the waterfall model (Royce 1970) due to the fact, that most of the development was achieved in Germany, while installation of the tools and a usability study took place in Kenya (Fig. 3).

During the proof-of-concept it became apparent, that the intended concept could not be realized due to various reasons: A complete data transfer including the recorded photographs via Python script to the database would have required substantially more programming or an ArcGIS Server license, the latter currently not being available at ACK. Therefore, Esri's File Geodatabase was used locally on a standard computer, also because a suitable server machine was lacking. Further,

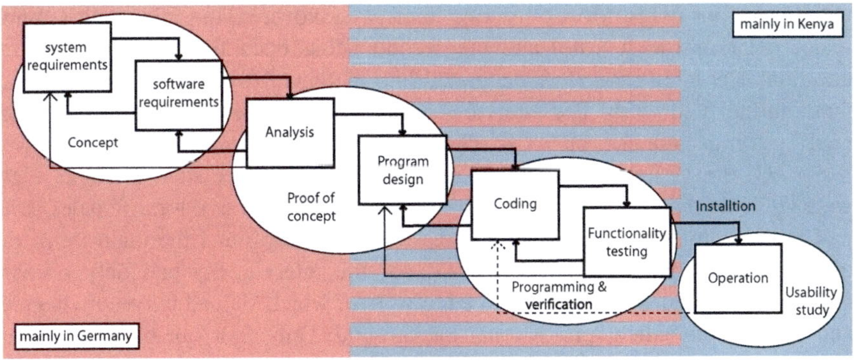

Fig. 3 Workflow of the chosen software development approach for making a mobile app monitoring system available for supporting the cheetah monitoring work of ACK

the frontend will need to be included in a rather restrictive content-based management system of an external service provider. This server cannot host for free large files like the tile packages. Therefore, it was opted to utilize ArcGIS Online for the Web frontend's map display (via OpenLayers) and photo upload (via JavaScript API for ArcGIS). The individually required tile packages need to be distributed together with the apps. The idea of the mobile apps' sightings map was dropped due to the need of extra programming efforts for not only showing current but also older sightings, i.e. those having been uploaded already and thus being locally deleted. Otherwise the scouts could fear that they would have lost data. Due to technical issues with the asynchronous functioning of JavaScript in conjunction with the used Qt runtime version, no automatic data upload will be available. Instead the user has to manually start the feature service layer uploads one after the other. Of advantage is here, that the user can decide on the time of upload.

2.4 Testing and Usability Study

For verification of the programming codes black-box and white-box testing were performed and single module, module integration and environment integration were tested (Franz 2014). Although the mobile apps were tested on site in Kenya, the overall success of the implementation was surveyed via a usability study at ACK study sites. The questionnaire evaluated the applications' user friendliness and gathered feedback for further improvements.

A usable software requires a good user experience (UX) which is tied to the users' expectations in regard to the functionalities and depends on the user interface (UI) design (Hsu et al. 2017). The use of qualitative methods is appropriate when the sample size is small and can ask for detailed statements (Brosius et al. 2016). The usability study used semi-structured interviews that allow for spontaneous questions to trigger more in-depth answers (Aeppli et al. 2014). 33 research questions with a phrased hypothesis for each resulted in 39 questions (indicators) with up to four items. Most often two items were covered, the first using a rating scale (polar or Likert-scale) and the second being open thus acting as control question. The interviews were integrated into members' work, e.g. accompanying them during field work. The interviewer was the developer, sometimes accompanied by a translator, i.e. an ACK office staff member who had been interviewed before. Of the 12 interviewees, four represented ACK office staff, the other eight were ACK field staff. Seven interviewees were males, and five were females, with half falling into the senior youth age class (26–35 years) and the majority of the others being above the age of 45. 25% of the interviewees had only primary education, 75% finished high school education, of which six had university degrees and two had an 'alien' professional background. Only four out of 39 questions seemed not to have been fully understood. Figure 4 provides the structure of the testing procedure that was followed with in total four interview parts dedicated to different user groups and interleaved by tool testing phases.

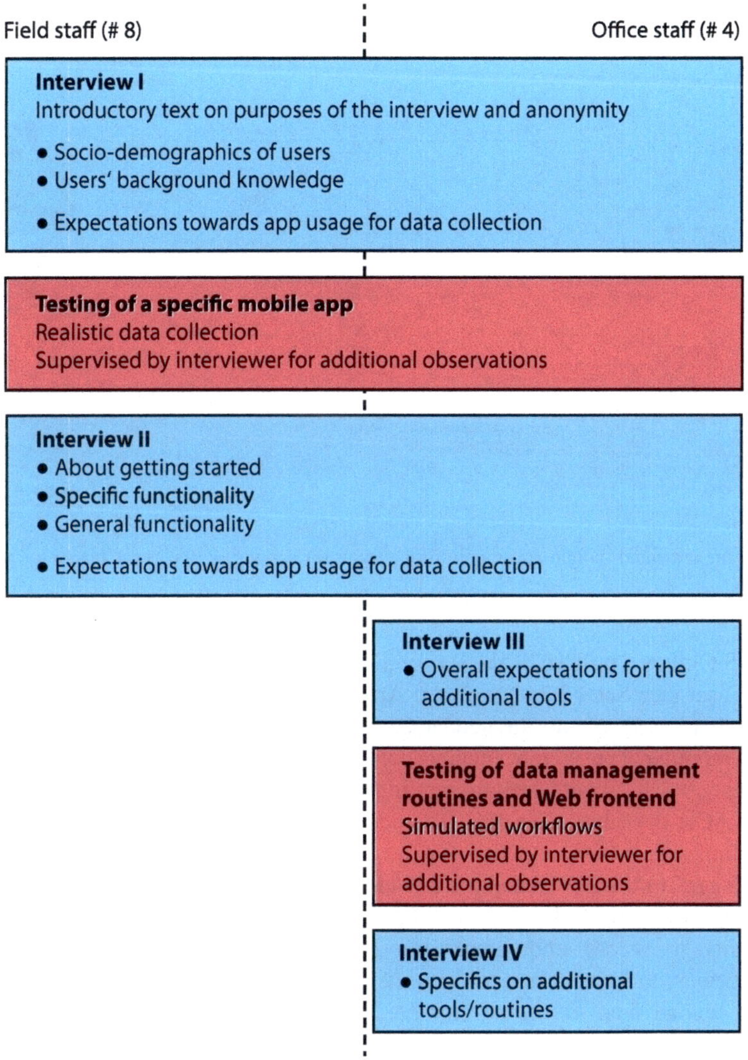

Fig. 4 Usability study procedure of interleaving interview questions and actual testing phases depending on the targeted user group

3 Results

3.1 Realized Set-up of the Mobile App Monitoring System

Figure 5 shows the realized mobile app monitoring system set-up. It covers the three mobile apps with the interview app now also including a map display. The custom map tiles have to be delivered to the app users using an external device.

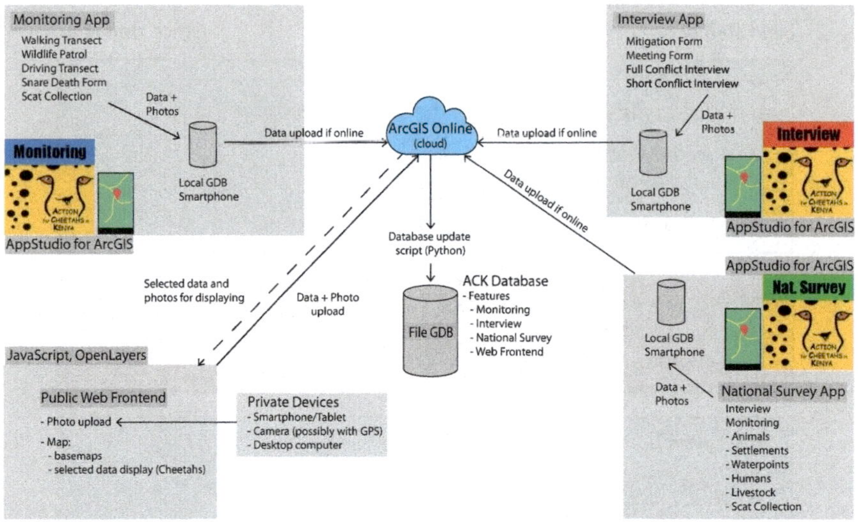

Fig. 5 Implemented mobile app monitoring system for supporting the cheetah monitoring work of ACK

ArcGIS Online is utilized as the server. Data upload to the cloud has to be triggered by the user (see Sect. 2.3 above). An ArcPy/Python script allows for downloading field data recordings into the geodatabase which is accessed and maintained on an ACK computer. The Web frontend accommodates the citizen science component of informing ACK about cheetah sightings and of visualizing monitoring success where ACK considers it helpful. The additional WebGIS functionality has not yet been implemented.

In regard to the functionalities included in the developed mobile apps, Table 1 provides an overview. The functionalities are mainly related to the offline map, the capability to record photographs, to handle additional textual messages and explanations, to write data records to the memory, to save them in a local database and to upload them for cloud storage. The table reveals what functionality was easily coded, while pointing to those features that required more or sometimes substantial additional coding efforts. The efforts included the accommodation of different devices (despite AppStudio's claim of being platform independent), the uploading of audio files similar to the method of uploading photographs, a higher versatility in regard to input options via forms, the length of the input forms covering up to 16 pages as well as the complex app structures with several input forms and differing functionality. Furthermore, the upload to the cloud happens not per observation but per feature service layer due to the complexity of the monitoring or interview schemes.

In Fig. 6 a screenshot each of the monitoring apps' mobile map and the Web frontend is presented. The apps' map reveals deficiencies in regard to the matching of line features, which is due to their various external sources. However,

Table 1 Functionalities implemented in the developed mobile apps, pointing here to additionally required coding efforts

App functionality	Implemented functionalities	Additional programming efforts required for
Offline map	Loading as raster tile package	Use of different device models and operating system versions
	User centered display and dynamic zoom	
	Switching between topographic feature map and VHR satellite imagery	
	Static legend and dynamic scale bar	Newly developed due to problem with QML scale bar object
	Pinpointing location by moving the map	Storing GNSS coordinates in addition
Taking photograph or audio file	Via device camera or from file system	Modification to also access audio files
Dynamic input form	Data entry via dynamic input forms	Flexibility in order to support radio buttons, checkboxes and dynamic page structures
	Navigation between consecutive pages	Several pages per input form
	Input restrictions, default values, mandatory fields	
Responsive design	Working with dimensions relative to the screen size	(QML-type display scale factor not working correctly, doing without)
User communication	Messages: warning for not losing data, mandatory fields still to be filled, data in local database still in need of uploading, upload is completed	Checking conditions and offering options for required actions
	Providing information pages and about pages	
Storing data in temporary memory	Supporting several feature service layers (one per input form)	Complex app structures with several input forms plus differing in functionalities
	Checking pages for last edits before storing in the local database	
Saving data in local database	Writing data from temporary memory to the database tables	
Uploading data to ArcGIS Online	Offering upload possibility when online: a button per feature service layer, to be triggered manually	Not per observation, but per feature service layer for avoiding data loss due to asynchronous JavaScript and in regard to attached files

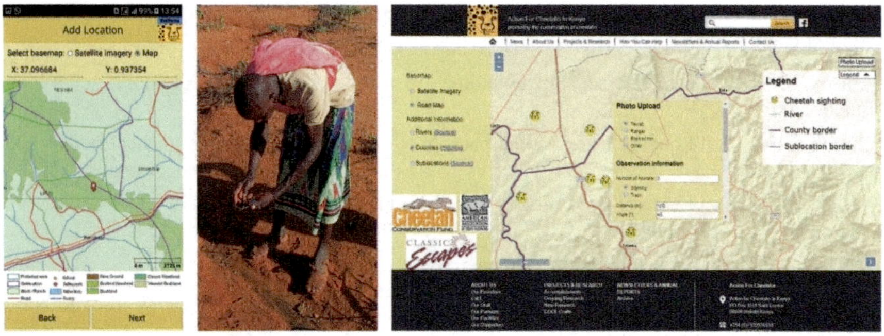

Fig. 6 Screenshots of the monitoring apps' mobile map (plus a photo from app use in the field) and the public Web frontend

free geodata of a higher quality is currently not available in Kenya. The Web frontend is not yet publicly accessible and there is still potential to enhance its graphical appearance.

3.2 Outcomes of the Usability Study

Table 2 summarizes the relevant outcomes of the usability study and their interpretation. In regard to background knowledge, the users judged their knowledge on mobile device experience, usage of maps (in general and on smartphones) and their familiarity with monitoring apps higher than expected. Reasons for that might be the overall higher education level of ACK staff plus the ACK training including the prior use of the iNaturalist platform. However, there was almost no experience with interpreting satellite imagery.

The concept of CloudGIS and cloud storage was only known to about half of the interviewees, who had not necessarily a clear idea about it. The expectations that the new apps would simplify the data collection process were high. After having tested and used a mobile app for one to four hours, the feedback on getting started and the general process flow was positive. This was expected as the mobile apps are closely based on the paper forms. The answers related to specific functionalities provided a more mixed picture: Users stated that they felt comfortable pinpointing positions, which is in contradiction to an uncertainty they felt about the usefulness of the basemap contents. Most of the other functionality seemed to be understood, apart from the data upload to the cloud. After a maximum of half a day of using the app, users were unable to remember the displayed messages. When being asked if all acted as expected and for elements distracting from the actual work, the interviewees pointed out general technical issues encountered. On repetition of the expectation question, still nine of the earlier eleven respondents judged the data recording process to be much easier, however, the reasons changed and provided a more realistic perception (see Sect. 4 Discussion).

Table 2 Outcomes of the usability study testing the developed mobile app monitoring system

Construct (Topic)	Relevant outcomes	Hypoth. confirm.	Interpretation	Impact on tool use
Checked before using the mobile apps				
User background knowledge	Confident in regard to experience with mobile devices; the majority using it already for monitoring (iNaturalist)	x	Due to efforts/ support by ACK (using iNaturalist for transition)	Will add to likelihood of sustainable app usage
	Map use seems common while half of the interviewees are not familiar with imagery (satellite/aerial)	(x)	Does not clarify on map reading abilities	Need of learning how to read/ interpret a satellite image
	Half of the people have a vague idea on the concept of cloud GIS and storage	(x)	Cloud storage being the more familiar concept	Confidence in using the cloud needs to be built
	Half of the people use maps on the smartphone (mentioning Google Maps)	(x)	Why not all (but 2) remembering map functionality in iNaturalist?	Skeptical in regard to the apps' map
Expectations on app use	High expectations on apps simplifying the data collection process	(x)	Technology-devout; benefits mainly listed on field work	Disappointment likely to impact
After having tested/used the mobile apps				
Getting started	A large majority judged the general setup (start/process flow) to be easy going	✓	Experience with iNaturalist; apps following paper forms	Not much training required

(continued)

Table 2 (continued)

Construct (Topic)	Relevant outcomes	Hypoth. confirm.	Interpretation	Impact on tool use
Specific functionality	Offline basemap: not a clear picture in regard to usefulness of map content, most users feel comfortable with pinpointing locations	x	Preference of basemap over imagery contradicts confidence level in pinpointing locations	Accurateness of pinpointed locations to be questioned
	Most of the functionalities easy to understand	✓	Using the apps seems to be straight forward	Adds to promising outlook for app usage
	More difficult to understand are: displayed messages, uploading data for cloud storage	(✓)	Helpfulness of warnings/reminders doubted; questioned if upload process understood	Thoroughness in understanding the upload process being crucial
General functionality	Expected actions, distracting elements: only time when technical issues were pointed out	x	No issue of not being informed or being distracted	Remaining technical issues should be solved
Judgement (difference in perception before/after app use)	Still a high percentage (9 of 11 earlier on) judged the data recording process to be much easier, but the reasons given change	✓	More realistic perception now: not less work, but beneficial for data transfer	App testing a needed step for leveling the expectations
Addressing ACK office staff only				
Overall expectations	Expectations referred to easier and more efficient data transfer mainly, not aware of technological difficulties, critical in regard to Internet speed when sending large files	(✓)	Positive attitude	Open-minded, i.e. will be supportive in making the transition work

(continued)

Table 2 (continued)

Construct (Topic)	Relevant outcomes	Hypoth. confirm.	Interpretation	Impact on tool use
Specifics on additional tools/routines (data management, Web frontend)	Usage of new database structure, Python script to download data from the cloud, benefit of Web frontend, its photo upload, tagging the photographs: easy to use, able to list some benefits	(✓)	Optimistic in handling add-ons; see benefits of database in data analysis mainly and of Web frontend in visual presentation of ACK work	Promising outlook for data handling around actual mobile app use
	Creating tile packages of basemap: don't see problems	x		

The questions posted to ACK office staff members only revealed an overall positive attitude, being happy about the easier and more efficient data transfer. Technology-wise they felt the greatest challenge being linked to the limited Internet speed. The office staff were optimistic in regard to handling the add-ons for data management, they see the benefits of the database mainly for data analysis, and of the Web frontend for visual presentation of ACK's work.

4 Discussion

The Esri software environment was chosen for development of the mobile app monitoring system due to the variety of options available and the previous, but limited, use of ArcGIS Desktop software by ACK office staff. AppStudio was selected over Collector for ArcGIS or Survey 123 due to limitations in customization and the complexity of ACK interview forms. Only AppStudio allowed the full freedom in app design, but it requires advanced programming efforts. By incorporating ArcGIS Online as server a version working for ACK's current ability was realized, which covers the required functionality. As a non-profit organization, ACK qualifies for a conservation grant to utilize Esri support for the software tools used.

However, not all of the anticipated functionality was supported in this way. For example, serving the mobile app users with custom basemap services depending on their current working area (e.g. when attending to the national survey) is not possible right now. Also the database structure had to be kept simple, which is of disadvantage considering the growing amount of data to which the new mobile apps will contribute. The concept for a state-of-the-art solution (Fig. 2) reveals the need for scaling at ACK, both in regard to server technology and skilled personnel for maintenance and development. With their own server the programming of a custom WebGIS for ACK's work, accessible anywhere in the country, would be the next step adding to the smooth functionality of the newly developed package of dedicated mobile apps.

ACK is still at the beginning of going digital. Using iNaturalist, although it serves a different purpose, clearly helped for the transition from paper forms to the new mobile apps, i.e. in training the field staff. Training is also needed for ACK office staff for moving away from the use of Excel for data handling to completely relying on the geodatabase. The Web frontend will best serve as a tool to visualize ACK monitoring efforts to potential donors and allowing others (mainly park rangers and naturalists) and the public to contribute in a simplified way.

According to Newman et al. (2016), mobile applications make the entering of monitoring data more efficient and facilitate the coordination of participants. The harmonized workflows of locating an observation, documenting it by a photograph, adding descriptive information, checking and editing the reported data, and saving the data is linked to both efficiency and coordination. Consistency in the data and ready-to-use data formats for immediate visualization and subsequent analysis are ensured through this technology. Although the online functionality promises easy data transmission from the field to the office, this seems to be the weak spot in the current implementation. Due to the need to work offline, automated uploading is not feasible with the current technology but workarounds had to be implemented whose responsible handling requires a certain understanding by the users. Here, ACK's office staff need to invest in training and to take extra cautions to ensure complete data upload to the cloud by means of supervision. Sticking to one type of smartphone would be advantageous with less programming hassle despite support of responsive design by AppStudio. Issues of the mobile apps interfering with the field staffs' privacy could arise (see Christin 2016) internally due to creating movement profiles or externally due to chances of gaining information about the people affiliated with the project.

The public Web frontend is still a rather basic version that should be further elaborated. Emphasis should go into an appealing graphical design to serve the purpose of attracting donors. Limiting the citizen science input (based on the Wildlife patrol form) is intentional. It is also not competing with iNaturalist, which is a portal supporting communication. The tool has been developed as a contributory project only (cp. Gray et al. 2017) where volunteers help to collect data. This is the most common type of citizen science projects with the benefit that the scientists still have the control. Among the advantages listed for citizen science, which are of relevance for the developed applications aiming at data for a large area, are the chance of involving a higher number of participants (Vann-Sander et al. 2016) and to monitor more unique occurrences (Levine and Feinholz 2015). Due to the technological advances it is possible to reach out to anyone. However, care has to be taken not to attract those causing mischief (see Mousa et al. 2015).

The usability study, which did not include external users, has shown that the developed apps and tools are well-received (see Table 2, last column). Overall it provides a promising outlook, but we are aware that there is still room for enhancements. According to Shneiderman et al. (2014) usability becomes more important if there is the possibility to choose an alternative. Therefore, qualitative testing was used to gather feedback for gaining insights how to improve the tested application (Krug 2010). The opinion of Hennig (2016) is that developers tend to overlook that the app users might be IT and GI laypersons.

Even though we believed that most field staff lacked experience with smartphones or knowledge about maps, the little knowledge available by some was sufficient to handle the new process of entering data after recording a few observations with ease. Even if inputting the required information in the forms stretched across as many as 16 pages, it was not an issue because it borrowed from the known application of filling lengthy paper forms (cp. Hennig 2016). Indirectly, the test answers revealed that data upload was not easily understood by the tested people. Showing the uploaded data within ArcGIS Online improved understanding and should be considered as a part of any training of the developed or similar mobile apps.

The greatest challenge observed in the field was using the map for pinpointing the location, as some test persons had obvious difficulties reading the map or the satellite image. Hennig (2016) points to the challenge of app users often falsely placing locations. The map included in the mobile apps does not necessarily show the required details for being accurate. However, using in addition the very-high resolution satellite imagery was not considered by the users. Here, training is required to create skills of reading maps and interpreting the satellite imagery.

Figure 7 shows the change in user expectations answering the same question before and after the actual testing of a mobile app. While before the testing users expected faster or less work besides an easier reporting to office staff, after several

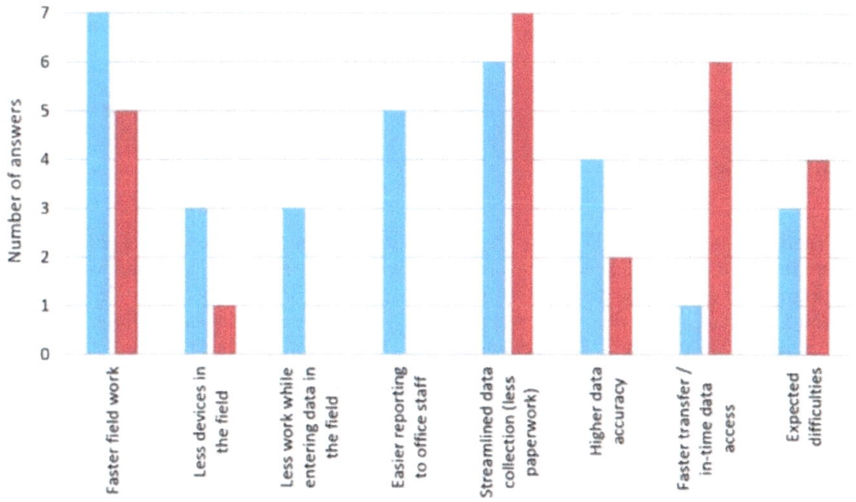

Fig. 7 Change in user expectations of mobile app impacts on data collection based on the same question but asked before (blue bars) and after the app testing (red bars)

working hours with an app answers became more realistic. As main benefits besides a faster and streamlined data collection, now faster data transfer and in-time data access were stated.

Although the usability testing revealed a bright future for the use of the apps in the field, reality in the subsequent months turned out differently. The scouts have to be closely guided over a longer period to assure the continuous use of the apps as well as a successful upload of observation data. Here, disappointment towards the technology not making the recording process less laboursome in combination with intangible results for the scouts might be still impacting. The Web frontend with its visualization of a constantly growing pool of cheetah recordings could help here spurring the motivation within ACK's field staff.

5 Conclusion

This paper described the process and discussed the successes and challenges of implementing a custom-designed mobile app monitoring system for supporting the monitoring of cheetahs in Kenya. For reaching a routine work level, there is still some way to go. But overall, signs point to the system becoming greatly beneficial in the long run. As ACK is preparing for a national survey this is required. The setup provides ACK with a system of apps and tools to support its conservation efforts. It allows monitoring the distribution of cheetahs, determining the populations and connections between them, and assessing their various threats. It further allows to record information of the local human population about their experience with and perceptions towards cheetahs, which helps in reducing threats. All data and information can now be collected in a digital format and transferred from the field via a cloud into a database at ACK. The added georeference per data record allows for spatial data analysis and species distribution modelling.

References

Adriaens T, Sutton-Croft M, Owen K, Brosens D, van Valkenburg J, Kilbey D et al (2015) Trying to engage the crowd in recording invasive alien species in Europe: experiences from two smartphone applications in northwest Europe. Manag Biol Invasion 6(2):215–225

Aeppli J, Gasser L, Gutzwiller E, Tettenborn A (2014) Empirisches wissenschaftliches Arbeiten, 3rd edn. UTB/Klinkhardt, Bad Heilbrunn

Andresen L, Everatt K, Somers MJ (2014) Use of site occupancy models for targeted monitoring of the cheetah. J Zool (Lond) 292(3):212–220

Brosius H, Haas A, Koschel F (2016) Methoden der empirischen Kommunikationsforschung, 7th edn. Springer Fachmedien, Wiesbaden

Brovelli MA, Minghini M, Zamboni G (2016) Public participation in GIS via mobile applications. ISPRS J Photogramm Remote Sens 114:306–315

Bunce RGH, Metzger MJ, Jongman RHG, Brandt J, de Blust G, Elena-Rossello R et al (2008) A standardized procedure for surveillance and monitoring European habitats and provision of spatial data. Landsc Ecol 23(1):11–25
Burgess HK, DeBey LB, Froehlich HE, Schmidt N, Theobald EJ, Ettinger AK et al (2017) The science of citizen science: exploring barriers to use as a primary research tool. Biol Conserv 208:15–28
Christin D (2016) Privacy in mobile participatory sensing: current trends and future challenges. J Syst Softw 116:57–68
Durant S (2004) Survival of the fastest—the cheetahs of Serengeti. Africa Geographic, June 2004, pp 30–33
Durant S, Mitchell N, Ipavec A, Groom R (2015) *Acinonyx jubatus*, Cheetah. In: IUCN (ed) The IUCN red list of threatened species 2015, e.T219A50649567. http://dx.doi.org/10.2305/IUCN.UK.2015-4.RLTS.T219A50649567.en
Durant S, Mitchell N, Groom R, Pettorelli N, Ipavec A, Jacobson AP et al (2017) The global decline of cheetah *Acinonyx jubatus* and what it means for conservation. Proc Natl Acad Sci USA 114(3):528–533
Ellwood ER, Crimmins TM, Miller-Rushing AJ (2017) Citizen science and conservation: recommendations for a rapidly moving field. Biol Conserv 208:1–4
Forrester TD, Baker M, Costello R, Kays R, Parsons A, McShea WJ (2017) Creating advocates for mammal conservation through citizen science. Biol Conserv 208:98–105
Franz K (2014) Handbuch zum Testen von Web- und Mobile-Apps—Testverfahren, Werkzeuge, Praxistipps, 2nd edn. Springer Vieweg, Groß-Gerau
Gray S, Jordan R, Crall A, Newman G, Hmelo-Silver C, Huang J et al (2017) Combining participatory modelling and citizen science to support volunteer conservation action. Biol Conserv 208:76–86
Haklay M (2013) Citizen science and volunteered geographic information: overview and typology of participation. In: Sui D, Elwood S, Goodchild M (eds) Crowdsourcing geographic knowledge. Volunteered geographic information (VGI) in theory and practice, pp 105–122. Springer, Dordrecht
Hennig S (2016) Zur Berücksichtigung von Nutzern, ihren (Usability-) Anforderungen und Kompetenzen in Bezug auf Online-Karten. In: Hennig S (ed) Online-Karten im Fokus: Praxis-orientierte Entwicklung und Umsetzung. Wichmann, Berlin, pp 53–70
Hsu C, Chen Y, Yang T, Lin W (2017) Do website features matter in an online gamification context? Focusing on the mediating roles of user experience and attitude. Telemat Inform 34 (4):196–205
Jedersberger J (2017) Conceptualizing and implementing mobile mapping tools to support cheetah monitoring in Kenya. Master's thesis, Faculty of Information Management and Media, Karlsruhe University of Applied Sciences, Germany
Jepson P, Ladle RJ (2015) Nature apps: Waiting for the revolution. Ambio 44(8):827–832
Kenya Wildlife Service (2010) Kenya national strategy for the conservation of cheetahs and wild dogs. In: KWS-Research. Kenya Wildlife Service, Nairobi
Krug S (2010) Web Usability – Rocket Surgery Made Easy. Addison Wesley, München
Kuloba BM (2011) Modeling Cheetah *Acinonyx jubatus* Fundamental Niche in Kenya. Master degree assignment, Faculty of Geo-Information Science and Earth Observation, University of Twente, The Netherlands
Levine AS, Feinholz CL (2015) Participatory GIS to inform coral reef ecosystem management: mapping human coastal and ocean uses in Hawaii. Appl Geogr 59:60–69
Masseloux J, Epps C, Duart A, Schwalm D, Wycstra M (in press) Using detection/non-detection surveys and interviews to assess carnivore site use in Kenya. Afr J Wildl Res 48(1)
Mihanyar P, Abd Rahman S, Aminudin N (2016) The effect of national park mobile apps on national park behavioral intention: Taman Negara national park. Procedia Econ Financ 37:324–330

Mousa H, Mokhtar SB, Hasan O, Younes O, Hadhoud M, Brunie L (2015) Trust management and reputation systems in mobile participatory sensing applications: a survey. Comput Netw 90:49–73

Mutoro N, Schaab G, Wykstra M (2016) Analysis of cheetah (*Acinonyx jubatus*) occupancy and gene flow in Kenya using GIS tools. In: Conference proceedings, 11th Esri eastern Africa user conference, Kisumu (Kenya), 2–4 Nov 2016, pp 18–22. https://www.esriea.com/user-conference/Proceedings%202016.pdf. Accessed 30 Jan 2018

Newman G, Chandler M, Clyde M, McGreavy B, Haklay M, Ballard H et al (2016) Leveraging the power of place in citizen science for effective conservation decision making. Biol Conserv 208:55–64

Ott G (2015) Hobbys—Private Quellen der Bürgerwissenschaft. In: Finke P (ed) Freie Bürger, freie Forschung. Oekom, München, pp 70–79

Royce W (1970) Managing the development of large software systems. In: ICSE '87 Proceedings of the 9th international conference on software engineering, 30 Mar–2 Apr 1987, pp 328–338. Monterey (USA)

Saag A, Randlane T, Leht M (2010) Keys to plants and lichens on smartphones: Estonian examples. In: Nimis PL, Vignes Lebbe R (eds) Tools for identifying biodiversity: progress and problems. EUT, Trieste, pp 195–199

Shneiderman B, Plaisant C, Cohen M, Jacobs S (2014) Designing the user interface: strategies for effective human-computer interaction, 5th edn. Pearson Education, Essex

van der Meer E, Broekhuis F, Chelysheva E, Wykstra M, Davies-Mostert H (2017) Citizen science in cheetah research. In: Nyhus P (ed) Cheetahs: biology and conservation. Elsevier Science Publishing, San Diego (CA), pp 471–483

Vann-Sander S, Clifton J, Harvey E (2016) Can citizen science work? Perceptions of the role and utility of citizen science in a marine policy and management context. Mar Policy 72:82–93

Wambua CM (2008) Wildlife density, distribution and abundance with emphasis on the cheetah prey in Machakos and Makueni Districts, Kenya. Master's thesis, Department of Biology, Addis Ababa University, Ethiopia

Introducing Spatial Variability to the Impact Significance Assessment

Rusne Sileryte, Jorge Gil, Alexander Wandl and Arjan van Timmeren

Abstract The concept of Circular Economy has gained momentum during the last decade. Yet unsustainable circular systems can also create unintended social, economic and environmental damage. Sustainability is highly dependent on a system's geographical context, such as location of resources, cultural acceptance, economic, environmental and transport geography. While in some cases an impact of the proposed change may be considered equally significant under all circumstances (e.g. increase of carbon emissions as a main contributor to the global climate change), many impacts may change both their direction and the extent of significance dependent on their context (e.g. land consumption may be positively evaluated if applied to abandoned territories or negatively if a forest needs to be sacrificed). The geographical context, (i.e. its sensitivity, vulnerability or potential) is commonly assessed by Spatial Decision Support Systems. However, currently those systems typically do not perform an actual impact assessment as impact characteristics stay constant regardless of location. Likewise, relevant Impact Assessment methods, although gradually becoming more spatial, assume their context as invariable. As a consequence, impact significance so far is also a spatially unvarying concept. However, current technological developments allow to rapidly record, analyse and visualise spatial data. This article introduces the concept of spatially varying impact significance assessment, by reviewing its current definitions in literature, and analysing to what extent the concept is applied in existing assessment methods. It concludes with a formulation of spatially varying impact significance assessment for innovation in the field of impact assessment.

R. Sileryte (✉) · J. Gil · A. Wandl · A. van Timmeren
Faculty of Architecture and the Built Environment,
Delft University of Technology, Delft, The Netherlands
e-mail: r.sileryte@tudelft.nl

J. Gil
e-mail: j.a.lopesgil@tudelft.nl

A. Wandl
e-mail: a.wandl@tudelft.nl

A. van Timmeren
e-mail: a.vanTimmeren@tudelft.nl

A. Mansourian et al. (eds.), *Geospatial Technologies for All*,
Lecture Notes in Geoinformation and Cartography,
https://doi.org/10.1007/978-3-319-78208-9_10

Keywords Impact significance assessment · Impact significance determination
Spatial decision support · Spatial differentiation

1 Introduction

Resource scarcity and rapid urbanisation both in light of rapidly changing demographics, power shifts and climate change create a snowballing challenge for sustainability. Fortunately, another, more positive, megatrend is the accelerating technological innovation that could provide important contributions to human well-being, improve labour efficiency, communication and education, and in that way rise society to the aforementioned challenges (Retief et al. 2016). Indeed the rapidly increasing computational power, means of sharing data and information, and digital literacy, are key drivers in the pursuit of sustainability.

In the past decade the concept of Circular Economy (CE), as a response to the aforementioned trends, has gained momentum with a rapidly increasing number of publications each year (Geissdoerfer et al. 2017). CE is an economic model based on renewability of all resources energy, materials, water, topsoil, land and air while retaining or creating value, promoting positive systemic impacts on ecology, economy and society, and preventing negative impacts (REPAiR D6.1 2017).

However, it is important to realise that the ultimate goal is not achieving circularity but sustainability. While these two terms tend to appear hand in hand, unsustainable circular systems also exist, which can cause unintended negative consequences (e.g., due to excessive use of transport and energy, unattractive working conditions or business abandonment due to failed adoption) (van Buren et al. 2016). Some previous studies upon conducting Life Cycle Assessment (LCA) have shown that closed loops are not always favourable from an environmental point of view (Haupt and Zschokke 2017). Therefore complex highly interdependent systems require a systems approach (Williams et al. 2017).

The shift towards circularity is going to require changes in design, production, logistics and consumer behaviour. The sustainability of these systems is highly dependent on their geographical contexts, such as location and availability of resources, presence of skilled labour force, economic, environmental and transport geography (Accorsi et al. 2015). Policies and shift supporting tools cannot be applied uniformly across the territory because the economic, social, environmental and institutional situations differ not only on a national level but also locally, on a community level. These instruments need to include place-based contextualised significance assessments of probable impacts, with Geographic Information Systems (GIS) as their basis.

This paper is linked with the H2020 Research and Innovation Action project REPAiR (Resource Management in Peri-urban Areas). The project aims to provide a Geodesign Decision Support Environment (GDSE) as a tool to assist local and regional authorities in creating and evaluating integrated spatial development strategies for Circular Economy. The strategies need to be specific for the place at hand, transdisciplinary, eco-innovative and promote the use of waste as a resource.

In the context of sustainability pursuit and transition towards CE, this paper proposes that both impact and its context assessments cannot be applied uniformly, and that the significance of impacts is a spatially varying measure. The paper is organised as follows. First, the general concept of impact significance is reviewed setting the theoretical framework of this study. Then, the need for spatial differentiation is discussed, defining the analytical framework that is later applied to four methods of impact assessment considered the most relevant in the context of this research. Recommendations for spatially differentiated impact significance assessment are given in the fifth section. Finally, conclusions are drawn followed by discussion on future work.

2 Theoretical Framework

"Impact Significance Assessment" or "Impact Significance Determination" is not commonly explored as a separate subject as a combined query in Scopus returns merely 11 distinct results (Query 1, Table 1). Reducing the query into "Impact Significance" results into a significantly larger number of 92 documents (Query 2). Analysis of keywords reveals that impact significance is most commonly associated with the topics of Environmental Impact Assessment (47/92 documents, Query 3) and Decision Making (10/92 documents, Query 4). Spatial Analysis or GIS are among the keywords in only 5 out of 92 documents (Query 5).

Impact significance assessment may serve two purposes (Zulueta et al. 2017): (1) identification of significant impacts to trigger authoritative actions after conducting an impact assessment of a certain project, and (2) impact significance assessment for the purpose of comparison between multiple alternatives as a support to the decision making process. The latter purpose is considered in context of this paper.

It differs notably how impact significance is assessed by different jurisdictions, as there is clearly an absence of a legal definition for the concept (Jones and Morrison-Saunders 2016). Wood (2008) describes impact significance as a dynamic, contextual, and political concept, characterised by uncertainty. The need for greater transparency, clarity and understanding of the significance determination process is recognized in the literature for decades. However, there is little apparent progress evident as the latest publications on the topic, such as Retief et al. (2016), Ehrlich and Ross (2015), Jones and Morrison-Saunders (2016), still mention the same issues related to significance assessment—i.e. lack of guidelines, vague terminology, high lexical and process uncertainty and low consistency and coherence.

The act of decision making is closely associated with social and political conflicts and deeply held values that reflect cultural, historical and social norms rendered acceptable by the community (Jones and Morrison-Saunders 2016). When the primary goal of significance assessment is sustainability, the focus shifts from minimising damage to maximising long-term gains (Gibson et al. 2005). The timespan

Table 1 A list of literature queries

No.	Query	Platform	Date
1	TITLE-ABS-KEY ("Impact Significance Assessment" OR "Impact Significance Determination")	Scopus	15 Sep 2017
2	TITLE-ABS-KEY ("Impact Significance")	Scopus	15 Sep 2017
3	TITLE-ABS-KEY ("Impact Significance") AND (LIMIT-TO (EXACTKEYWORD, "Environmental Impact Assessment") OR LIMIT-TO (EXACTKEYWORD, "Environmental Impact") OR LIMIT-TO (EXACTKEYWORD, "Environmental Impact Assessments") OR LIMIT-TO (EXACTKEYWORD, "EIA") OR LIMIT-TO (EXACTKEYWORD, "Environmental Impact Assessment (EIA)") OR LIMIT-TO (EXACTKEYWORD, "Environmental Assessment") OR LIMIT-TO (EXACTKEYWORD, "Environmental Impact Significance Assessment")	Scopus	22 Nov 2017
4	TITLE-ABS-KEY ("Impact Significance") AND (LIMIT-TO (EXACTKEYWORD, "Decision Making")	Scopus	22 Nov 2017
5	TITLE-ABS-KEY ("Impact Significance") AND (LIMIT-TO (EXACTKEYWORD, "GIS") OR LIMIT-TO (EXACTKEYWORD, "Geographic Information Systems") OR LIMIT-TO (EXACTKEYWORD, "Spatial Analysis")	Scopus	22 Nov 2017
6	"GIS AND" multi criteria "AND" decision support "AND (collaborative OR participatory OR cooperative) AND sustainability AND urban YEAR > 2015"	Google Scholar	1 March 2017

considered is longer, to include future generations, and more attention is given to assessing cumulative impacts (Lawrence 2007c). Both negative and positive impacts are addressed in contrast with assessments targeted solely at project approval. An impact of a proposed action is considered negatively significant if it inhibits sustainability. It is considered positively significant if it makes a durable contribution to achieving sustainable visions and strategies as compared to the baseline scenario (Barrow 2000).

To investigate what supplements impact magnitude to determine impact significance, a number of scientific publications have been reviewed. Besides publications returned by Query 1, additional studies have been chosen based on the summary made by Cloquell-Ballester et al. (2007), namely Table 1: Criteria to determine the significance of environmental impacts according to different authors (pg. 64); and some related citations in recent publications (Table 2).

One statement researchers and reviewers seem to agree on is that impact magnitude and impact significance are essentially different concepts that must not be confused (Thompson 1990; Lawrence 2007a; Wood 2008; Ehrlich and Ross

Table 2 A list of literature used for the review on impact significance assessment

List of references	
Duinker and Beanlands (1986)	Wood (2008)
Thompson (1990)	Ijäs et al. (2010)
Canter and Canty (1993)	Gangolells et al. (2011)
Antunes et al. (2001)	Briggs and Hudson (2013)
Bojórquez-Tapia et al. (2002)	Zulueta et al. (2013)
Cloquell-Ballester et al. (2007)	Ehrlich and Ross (2015)
Lawrence (2007a)	Jones and Morrison-Saunders (2016)
Lawrence (2007c)	Zulueta et al. (2017)
Lawrence (2007b)	

2015). Furthermore, there is general agreement that subjectivity cannot be avoided in the process, although it can be well informed by science and maximally transparent (Briggs and Hudson 2013). Thus, all reviewed publications seem to agree that there are two sides of impact significance—the rather objective side related with the impact's assessment, and the rather subjective one related to the values of importance given to that impact. Table 3 gives an overview of how different authors define significance and its two major components.

In its essence, impact significance determination is a multicriteria problem (Cloquell-Ballester et al. 2007). What the different authors (as well as official regulations) do not seem to agree on is which factors exactly characterise impacts, and which ones characterise importance. Generally, there is a lot of inconsistency in how the arguments are classified by authors. E.g. Bojórquez-Tapia et al. (2002), Cloquell-Ballester et al. (2007) regard synergic and cumulative effects as properties of the impact intensity, while Antunes et al. (2001), Lawrence (2007b), Wood (2008) regard cumulative effects as properties of the impact receiving context. Institutional arrangements are often viewed as constraints or background of the significance determination procedures (Briggs and Hudson 2013; Ehrlich and Ross 2015) rather than context properties (Lawrence 2007a; Wood 2008). Ijäs et al. (2010) classify impact permanence and reversibility on the same side as context susceptibility and Ehrlich and Ross (2015) regards everything as impact properties, while decision makers are responsible for setting a subjective threshold value to determine how all of these properties qualify for significance.

Moreover, there does not seem to be consensus between the authors on who is responsible for providing value judgements to determine the significance. While some authors attribute this responsibility to the experts and scientists (Antunes et al. 2001; Cloquell-Ballester et al. 2007; Zulueta et al. 2017), others suggest to ask public opinion (Antunes et al. 2001; Gibson et al. 2005; Gangolells et al. 2011; Briggs and Hudson 2013) or to leave it in the hands of decision-makers as advocates of society (Duinker and Beanlands 1986; Ehrlich and Ross 2015).

Table 3 Variables of impact significance according to different authors

Publication	Objective (impact) measure	Subjective (judgement) measure
Duinker and Beanlands (1986)	Magnitude and spatiotemporal distribution of change, reliability of prediction	Importance of environmental attribute to project decision makers
Canter and Canty (1993)	Impact intensity	Impact Context
Antunes et al. (2001), Wood (2008)	Impact magnitude	Context sensitivity
Bojórquez-Tapia et al. (2002)	Interaction intensity	Environmental vulnerability
Lawrence (2007a)	Impact characteristics	Characteristics of the receiving environment
Cloquell-Ballester et al. (2007)	Project activities	Environmental factors
Ijäs et al. (2010)	Scale of importance, magnitude of change	Permanence, reversibility, cumulativity, context susceptibility
Gangolells et al. (2011)	Impact severity	Concerns of interested parties
Zulueta et al. (2013, 2017)	Impact characteristics	Expert judgement
Briggs and Hudson (2013)	Impact on a receptor	Value of the receptor
Ehrlich and Ross (2015)	Impact adversity	Threshold of acceptability
Jones and Morrison-Saunders (2016)	Impact characterisation	Impact importance

This article's focus is on adding a spatial dimension to the objective procedure of impact assessment and to the subjective procedure of judgement. To offer a clear definition of the two, the arguments collected during the literature review were sorted into two groups (Table 4), one for the arguments given on the basis of impact characteristics and the other for the arguments given on the basis of the impact receiving context, based on the following definitions:

Impact Characteristics refer to all characteristics that would be computed using the same formula, if the same intervention was moved to a different context. E.g. if odour from a new facility affects 1000 m radius around the facility, then moving the facility to a new location would not change the radius.

Context Characteristics refer to all characteristics that would be computed with the same formula if an intervention with different impact would be placed in the same context. E.g. if habitat is negatively affected by odour, then placing a facility with smaller odour radius would not change habitat's sensitivity.

Based on the literature review, it has been concluded that Impact Significance can be defined as a function between Impact Characteristics and Context Importance (Eq. 1), where impact characteristics are provided by an objective assessment procedure and context importance is provided by a subjective judgement.

Table 4 Arguments for significance determination, based on impact characteristics and context characteristics

Arguments based on impact characteristics	Examples	References
Magnitude or intensity	Noise levels, odour intensity, amount of pollutants, amount of required resources, amount of employment	All
Extent of potentially affected factors	Amount of affected population, volume of polluted water, "the greatest good for the greatest number"	Duinker and Beanlands (1986), Canter and Canty (1993), Antunes et al. (2001), Lawrence (2007a), Ijäs et al. (2010), Briggs and Hudson (2013), Zulueta et al. (2017)
Economic considerations	Costs for certain institutions, revenue potential	Wood (2008)
Spatial patterns	Spreading distance, density, affected area, fragmentation, inclusion	Duinker and Beanlands (1986), Bojórquez-Tapia et al. (1998), Antunes et al. (2001), Lawrence (2007a), Wood (2008)
Temporal patterns	Duration, frequency, periodicity, swiftness	Duinker and Beanlands (1986), Canter and Canty (1993), Bojórquez-Tapia et al. (1998), Antunes et al. (2001), Lawrence (2007a), Wood (2008), Ijäs et al. (2010), Briggs and Hudson (2013), Zulueta et al. (2017)
Reversibility	Depletion of fossil fuels, erosion of tropical forests, human toxicity	Canter and Canty (1993), Antunes et al. (2001), Ijäs et al. (2010), Briggs and Hudson (2013), Zulueta et al. (2017)
Reliability	Certainity, probability, predictability	Duinker and Beanlands (1986); Canter and Canty (1993)
Social and ethical importance	Child labour, public controversy, public priority, "the greatest good for the least advantaged"	Duinker and Beanlands (1986), Canter and Canty (1993), Bojórquez-Tapia et al. (1998), Lawrence (2007a), Wood (2008)
Ecological sensitivity	Species extinction potential, resilience, recovery capacity	Canter and Canty (1993), Bojórquez-Tapia et al. (1998), Wood (2008)

(continued)

Table 4 (continued)

Arguments based on impact characteristics	Examples	References
Cultural sensitivity	Proximity to scientific, cultural or historic resources, aesthetic effect in scenic landscapes	Canter and Canty (1993)
Competition for resources	Groundwater depletion, agricultural land use	Duinker and Beanlands (1986)
Socioeconomic sensitivity	Accessibility, employment, agricultural production	Antunes et al. (2001), Canter and Canty (1993)
Institutional arrangements	Legal noise thresholds, target recycling rates, political targets	Duinker and Beanlands (1986), Canter and Canty (1993), Lawrence (2007a), Wood (2008)
Cumulative effects	Current pollution rates, synergy, spatiotemporal crowding of effects, induction potential, precedent setting, feedback resistance, biomagnification	Canter and Canty (1993), Bojórquez-Tapia et al. (2002), Lawrence (2007a), Wood (2008), Ijäs et al. (2010), Zulueta et al. (2017)

$$IS = f(I, C) \tag{1}$$

where:

IS Impact Significance,
I Impact Characteristics,
C Context Importance.

3 Spatial Variability

It has been noticed almost three decades ago "that methodologies which proceed through full aggregation of impacts to a 'final score', should not be used as an assessment technique, the results of which are intended for use by the decision-maker. Such an approach would remove the decision from those appointed or elected for that purpose and place it in the hands of the study-team" (Thompson 1990).

Based on the reviewed literature, it seems that although 'final score' is avoided for the clarification of diverse impacts, the significance of impacts is still spatially invariable. The spatial extent and spatial patterns are used only as one of the impact defining characteristics. E.g. the Spatial Impact Assessment Methodology (SIAM) proposed by Antunes et al. (2001) is mainly aimed at performing an aggregation of impacts in the spatial dimension. However, the spatial differences between alternatives are not communicated back to the decision makers.

There are multiple reasons why impact significance should not be a spatially uniform measure. First, by stripping the spatial dimension local impacts either get completely absorbed by the impacts at the larger scale or are wrongly given the same weight (Antunes et al. 2001). Second, impacts of different nature can accumulate in space and time and that way synergistically affect not only environmental but also social or economic sustainability. Third, impact assessment practices "will increasingly have to deal with significance judgements in relation to new proposals where existing thresholds, even without the proposal, have already been exceeded for various valued components" (Retief et al. 2016).

Furthermore, the concerns of the affected communities may differ from place to place (Gangolells et al. 2011). Therefore, using values of one community may not fit the judgements of the neighbouring one. In case of large scale changes that involve national or regional policies, each of the multiple affected communities would take the changes differently. E.g. a small development proposal in an ecologically sensitive environment may have a more significant impact than a far larger development located in a more robust setting. Similarly, a community dominated by high unemployment may be more supportive of controversial development proposals than comparable areas with full employment (Wood 2008).

Finally, two conditions must be controlled to accept a judgement as well-founded: consistency and consensus (Cloquell-Ballester et al. 2007). While consistency refers to the standard deviation of individual judgements, a study by Janssen et al. (2015) has demonstrated that associating individual stakeholder values with particular locations helped to arrive to a consensus which could not be reached otherwise.

Having spatial variability in impact significance assessment requires a spatially explicit model. Goodchild (2001) suggests four tests to determine if a model is (or should be) spatially explicit:

The Invariance Test considers a model spatially explicit if its outcomes (rankings or orderings of decision alternatives) are not invariant under relocation of the feasible alternatives. This implies that a change in the spatial pattern of feasible alternatives result in the changes of their rankings.

The Representation Test requires decision alternatives to be geographically defined. Such alternatives consist of, at least, two elements: action (what to do?) and location (where to do it?).

The Formulation Test declares a model spatially explicit if it contains spatial concepts such as location, distance, contiguity, connectivity, adjacency, or direction.

The Outcome Test checks if the spatial form of outputs is different than the spatial form of its inputs. E.g. the input values of spatial decision problems may be assigned to various spatial objects, while the output maps would represent the overall values associated with each location using raster data format.

4 Analysis of Impact Significance Assessment Methods

Although rarely considered as a subject on its own, impact significance assessment is an intrinsic part of Impact Assessment methods and Decision Support Systems. Based on the review in Sect. 2, impact significance assessment is a procedure that can rank or classify impacts taking into account both impact characteristics and the importance of the context where they occur. To determine current state-of-the-art of spatial variability in impact significance assessment, four methods have been selected as the most relevant in context of transitioning towards CE: Environmental Impact Assessment (EIA), Life Cycle Assessment (LCA), impact assessment in Geodesign and Spatial Decision Support Systems (SDSS). These methods were evaluated using spatial variability tests (Goodchild 2001). The analysis results (Tables 5, 6, 8 and 9) have shown that the spatial variability of impact significance corresponds to one of the two equations (Eqs. 2 and 3).

$$IS_{(x,y)} = f(I_{(x,y)}, C) \tag{2}$$

where:

$IS_{(x,y)}$ Impact Significance at location (x, y),
$I_{(x,y)}$ Impact Characteristics at location (x, y),
C Context Importance.

Table 5 Spatial variability of impact significance assessment in EIA

Spatial variability test	Impact Characteristics	Context Importance
Invariance	±	–
	Subject to change based on the project relocation	No requirement for spatially differentiated environmental sensitivity or public judgement values
Representation	–	–
	Decision alternatives may not be associated with project relocation	No requirement for spatially differentiated environmental sensitivity or public judgement values
Formulation	+	–
	Project and its impacts must be associated with particular geographical location	No requirement for geographic definition of environmental sensitivity or public opinion
Outcome	±	–
	Spatial extent must be provided, but there is no defined format	No required format for the description of environmental sensitivity

$$IS_{(x,y)} = f(I, C_{(x,y)}) \tag{3}$$

where:

$IS_{(x,y)}$ Impact Significance at location (x, y),
I Impact Characteristics,
$C_{(x,y)}$ Context Importance at location (x, y).

4.1 Environmental Impact Assessment

Environmental Impact Assessment (EIA) is a procedure used to provide an analysis of the potential significant environmental effects associated with major development proposals and to communicate this information to decision-makers and the broader public (Wood 2008). As a vast amount of different methodologies exist for impact identification and assessment, it is characterized by diversity in its practice, and by associated ambivalence (Pope et al. 2013). The latest review on EIA state-of-the-art by Zelenakova and Zvijakova (2017) describes EIA as a seven step procedure: scoping, impact identification, description of environment, impact prediction, impact assessment, decision making and communication of results. Although, impact significance assessment is not explicitly mentioned as a separate step, it should intrinsically be part of decision making.

The analysis of spatial variability has been made on the basis of Directive 2011/92/EU as amended by Directive 2014/52/EU (known as the "EIA Directive"). The main principle of the EIA Directive is to ensure that plans, programmes and projects likely to have significant effects on the environment are assessed and their implications made public prior to their approval or authorisation (European Commission 2014). The Directive indicates the rules for reporting the carried EIA, however it does not appoint a single method of assessment. Nevertheless, the Directive provides a list of impact characteristics that need to be considered, among which is spatial extent. A description of the location of the project, with particular regard to the environmental sensitivity of geographical areas likely to be affected is also required.

According to the EIA Directive "Member States may set thresholds or criteria to determine when projects need not undergo [...] environmental impact assessment" European Commission (2014). Also the public interested in environmental decision-making needs to be informed and allowed to express comments and opinions. However, the Directive does not require project developers to collect either the importance judgement of the public or institutional judgements, which would later be juxtaposed with the predicted impacts.

Based on the analysis results in Table 5, it appears that according to the EIA Directive, Impact Significance in a particular location is determined by the Impact Characteristics in that location and spatially non-differentiated values of Context Importance as in Eq. 2.

4.2 Life Cycle Assessment

LCA is especially relevant in the context of transitioning towards the CE as it can tell whether the achieved circularity of a certain resource would actually enhance the overall sustainability or not (Haupt and Zschokke 2017). LCA is "primarily a steady-state-tool" that does not consider temporal or spatial information and mostly has no relation with the context. In fact, often this information becomes lost due to aggregation (Udo de Haes 2006). The comparison between impacts is instead done by employing a functional unit (e.g. treatment of household waste produced in the city of Amsterdam during one year) and aggregating all the emissions into indicators that can be compared directly, or at midpoint or endpoint levels. While LCA is able to provide a complete picture of all impacts associated with a product or process, the communication of results usually requires an expert audience (Elia et al. 2017).

Although LCA was developed as a spatially independent approach, spatial LCA attempts associated with every stage can be found in the literature (Nitschelm et al. 2016). The significance of impacts in LCA is typically determined by the impact indicators and characterisation factors. Both impact inventory and characterisation factors may be spatially differentiated. The spatial variability of impact significance assessment is analysed based on the selection of recent publications (Table 7).

Based on the analysis results in Table 6, it seems that impact significance in a particular location is typically determined according to the Eq. 1, although Eqs. 2 and 3 are also possible in case of spatial LCA.

Table 6 Spatial variability of impact significance assessment in LCA according to the selection of literature as in Table 7

Spatial variability test	Impact characteristics	Context importance
Invariance	±	±
	May be subject to change on relocation of alternatives in both spatial and non-spatial LCA	Typically not spatially differentiated, although precedents exist
Representation	±	±
	The decision alternatives may have both a choice of actions and locations, although typically on a coarse granularity	Typically not spatially differentiated, although precedents exist
Formulation	–	–
	Spatial concepts are not included in impact assessment	Spatial concepts are not included in characterisation
Outcome	±	–
	Impacts may be geolocated based on processesses as objects in different spatial form (e.g. grid cell assignment)	Spatially differentiated characterisation factors typically do not change spatial form

Table 7 A list of literature used for the review on Life Cycle Assessment

List of references	
Haupt and Zschokke (2017)	Nitschelm et al. (2016)
Hiloidhari et al. (2017)	Kim et al. (2015)
Maier et al. (2017)	Smetana et al. (2015)
Escamilla and Habert (2016)	Hellweg and Mila i Canals (2014)

4.3 Geodesign

Geodesign has been chosen as a leading methodology for the decision support environment in the REPAiR project (REPAiR 2016) as it is a design and planning method that tightly couples the creation of design proposals with impact simulations informed by geographical context. Impact Assessment is the 4th step of the geodesign methodology (Steinitz 2012) and refers to the question "What differences might the changes cause"? The impacts are then assessed by experts and stakeholders using simple assessment matrices, that assign values from "very bad" to "very good" to each scenario of change for each of the valued factors. Impact significance is determined based on a consensus between the workshop participants considering their judgement and expertise.

Analysis results in Table 8 reveal that impact significance in geodesign is generally not spatially differentiated because context importance is not spatially explicit. Moreover, although impact characteristics are of spatial nature and determined by

Table 8 Spatial variability of impact significance assessment in geodesign methodology

Spatial variability test	Impact characteristics	Context importance
Invariance	+	–
	All alternatives are of a spatial nature, thus the ranking of impacts directly depends on them	The stakeholder values are not spatially defined
Representation	+	–
	The decision alternatives consist of actions and geographical locations	Stakeholder values are associated with actions but not particular locations
Formulation	–	–
	Impacts are not characterised by spatial concepts	Stakeholder values are not characterised by spatial concepts
Outcome	±	–
	Output is not presented in spatial format, but as a matrix	Output is not presented in spatial format, but as a matrix

Table 9 Spatial variability of impact significance assessment in SDSS according to the selected literature as in Table 10

Spatial variability test	Impact characteristics	Context importance
Invariance	–	+
	Uniform throughout the study area	Expressed per spatial unit in means of sensitivity, vulnerability or potential
Representation	–	+
	Location varies among alternatives, but actions and thus their impacts remain spatially constant	Decision alternatives are associated with context characteristics that define its importance
Formulation	–	±
	Not spatially defined	Mostly limited to location, but may also include distance, adjacency, direction, etc.
Outcome	–	+
	Not spatially defined and therefore not output in spatial format	May be based on different spatial form than decision alternatives

Table 10 A list of literature used for the review on Spatial Decision Support Systems

List of references	
Meerow and Newell (2017)	Corral et al. (2016)
Bonzanigo et al. (2016)	Janssen et al. (2015)
Jeong and Garcia-Moruno (2016)	Dapueto et al. (2015)
Rovai et al. (2016)	Bojesen et al. (2015)
Ottomano Palmisano et al. (2016)	van Niekerk et al. (2015)
Grêt-Regamey et al. (2016)	Erfani et al. (2015)

the spatial alternatives, impact significance is assessed uniformly for the whole study area. This would lead to Eq. 2 as the most suitable to describe impact significance determination in geodesign. However, workshop participants may implicitly assume spatial variability and accordingly adjust their ratings of the alternatives without expressing them formally.

4.4 Spatial Decision Support Systems

An SDSS can be defined as an interactive, computer-based system designed to support a user or group of users in achieving higher effectiveness in decision making while solving a semi-structured problem that has spatial consequences (Malczewski

1999). Decision Support Systems are meant to rather support than replace human judgements and improve effectiveness rather than efficiency of a process (Uran and Janssen 2003). This means that a user is expected to utilise the system as an advisory unit that is simply more capable to digest large amounts of data and perform quick computations.

There is an increasing amount of SDSS related scientific articles being published every year on solving an increasing variety of spatial decision problems that follow rather distinct methodologies (Ferretti and Montibeller 2016). In order to investigate the current practices and how they approach impact significance assessment, a small set of 12 relevant publications has been chosen based on Query 6 (Table 10).

Evidently, none of the studies have performed an actual impact assessment. Instead impact significance has been decided purely based on the context importance. E.g. presence of ecosystem services increases access to green spaces. Therefore ecosystem services should be located in a cell where the access to green spaces is the lowest (Meerow and Newell 2017). In some studies impacts refer not to the impacts a project would cause to the environment but to the impacts environment would have on project's success. E.g. more transport infrastructure is better for urban development. Therefore urban development should be located where transport infrastructure is the best (Grêt-Regamey et al. 2016). Equation 3 is the most suitable to describe how impact significance in a particular location is determined in SDSS.

5 Recommendations for Spatially Differentiated Impact Significance

According to Eqs. 2 and 3, for Impact Significance to be spatially differentiated it is sufficient that either Impact Characteristics or Context Importance is spatially differentiated. However, if only one variable in the equation is spatially differentiated and the other is spatially constant, the value of impact significance does not account equally for both impact characteristics and context importance. Instead, it aligns with the variability of the spatially differentiated one. Spatial variations of both impact characteristics and context importance should be taken into account in order to conduct a spatially differentiated impact significance assessment, as per Eq. 4.

$$IS_{(x,y)} = f(I_{(x,y)}, C_{(x,y)}) \tag{4}$$

where:

$IS_{(x,y)}$ Impact Significance at location (x, y),
$I_{(x,y)}$ Impact Characteristics at location (x, y),
$C_{(x,y)}$ Context Importance at location (x, y).

Several recommendations are provided for achieving spatially differentiated impact significance that reuse elements from existing methodologies, following the four tests defined by Goodchild (2001).

The Invariance Test on Impact Characteristics. Impact characteristics should be subject to change if the location of an object or action is changed. E.g. if a decision needs to be made upon which neighborhood to place a compost park, and one of the considered impacts is "increased accessibility to green spaces", then the number of people able to access the new park needs to be calculated for each of the neighborhoods.

The Invariance Test on Context Importance. The values of context importance should as well be varying between different locations. E.g. following the same example of locating a compost park, context importance may be dependant on the neighbourhood demographics with higher preference for young families and lower for students, which will be varying from neighborhood to neighborhood.

The Representation Test on Impact Characteristics. If decision alternatives involve both choice of actions and their locations, the characteristics of impacts need to change accordingly. E.g. if a choice needs to be made between locating a compost park in an existing green space or in a newly created one, then impact assessment should describe the impact of the new and adapted park dependent on the location characteristics, as some of them might be more favourable for adaptation while the others for a new green space.

The Representation Test on Context Importance. When decision alternatives involve both choice of actions and their locations, the importance needs to be given not only on basis of the preferred action but also considering the different location possibilities. E.g. acceptability and usage of a compost park may depend on the social composition of a particular neighborhood, while a need for greater green space accessibility may depend solely on neighborhood demographics.

The Formulation Test on Impact Characteristics. Those impact characteristics that change depending on the context characteristics, should be formulated with spatial concepts. While impact characteristics such as reversibility or duration may be dependent only on the chosen action and not vary in different contexts, impact magnitude may be well associated with the context characteristics. E.g. possible odour from the composting facilities may affect different areas by different intensities depending on the wind patterns.

The Formulation Test on Context Importance. Distance, adjacency, connectivity or direction may also serve for defining context importance. The importance does not always to have to be bounded to specific cells but expressed as adjacency to certain facilities or sensitive habitats, a function of distance from risk inducing object, accessibility over a network or gradually decreasing while moving north or south due to climate or cultural variations.

The Outcome Test on Impact Characteristics. In order to evaluate the impact on each valued component, it is necessary to identify the receptors and to describe the impact pathways affecting those receptors (Antunes et al. 2001). The receptors will eventually have a spatial dimension (e.g. population density, species distribution, location of resources). However, the spatial form of an impact may be different than that of the receptor.

The Outcome Test on Context Importance. Similar to impact characteristics, context importance can be expressed in a different spatial form than the significance assessment. Context importance may be based on e.g. topography, network centrality or administrative boundaries, while impact significance may be assessed per individual neighborhoods.

The four tests help to determine whether the assessment is or could be spatially differentiated and on what grounds. Passing one of the four tests is sufficient to qualify for the spatially differentiated impact significance assessment, however a balance between spatial differentiation in impact characteristics and context importance needs to be retained, i.e. if Impact Characteristics are spatially explicit, then Context Characteristics must also be spatially explicit.

The need for spatial differentiation in impact significance should also be critically evaluated based on its added value. As Nitschelm et al. (2016) have noted "the debate about whether spatialized LCA reduces uncertainties in LCA studies remains open. The amount of local data needed for spatialized LCA studies can indeed increase uncertainties in the LCI phase." The same observation stands true not only for LCA but impact assessment and decision support methods in general. However, evidence from SDSS demonstrates that judgement of context characteristics is spatially varying, while Impact Assessment studies prove the same about impact characteristics. This suggests that accounting for both components of the significance assessment should lead to a more informative and just result.

6 Conclusions and Future Work

The literature review on impact significance assessment has revealed that although the process is commonly performed during impact assessment and decision making, there is no single method that could be followed. Significance assessment is required by legal documents such as the EIA Directive, but there is a lack of legal definition or standardised method. What different authors agree on is that impact significance assessment is a double-sided procedure that involves objective assessment of impacts and subjective judgement of their importance. However, there is no consensus on what exactly characterises impacts, and who needs to provide judgement of importance and how. The review provides an overview of how different authors describe the two components of impact significance and what arguments are used to support the judgement.

As a result, this research suggests to regard impact significance assessment as a function between impact characteristics and the importance of the context that the impact occurs in. While impact characteristics can be estimated using objective measures, context importance requires judgement of importance that may be provided by stakeholders, decision makers, public opinion or institutionally.

It has been observed that up to now publications on impact significance regard spatial aspects only as possible impact characteristics and not a separate dimension of assessment. However, when decision making involves local impacts whose significance highly depends on context characteristics, the assessment requires spatial differentiation. Following this assumption, three main challenges need to be overcome: (1) probable impacts need to be characterised according to their geographical context; (2) the geographical context needs to be evaluated for its relative importance; and (3) finally, the values need to be combined to represent impact significance that may have spatial variability dependent on both components.

Environmental Impact Assessment, Life Cycle Assessment, Geodesign and Spatial Decision Support Systems, all employ impact significance assessment prior to comparison of decision alternatives. Although the alternatives often have spatial form and cause impacts that can be represented spatially, the four spatial tests by Goodchild (2001) have revealed that spatial differentiation is mostly based on either impact characteristics or context importance but not both of them simultaneously. As a result of this study, recommendations have been provided to overcome this gap in future impact significance determinations.

The recommendations drawn from the analysis will be further tested and refined in practice during the development of a Geodesign Decision Support Environment. They could, when supplemented by further related analyses, contribute to more systematic and spatially explicit significance determination approaches. In order to do so future work still includes providing clear unambiguous definitions of the used terms (e.g. context vs. impact) and demonstrations how the devised theory can be implemented in decision support. The created frameworks and tools aim to be sustainable and exceed the specifics of a single case study (Circular Economy). Finally, the same or very similar principles could be applied for temporal dimension to provide temporally differentiated significance assessment.

Acknowledgements This research has received funding from the European Union's Horizon 2020 research and innovation programme under Grant Agreement No 688920 and has been supported by the Amsterdam Institute of Advanced Metropolitan Solutions.

References

Accorsi R, Manzini R, Pini C, Penazzi S (2015) On the design of closed-loop networks for product life cycle management: economic, environmental and geography considerations. J Transp Geogr 48:121–134. https://doi.org/10.1016/j.jtrangeo.2015.09.005

Antunes P, Santos R, Jordão L (2001) The application of Geographical Information Systems to determine environmental impact significance. Environ Impact Assess Rev 21(6):511–535. https://doi.org/10.1016/S0195-9255(01)00090-7

Barrow CJ (2000) Social impact assessment: an introduction. Oxford University Press

Bojesen M, Boerboom L, Skov-Petersen H (2015) Towards a sustainable capacity expansion of the Danish biogas sector. Land Policy 42:264–277. https://doi.org/10.1016/j.landusepol.2014.07.022
Bojórquez-Tapia LA, Ezcurra E, Garcıa O (1998) Appraisal of environmental impacts and mitigation measures through mathematical matrices. J Environ Manage 53(1):91–99
Bojórquez-Tapia LA, Juarez L, Cruz-Bello G (2002) Integrating fuzzy logic, optimization, and GIS for ecological impact assessments. Environ Manage 30(3):418–433. https://doi.org/10.1007/s00267-002-2655-1
Bonzanigo L, Giupponi C, Balbi S (2016) Sustainable tourism planning and climate change adaptation in the Alps: a case study of winter tourism in mountain communities in the Dolomites. J Sustain Tour 24(4):637–652. https://doi.org/10.1080/09669582.2015.1122013
Briggs S, Hudson MD (2013) Determination of significance in Ecological Impact Assessment: Past change, current practice and future improvements. Environ Impact Assess Rev 38:16–25, https://doi.org/10.1016/j.eiar.2012.04.003
van Buren N, Demmers M, van der Heijden R, Witlox F (2016) Towards a circular economy: The role of Dutch logistics industries and governments. Sustainability (Switzerland) 8(7):1–17. https://doi.org/10.3390/su8070647
Canter LW, Canty GA (1993) Impact significance determination-Basic considerations and a sequenced approach. Environ Impact Assess Rev 13(5):275–297. https://doi.org/10.1016/0195-9255(93)90020-C
Cloquell-Ballester VA, Monterde-Díaz R, Cloquell-Ballester VA, Santamarina-Siurana MC (2007) Systematic comparative and sensitivity analyses of additive and outranking techniques for supporting impact significance assessments. Environ Impact Assess Rev 27(1):62–83. https://doi.org/10.1016/j.eiar.2006.08.005
Corral S, De Lara DRM, Salguero MT, Mendoza CCJ, De La Nuez DL, Santos MD, Peña FD (2016) Assessing Jatropha crop production alternatives in abandoned agricultural arid soils using MCA and GIS. Sustainability 8(6). https://doi.org/10.3390/su8060505
Dapueto G, Massa F, Costa S, Cimoli L, Olivari E, Chiantore M, Federici B, Povero P (2015) A spatial multi-criteria evaluation for site selection of offshore marine fish farm in the Ligurian Sea, Italy. Ocean Coast Manage 116:64–77. https://doi.org/10.1016/j.ocecoaman.2015.06.030
Duinker PN, Beanlands GE (1986) The significance of environmental impacts: an exploration of the concept. Environ Manage 10(1):1–10
Ehrlich A, Ross W (2015) The significance spectrum and EIA significance determinations. Impact Assess Project Apprais 5517(January):37–41. https://doi.org/10.1080/14615517.2014.981023
Elia V, Gnoni MG, Tornese F (2017) Measuring circular economy strategies through index methods: a critical analysis. J Clean Prod 142:2741–2751. https://doi.org/10.1016/j.jclepro.2016.10.196
Erfani M, Afrougheh S, Ardakani T, Sadeghi A (2015) Tourism positioning using decision support system (case study: Chahnime-Zabol, Iran). Environ Earth Sci 74(4):3135–3144. https://doi.org/10.1007/s12665-015-4365-z
Escamilla EZ, Habert G (2016) Method and application of characterisation of life cycle impact data of construction materials using geographic information systems. Int J Life Cycle Assess. https://doi.org/10.1007/s11367-016-1238-y
European Commission (2014) Directive 2011/92/EU of the European Parliament and of the Council, as amended by: directive 2014/52/EU of the European Parliament and of the Council
Ferretti V, Montibeller G (2016) Key challenges and meta-choices in designing and applying multi-criteria spatial decision support systems. Decis Support Syst 84:41–52. https://doi.org/10.1016/j.dss.2016.01.005
Gangolells M, Casals M, Gasso S, Forcada N, Roca X, Fuertes A (2011) Assessing concerns of interested parties when predicting the significance of environmental impacts related to the construction process of residential buildings. Build Environ 46(5):1023–1037. https://doi.org/10.1016/j.buildenv.2010.11.004

Geissdoerfer M, Savaget P, Bocken NM, Hultink EJ (2017) The circular economy a new sustainability paradigm? J Clean Prod 143:757–768. https://doi.org/10.1016/j.jclepro.2016.12.048
Gibson R, Hassan S, Holtz S, Tansey J, Whitelaw G (2005) Sustainability assessment: criteria. Process Appl Earthscan
Goodchild M (2001) Issues in spatially explicit modeling. Agent-based models of land-use and land-cover change, pp 13–17
Grêt-Regamey A, Altwegg J, Sirén EA, van Strien MJ, Weibel B (2016) Integrating ecosystem services into spatial planningA spatial decision support tool. Landscape Urban Plan. https://doi.org/10.1016/j.landurbplan.2016.05.003
Haupt M, Zschokke M (2017) How can LCA support the circular economy?-63rd discussion forum on life cycle assessment, Zurich, Switzerland, November 30, 2016. Int J Life Cycle Assess 22(5):832–837. https://doi.org/10.1007/s11367-017-1267-1
Hellweg S, Mila i Canals L (2014) Emerging approaches, challenges and opportunities in life cycle assessment. Science 1109–1114
Hiloidhari M, Baruah D, Singh A, Kataki S, Medhi K, Kumari S, Ramachandra T, Jenkins B, Shekhar Thakur I (2017) Emerging role of geographical information system (GIS), life cycle assessment (LCA) and spatial LCA (GIS-LCA) in sustainable bioenergy planning. Bioresour Technol https://doi.org/10.1016/j.biortech.2017.03.079
Ijäs A, Kuitunen MT, Jalava K (2010) Developing the RIAM method (rapid impact assessment matrix) in the context of impact significance assessment. Environ Impact Assess Rev 30(2):82–89. https://doi.org/10.1016/j.eiar.2009.05.009
Janssen R, Arciniegas G, Alexander Ka (2015) Decision support tools for collaborative marine spatial planning: identifying potential sites for tidal energy devices around the Mull of Kintyre, Scotland. J Environ Plan Manage 58(4):719–737. https://doi.org/10.1080/09640568.2014.887561
Jeong JS, Garcia-Moruno L (2016) The study of building integration into the surrounding rural landscape: Focus on implementation of a Web-based MC-SDSS and its validation by two-way participation. Land Policy 57:719–729. https://doi.org/10.1016/j.landusepol.2016.07.005
Jones M, Morrison-Saunders A (2016) Making sense of significance in environmental impact assessment. Impact Assess Proj Apprais 5517(January):1–7. https://doi.org/10.1080/14615517.2015.1125643
Kim J, Yalaltdinova A, Natalia S, Baranovskaya N (2015) Integration of life cycle assessment and regional emission information in agricultural systems. Sci Food Agric (March). https://doi.org/10.1002/jsfa.7149
Lawrence DP (2007) Impact significance determination-back to basics. Environ Impact Assess Rev 27(8):755–769. https://doi.org/10.1016/j.eiar.2007.02.011
Lawrence DP (2007b) Impact significance determination-Designing an approach. Environ Impact Assess Rev 27(8):730–754. https://doi.org/10.1016/j.eiar.2007.02.012
Lawrence DP (2007c) Impact significance determination-pushing the boundaries. Environ Impact Assess Rev 27(8):770–788. https://doi.org/10.1016/j.eiar.2007.02.010
Maier M, Mueller M, Yan X (2017) Introducing a localised spatio-temporal LCI method with wheat production as exploratory case study. J Clean Prod 140:492–501. https://doi.org/10.1016/j.jclepro.2016.07.160
Malczewski J (1999) GIS and multicriteria decision analysis. Wiley
Meerow S, Newell JP (2017) Spatial planning for multifunctional green infrastructure: growing resilience in detroit. Landscape Urban Plan 159:62–75. https://doi.org/10.1016/j.landurbplan.2016.10.005
van Niekerk A, du Plessis D, Boonzaaier I, Spocter M, Ferreira S, Loots L, Donaldson R (2015) Development of a multi-criteria spatial planning support system for growth potential modelling in the Western Cape, South Africa. Land Policy 50:179–193, https://doi.org/10.1016/j.landusepol.2015.09.014

Nitschelm L, Aubin J, Corson MS, Viaud V, Walter C (2016) Spatial differentiation in Life Cycle Assessment LCA applied to an agricultural territory : current practices and method development 112:2472–2484. https://doi.org/10.1016/j.jclepro.2015.09.138

Ottomano Palmisano G, Govindan K, Boggia A, Loisi RV, De Boni A, Roma R (2016) Local Action Groups and Rural Sustainable Development. A spatial multiple criteria approach for efficient territorial planning. Land Policy 59:12–26. https://doi.org/10.1016/j.landusepol.2016.08.002

Pope J, Bond A, Morrison-Saunders A, Retief F (2013) Advancing the theory and practice of impact assessment: setting the research agenda. Environ Impact Assess Rev 41:1–9, https://doi.org/10.1016/j.eiar.2013.01.008

REPAiR (2016) REPAiR—Resource Management in Peri-uran Areas: Going Beyond Urban Metabolism

REPAiR D61 (2017) D6.1 Governance and Decision-Making Processes in Pilot Cases. Technical report, H2020 project deliverable

Retief F, Bond A, Pope J, Morrison-Saunders A, King N (2016) Global megatrends and their implications for environmental assessment practice. Environ Impact Assess Rev 61:52–60. https://doi.org/10.1016/j.eiar.2016.07.002

Rovai M, Andreoli M, Gorelli S, Jussila H (2016) A DSS model for the governance of sustainable rural landscape: a first application to the cultural landscape of Orcia Valley (Tuscany, Italy). Land Policy 56:217–237. https://doi.org/10.1016/j.landusepol.2016.04.038

Smetana S, Tamásy C, Mathys A, Heinz V (2015) Sustainability and regions: sustainability assessment in regional perspective. Reg Sci Policy Pract 7(4):163–186. https://doi.org/10.1111/rsp3.12068

Steinitz C (2012) A framework for geodesign: changing geography by design. ESRI Press, Redlands, CA

Thompson MA (1990) Determining impact significance in EIA: a review of 24 methodologies. Journal of Environmental Management 30(3):235–250, https://doi.org/10.1016/0301-4797(90)90004-G

Udo de Haes H (2006) How to approach land use in LCIA or, how to avoid the Cinderella effect? The International Journal of Life Cycle Assessment 11(4):219–221, https://doi.org/10.1065/lca2006.07.257

Uran O, Janssen R (2003) Why are spatial decision support systems not used? Some experiences from the Netherlands. Computers, Environment and Urban Systems 27(5):511–526, https://doi.org/10.1016/S0198-9715(02)00064-9

Williams A, Kennedy S, Philipp F, Whiteman G (2017) Systems thinking: A review of sustainability management research. Journal of Cleaner Production 148:866–881, https://doi.org/10.1016/j.jclepro.2017.02.002

Wood G (2008) Thresholds and criteria for evaluating and communicating impact significance in environmental statements: 'See no evil, hear no evil, speak no evil'? Environmental Impact Assessment Review 28(1):22–38, https://doi.org/10.1016/j.eiar.2007.03.003

Zelenakova M, Zvijakova L (2017) Environmental Impact AssessmentState of the Art. In: Using Risk Analysis for Flood Protection Assessment, Springer International Publishing, chap 1, pp 1–72, https://doi.org/10.1007/978-3-319-52150-3_1

Zulueta Y, Rodríguez D, Bello R, Martínez L (2013) A linguistic fusion approach for heterogeneous Environmental Impact Significance Assessment. Applied Mathematical Modelling 40:1402–1417, https://doi.org/10.1016/j.apm.2015.07.016

Zulueta Y, Rodríguez R, Bello R, Martínez L (2017) A Hesitant Heterogeneous Approach for Environmental Impact Significance Assessment. Journal of Environmental Informatics 29(2, June):74–87, https://doi.org/10.3808/jei.201700363

Development of System for Real-Time Collection, Sharing, and Use of Disaster Information

Toshihiro Osaragi and Ikki Niwa

Abstract In large-scale earthquakes, it is important to quickly collect and utilize disaster information such as building collapse, street blockage, and fire outbreaks to mitigate disasters. In this paper, we develop a Web application for users to gather and share disaster information in real time. With this system, it is possible to not only share disaster information among users, but also to execute a simulation such as a fire. Next, conduct a demonstration experiment by local volunteers and investigate the usefulness and effectiveness of the system that collects disaster information. Furthermore, we analyze delay in sharing information under bandwidth-limited network environment and demonstrate the effectiveness of the system.

Keywords Disaster information · Information collection · Web application
Real time · Demonstration experiment

1 Introduction

In the aftermath of a major earthquake, it is essential to initiate rescue and fire extinguishing activities promptly in order to reduce the human and physical losses. In the past earthquake disasters, the lack of immediately available information about the kinds and locations of the property damage had resulted in delay of the initial responses and exacerbated losses. If it was possible to learn the locations of fires and trapped people quickly and precisely, the limited numbers of firemen and rescue teams available could be dispatched in the most effective manner.

T. Osaragi (✉) · I. Niwa
Department of Architecture and Building Engineering, School of Environment and Society, Tokyo Institute of Technology, 2-12-1-M1-25 Ookayama, Meguro-ku, Tokyo 152-8550, Japan
e-mail: osaragi.t.aa@m.titech.ac.jp

I. Niwa
e-mail: niwa.i.ac@m.titech.ac.jp

A. Mansourian et al. (eds.), *Geospatial Technologies for All*,
Lecture Notes in Geoinformation and Cartography,
https://doi.org/10.1007/978-3-319-78208-9_11

Additionally, if it was possible to learn the locations of street blockages due to obstacles such as collapsed buildings immediately following the disaster, routes could be chosen to avoid these, thereby facilitating travel to the teams' destinations. In summary, if disaster information immediately after the major earthquake could be collected, shared, and used in a quick and precise manner, it could contribute significantly to reducing human and physical losses. Numerous studies have been conducted addressing methods for early gathering of disaster information. We will analyze some of these previous studies in the later part.

In addition to patrols by fire companies, other methods of obtaining disaster information are visual surveillance from cameras in high locations or helicopters. However, there are concerns regarding not only the insufficient number of people who can oversee the gathering of disaster information but also the geographical scope and the precision or accuracy of the information obtained by such methods (Murai et al. 2008). Various conflicting information pours in during a disaster, and it is difficult to consolidate the data from multiple sources into a coherent picture.

Ngamassi et al. (2017) provided a synthesis of extant research on social media visual analytic and visualization toolkits for disaster management, and showed the past decade has seen a significant increase in the use of social media for disaster management (Petersen et al. 2017; Ngamassi et al. 2016; Denis et al. 2014; Hiltz et al. 2014; Hughes 2014; Yates and Paquette 2011). Relating to this research, there have also been some attempts in recent years to exploit the capabilities of handheld devices such as smart-phones and tablets as "data sensors" for the collection of disaster information (Table 1). For example, Kubota et al. (2013) have proposed a system in which images of disasters captured using smart-phones are emailed along with text describing the damage, and the disaster information is then automatically placed on a map by using the location data attached to the Exchangeable Image File Format (EXIF) information.

Hiruta et al. (2012) proposed an exclusively smart-phone-based system for sharing disaster information that requires no communications infrastructure or dedicated servers and is highly robust against disasters. Both concepts are seen as advances over the current systems, but it is difficult to update information in real time during a disaster because the situation changes from moment to moment, and because the information in the system is likely to degrade with time.

Other studies have addressed gathering disaster information promptly from wide regions by collecting information submitted to social networking services (SNS). For example, Stuart et al. (2014) extracted keywords related to disaster information from Twitter tweets in real time and linked them to geographical information in a proposed system for sharing disaster information on a map. In experiments involving the use of information and communications technology (ICT), such approaches have shown potential, but raise issues such as how to improve the accuracy of information contained in SNS posts by excluding false, mistaken, and obsolete information.

In response to a major disaster, such as the Haiti Earthquake (2010), the Great East Japan earthquake (2011), the Izu-Oshima rainstorm (2013), and the Kumamoto Earthquake (2016), an information sharing method called "crisis mapping", in

Table 1 Examples of disaster-information sharing system proposed in the previous research

Researches	A Real time property	B Robustness in disaster	C Capability of information sharing	D Independence from devices	E Easiness of utilization	F Secondary use of data	G Collecting information and assistance
Kubota et al. (2013)	×	○	×	○	◎	×	×
Hiruta et al. (2012)	×	◎	○	○	×	×	×
Stuart et al. (2014)	◎	×	○	○	○	×	×
Matthew et al. (2010)	×	○	○	×	◎	×	×

◎ = sufficiently satisfied, ○ = partly satisfied, × = not considered which inhabitants and on-site volunteers collect and submit information that is then visualized on an online map by information technology (IT) engineers and other personnel has been proposed. In recent years, volunteer-based geographical information systems (GIS) have become popular in a variety of fields and have shown a certain amount of success, but are commonly used some time after a disaster has occurred to get a better grasp of prevailing conditions. However, there are almost no systems capable of using volunteer-based support for firefighting or rescue activities immediately after a disaster.

In view of the above background, the authors have developed a practical, second-generation real-time system for collecting, sharing, and using disaster information. This system is intended specifically to support volunteer-based efforts to coordinate response by the authorities by collecting and sharing information obtained from multiple users during a disaster. This system was used in a field experiment by local citizens practicing as disaster mitigation volunteers and was found to be effective for gathering data. An additional experiment was also carried out to evaluate the system, assuming a bandwidth-limited network environment such as could be expected to occur during a disaster.

2 Development of a System for Gathering and Use of Disaster Information

2.1 System Components

Figure 1 shows the components of the proposed system. This system was constructed on a foreign-based cloud server (Amazon Web Services EC2) (Amazon.-com 2017) and was designed to minimize the risk of sustaining physical damage and failing in the course of a disaster. It was also designed as a web application in order to reduce the labor and time involved in pre-disaster installation and to facilitate maximal usability by limiting dependence on the user's platform.[1] Since even a first-time user can access the system over the web browser of the device he or she uses every day, whether a smart-phone or something similar, without going through any other facilities, he or she can easily submit photographs and other information.[2]

[1]Users in this research indicate information providers who collect disaster information using the system. On the other hand, firefighters or decision makers for disaster mitigation will be information seekers. We don't discuss the details of their roles due to the limitation of spaces.

[2]We do not limit data collection to employees of public bodies. We also anticipate numerous volunteers from the general population. Since we might see lower accuracy of the information obtained from large numbers of unspecified people, we expect it to be necessary to establish password-protected login accounts and filtering, or to use some other means of managing the submitted information as appropriate.

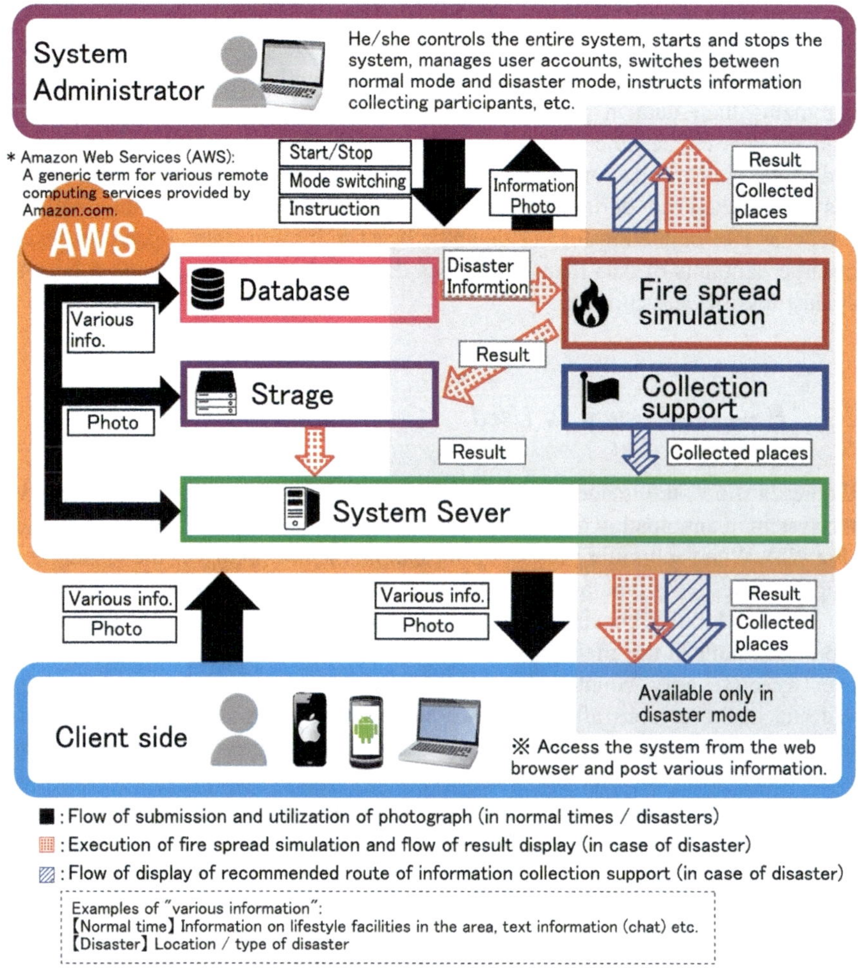

Fig. 1 The components of the proposed system

One of this system's greatest advantages is its real-time reporting,[3] which we achieve by employing Node.js (Japan Node.js Association 2017) and WebSocket (McKelvey 2017), thereby obtaining data synchronization with extremely short delays. Designing the system server to be event-driven[4] vastly increases the number

[3]"Real-time" in this study means the property that the required processing can be completed within a certain time in response.

[4]"Event-driven" means that other software executes the program. In other words, processing is initiated by the user or by an operation of another program (event). This is in contrast with the concept of "flow-driven" processing (in which the system is controlled by the execution flow of a program).

of connections it can handle simultaneously, and thus allows quick responses to each user. In addition, when the information in the database is updated, it is push-distributed by the server to all the users currently on the system, thus synchronizing their data in real time. Since users need not issue data transmission requests, this reduces the load on the server. As a result, the data remains current and stable synchronization can be expected, even on a transmission network that has been weakened during a disaster. We selected MongoDB (MongoDB Inc. 2017)[5] as the foundation for our database because it allows high-speed processing of larger amounts of data than in previous databases, greatly increasing robustness against loss of data currency.

2.2 How the System Is Used

Figure 2a shows the image seen by the user. A variety of functions are included in the system in anticipation of the conditions that can be expected following a disaster (Fig. 2b). When submitting disaster information, the user indicates the location on a map and selects the type of event (building collapse, street blockage, building fire, etc.) (Fig. 2(1)). It is conceivable that in an emergency, there would be little time to input the details of the affected location, so it is permissible to edit items besides the event type (severity, countermeasures being taken, miscellaneous comments, etc.) and attach photographs after the initial submission. The map image displays different markers for different event types showing what information the initial and other users have submitted about any event.

Selecting a marker opens an information window providing access to the submitted disaster information and photographs. When data is registered or updated, it is shared to all other users presently using the system, thereby preventing delay and degradation of the information. In addition to being provided on a map, the collected information is available for general use. It can be downloaded from the database in the comma separate value (CSV) file format (Fig. 2(6)) and sent to other systems. Other user support functions are also provided, including display of the user's present location (Fig. 2(2)), image refresh (Fig. 2(3)), chat (Fig. 2(4)), and location search by name (Fig. 2(5)). One can also switch to other user accounts by logging out and back in (Fig. 2(7)).

In order to improve the efficiency of information collection during a disaster, it is essential to use the system not only after a disaster strikes, but also during normal times in order to get accustomed to its operation. Therefore, this system was designed to be employed during normal times and to allow submission, sharing, and viewing of information while varying the types and numbers of events.

[5]The input/output processing involved in storage is distributed processing in MongoDB, which is a NoSQL database employing parallel processing that is compatible with Node.js. Many packages have been developed for operating this database format, which is designed to facilitate system server data searches, retrieval, and insertion.

(a) The image seen by the user

(b) Function list

No.	Function	Explanation
(1)	Submit disaster information	Specify the disaster occurrence location on the map and select the event type
(2)	Present location	Display user's present location based on GPS information
(3)	Image refresh	Re-read and refresh the map
(4)	Chat	Chat among users using the system
(5)	Location search	Location search by place name
(6)	CSV output	Output accumulated disaster information
(7)	Login / Logout	Control permissions for each account
1	Fire spread simulation	Display results of fire spread simulation
2	Edit disaster information	Edit information and attach photo
3	Move location of disaster	Location correction of posted disaster information
4	Delete disaster information	Delete posted disaster information

Fig. 2 The image seen by the user and function list of the system

For example, it is intended to allow other uses such as information sharing about daily facilities, etc. and chatting among multiple users in real time. However, when a disaster occurs, the managers switch the system from the ordinary to disaster mode, and the users can immediately begin participating in the collection of disaster information.

2.3 Linkage with Fire Spread Simulations

In addition to collecting disaster information, this system allows secondary usage of fire-related information submitted to the database in a fire spread simulation that estimates and visualizes the extent of damage as of certain time periods after a fire breaks out (Fig. 2 [1]).

Figure 3 provides images of the simulated spread of a fire. The spatial distribution of building, which was part of the Tokyo City Planning GIS data (collected from Tokyo's 23 wards in 2011, and from Tama City in 2012), was used with the fire spread speed equation of the Tokyo Fire Department (2001) in order to simulate spread from building to building (Hirokawa and Osaragi 2016a, b). Simulation input conditions (season, time, epicenter, earthquake intensity, etc.) can be altered via a simple process. Additionally, the simulated results for up to 12 h after the disaster can be stored in the GeoJSON format, thus eliminating the need to rerun the simulation every time it is desirable to view the results. The results can be

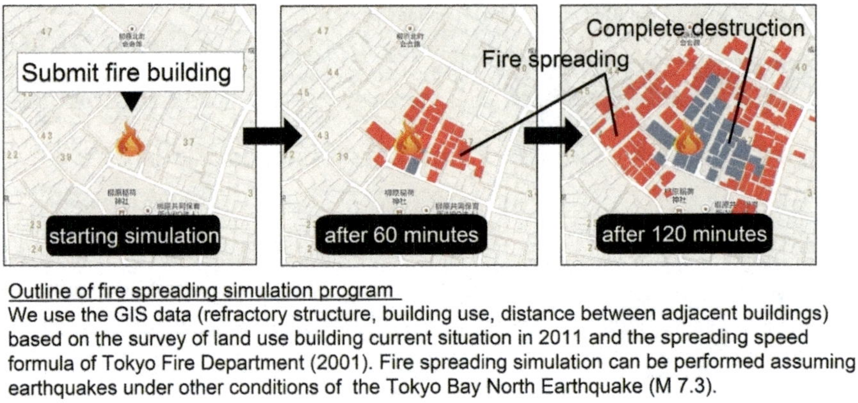

Fig. 3 Images of the simulated spread of a fire

visualized at any desired time interval. The application also allows assumption of a single or multiple buildings as the origins for the fire and its spread.

The ability of this function to predict the spread of fire for several hours after an outbreak can support individuals desiring to evacuate while avoiding the fire dangers, and firemen as they fight fires. Fire spread simulations are an example of a secondary usage of disaster information. Simulations have demonstrated that submitting information about street blockages contributes to the speedy arrival of firefighters at fires (Osaragi et al. 2015; Osaragi and Hirokawa 2017). There are many other possibilities for the secondary use of collected disaster information, which the authors will describe in future reports.

2.4 Information Collection Support Function

In order to collect information efficiently, it is more desirable to direct users to locations that need to be checked with the highest priority rather than to allow damage searches based only on individuals' judgments. Few of the existing systems contain functions supporting data collection (Table 1(G)). Therefore, systems supporting regional patrols by citizens (Kimura et al. 2016a, b) were examined and information collection support functions suggesting examples of what kinds of users should check what places were incorporated. This is expected to enable more efficient data collection while preventing overlapping efforts by multiple users. Another enhancement of efficiency is that users with no familiarity with the local area can be directed to locations of concern.

3 Validation Experiment with Local Citizens as Disaster Prevention Volunteers

3.1 Experimental Method and Assumptions

A field experiment was conducted with this system for collecting, sharing, and using disaster information in the following procedure, and the efficiency of data collection and its expected effectiveness were observed:

(1) The number and locations suffering physical damage (building collapse, street blockage, building fires) were set using the simulation results and were displayed (on a smart-phone screen) with "virtual damage" markers (Fig. 4a). By displaying the virtual damage on the screens of the users' devices, it was possible to reduce the time and labor involved in placing signboards and dispatching personnel in typical field experiments, and the experiment could be carried out over a wide region. This experiment consisted of a total of five runs

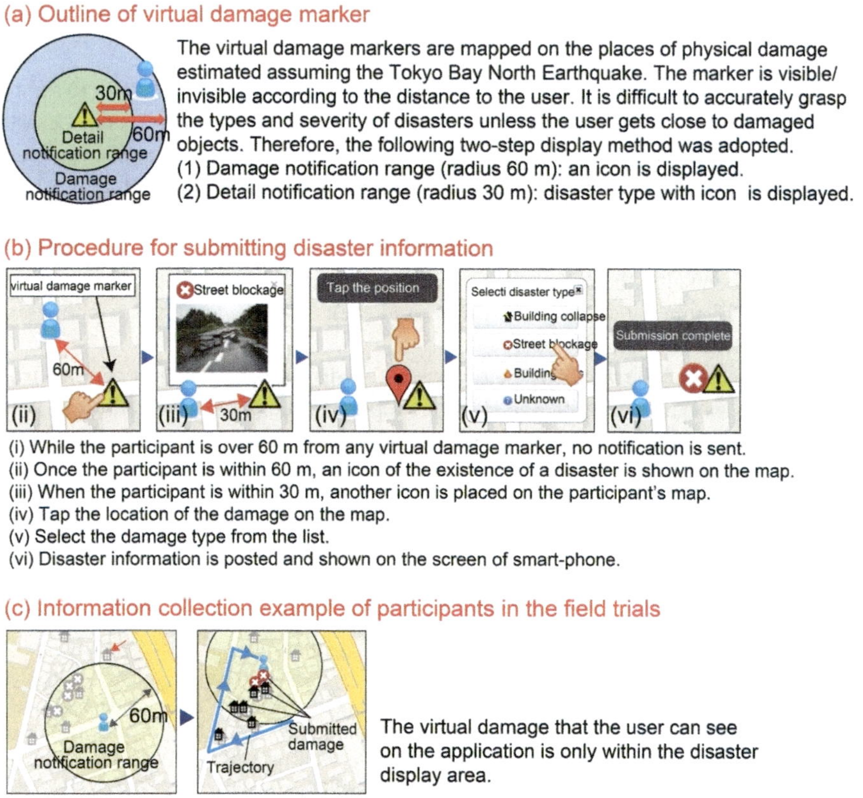

Fig. 4 Outline of experiment using virtual damages

assuming differing disasters, in order to observe how the participants in multiple tests learned in different and unfamiliar locations.

(2) When one is collecting data in an actual emergency and hears of damage in some location that is not immediately nearby, it is difficult to be sure of any details (the type of event and how serious it is) until one goes at least somewhat closer to the actual location. Therefore, the Global Positioning System (GPS) information on each participant's device was monitored and used as follows: (i) While the participant was over 60 m from any location flagged with a virtual damage marker, no notification was sent; (ii) once the participant had crossed within 60 m of such a location (damage notification range), an icon showing the existence of some kind of a disaster was placed on the participant's map; and (iii) when the participant approached even closer, to within 30 m (detail notification range), another icon was placed on the participant's map (Fig. 4b) with detailed information about the damage type (building collapse, street blockage, building fire).

(3) The participant walked around freely to look for the virtual damage (Fig. 4c). It is anticipated that since only damage that is physically close to one can be examined, the more proactive one is about moving around, the more disaster information one will be able to collect (in this experiment, the information collection support function was not used).

Setagaya City, one of Tokyo's 23 wards, was selected as the region for the experiment. An earthquake centered under northern Tokyo (Fire and Disaster Management Agency 2017) was assumed, and a physical damage simulation (Osaragi et al. 2015) was carried out. The results are shown in Table 2(1). The experiment was performed on February 10, 2015 (Tue) in five runs (three in the morning and two in the afternoon) with the generous participation of a total of 118 people on behalf of the Setagaya Council of Social Welfare. Each run was conducted for 15 min[6] and the participants used their own devices (smart-phones) to gather information.

The objects of the analysis below were the disaster information reports from participants who had submitted at least one report. Participants who had logged into the system but had only looked without submitting collected information were omitted. A total of 73 participant reports were employed. More women than men participated in each run, and most were in their 50 s or older (Fig. 5).

[6]According to Tokyo Fire Department (2015), from the point of view of reducing damage due to the spread of fires throughout the city, firefighters should not be dispatched as soon as they are notified of a fire somewhere. Instead, they should wait for some time (15 min) after the occurrence of an earthquake in order to gather and organize information. The most effective response is for them to be dispatched to the fires that have been prioritized on the basis of their risk of spreading. That is why disaster information is gathered for 15 min in this study.

Table 2 Estimated value of physical damage and the number of submissions by participants

(1) Estimated physical damage in Setagaya City at the time of the Northern Tokyo Bay Earthquake

Experiment number	Building collapse	Street blockage	Building fire	Total
1st	6,960 (4.05%)	3,192 (6.72%)	74 (0.04%)	10,226
2nd	6,972 (4.06%)	3,318 (6.98%)	74 (0.04%)	10,364
3rd	6,984 (4.07%)	3,190 (6.71%)	70 (0.04%)	10,244
4th	7,054 (4.11%)	3,275 (6.89%)	84 (0.04%)	10,413
5th	7,042 (4.10%)	3,224 (6.79%)	66 (0.04%)	10,332
Damage estimated by Tokyo Metropolitan Government	6,020 (3.51%)	–	62 (0.04%)	–

*Percentage in parenthesis = Estimated value of damage/Number of all buildings in Setagaya Ward (or number of all roads) (%)

(2) Number of submissions by damage type during the demonstration experiment

Experiment number	Building collapse	Street blockage	Building fire	Total
1st	42 (0.60%)	13 (0.41%)	2 (2.70%)	57 (0.56%)
2nd	46 (0.66%)	20 (0.60%)	1 (1.35%)	67 (0.65%)
3rd	68 (0.97%)	11 (0.34%)	0 (0.00%)	79 (0.77%)
4th	75 (1.06%)	28 (0.85%)	1 (1.19%)	104 (1.00%)
5th	50 (0.71%)	39 (1.21%)	0 (0.00%)	89 (0.86%)

*Percentage in parentheses = Number of submissions/Estimated value of damage (%)

(a) Outline of experiment participants

No.	Time	Person	Male	Female
1st	10:00 - 10:15	14(27)	3(9)	11(18)
2nd	10:30 - 10:45	12(23)	4(8)	8(15)
3rd	11:00 - 11:15	14(21)	4(10)	10(11)
4th	14:00 - 14:15	15(24)	6(10)	9(14)
5th	14:30 - 14:45	18(23)	5(7)	13(16)

The number represents the number of people to be analyzed, and the number in parentheses represents the number of participants in the experiment.

(b) Age group of experiment participants

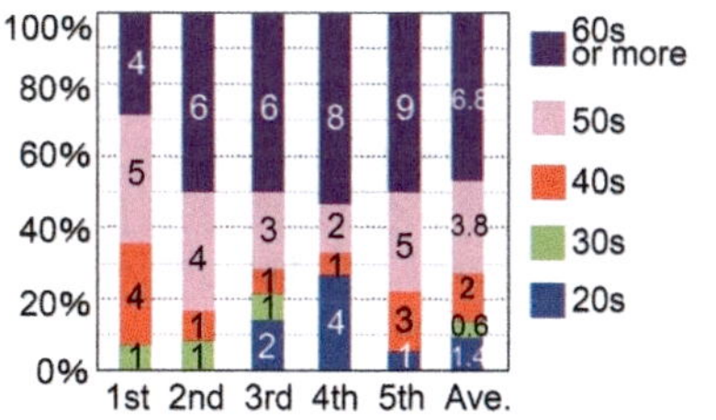

Fig. 5 Outline of experiment participants

3.2 Information Submission Rates

Table 2(2) shows the number of information submissions and collection rate in each run (numbers of reports of physical damage throughout Setagaya City). A mean of 15 participants collected information over the short 15-min period, succeeding in collecting about 0.7% of the virtual damage distributed throughout the entire ward. Examination of the correctly submitted reports[7] over time after the beginning of the experiment (Fig. 6) indicates an increase in proportion to time, and the rates of submissions did not show any signs of slackening as the 15-min limit approached. We can also expect the collection rate to increase in proportion with the number of participants. For example, if we assume that the real-time information sharing unique to this system is fully realized, that there are no redundant reports from the same location, and that the number of individuals participating in information collection corresponds to 0.2% of the population of Setagaya City (i.e., approximately 1,800, which is about 1.7 times the number of members of the Setagaya Volunteer[8] Fire Corps), then the collection rate reaches about 84% (=0.7% × 1800/15). If 84% of the street blockages were known, then according to Osaragi et al. (2015), the time required for a fire truck to reach a fire from the fire station could be reduced to about 67% of the time required if the fire department is completely ignorant of street blockages. Thus, as long as a certain number of users using this system are available to gather damage information, it has the potential to significantly contribute to reducing the scale of a disaster.

3.3 Accuracy Rate of Submitted Information

The number of information submissions and accuracy rate (see footnote 7) are tabulated in Fig. 7 based on the age group of the participant and his or her length of experience using a smart-phone. The age group tabulation shows a consistent rate of 90% (except for the group of participants in their thirties, in which the number of participants was low). The findings in this test showed no dependence on how long the participant had used a smart-phone. Thus, it was concluded that the system itself was easy for anyone to use, regardless of the type of user and, of particular note, not just for those of age groups typically skilled at using portable devices.

However, some examples of mistaken submissions were found that could have been caused by the display scales on the maps shown on the users' smart-phones.

[7]In this study, submitted damage types that differed from the types displayed in the icons of the virtual damage markers, and any virtual damage that had been misidentified due to submission from a location too distant from the virtual damage marker were treated as erroneous submissions.

[8]Volunteers in this research are assumed to be well trained and educated volunteers who will make action for disaster mitigation. Actually, around 0.12% of general residents are registered for such volunteers in Setagaya City.

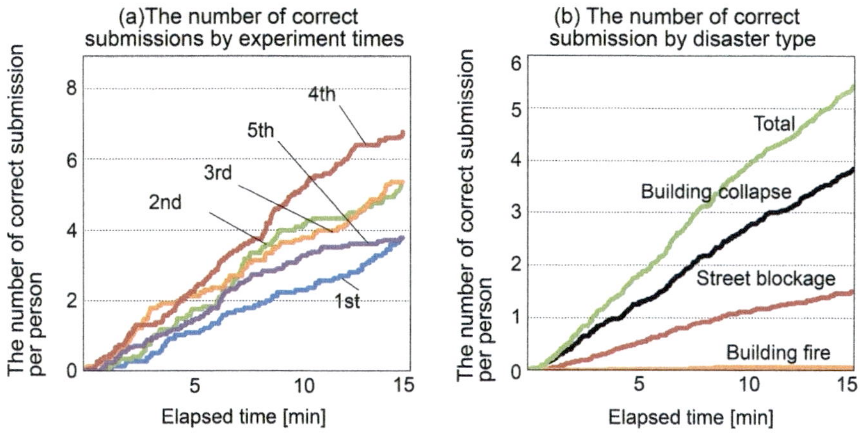

Fig. 6 Transition of the number of submissions by elapsed time of experiment

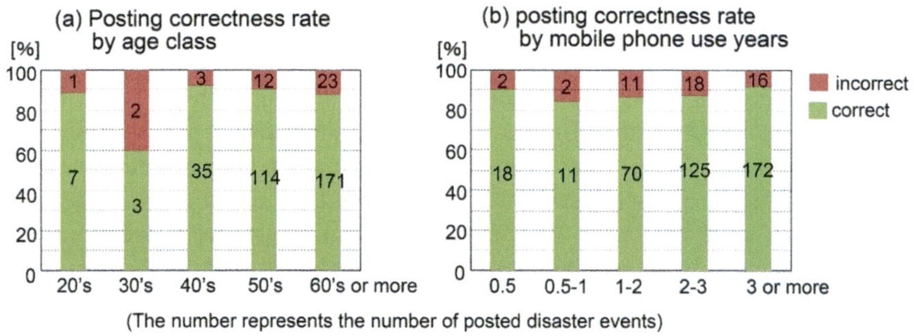

Fig. 7 The number of information submissions and accuracy rate by age group and length of experience using a smart-phone

Specifically, calculations were made as follows. With (a) the reported location of the reported damage, (b) the actual location of the virtual damage, and (c) the location of the participant at the time he or she submitted the report, the mean distances (a)–(b), (a)–(c), and (b)–(c) were evaluated by the correctness or incorrectness of the submitted information. Figure 8 presents the results. Here, the reader can see that the mean distances when the mistaken reports were submitted were over double the distances when the accurate reports were submitted. Thus, the distance tends to have a strong inverse relationship with the precision of submitted information. This issue might be resolvable by adding an automatic zoom function to the displayed map, which would allow the user to verify his or her location before submitting a report. Another way of checking the quality of submitted information would be to reduce the reliability of the submitted information when the distance between (a) the reported location of the reported damage and (c) the location of the participant at the time he or she submitted the report exceeds some threshold.

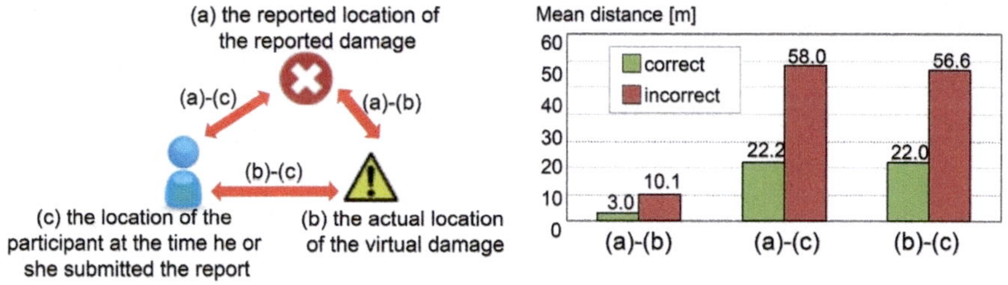

Fig. 8 Relationship between correct answer rate of posted information and distance between locations of user, posted disaster, and actual disaster

3.4 Efficiency of Information Collection

If the efficiency of information collection can be raised, it may be possible to gather more disaster information with fewer people. Therefore, some methods for improving this efficiency level were examined. Figure 9 shows the number of damage locations within the visual range (equal to the damage notification range: a 60 m radius from a location indicated with a virtual damage marker) of one of the participants and the number of reports submitted. During each run, each person received a mean of 12 damage events (an event notification approximately once each minute and 15 s) that were within his/her damage notification range. Nevertheless, the reader can see that only about half of these events were reported. In other words, it seems to be more efficient to make sure one actually submits information about events in their immediate vicinity than to walk around at random.

According to a questionnaire survey of the participants conducted after the experiment was completed, they were biased toward checking the following in order to collect data: (1) Their homes and the areas in which family members were; and (2) the locales they happened to be in when the experimental run began. Thus, when the system enters actual operation, it will be essential to use the information

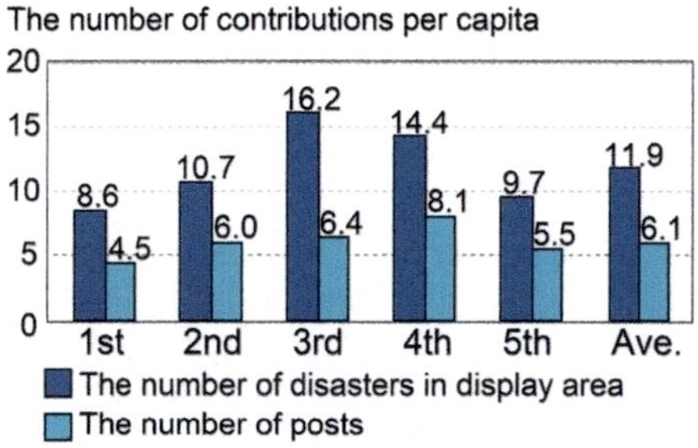

No.	Submission rate
1st	52.5%
2nd	56.3%
3rd	39.7%
4th	56.0%
5th	56.9%
Ave.	52.3%

Fig. 9 Disaster information submission rate within view range

collection support functions to obtain damage information about the facilities and roads most important for maintaining public safety and conducting firefighting and rescue activities in the most efficient manner possible.

4 Experiment to Evaluate Information Collection Support Functions

4.1 Background of Experiment and Methods

In the experiment in Sect. 3, the participants walked about freely to look for virtual damage and submit reports about it. This section presents a comparative experiment in which an information collection support function was used in order to see how much it improved information collection efficiency. Specifically, a total of five participants searched for and reported virtual damage (132 street blockages in the test region) over a 15-min period, beginning in that region (the neighborhood of Okusawa Station).

(a) Experiment 1 [without information collection support]: As in the experiment in the previous section, the participants moved freely in the region to look for and submit reports of street blockages.
(b) Experiment 2 [with information collection support]: Before starting, the participants were re-dispatched to different initial locations in the test region. Upon start of the experiment, they referred to the map displays provided by the system indicating prioritized locations where they should collect information.

The prioritized locations were the results of a street blockage simulation developed by Osaragi et al. (2015) (conducted for 1,000 cycles), tabulated on a 100 m mesh; 49 cells were selected from the top 3% calculated to have the highest rate of street blockage (the ratio of blocked roads with respect to the total road mileage in the grid cell). Thus, these prioritized locations were deemed to be locations with the highest potential to suffer street blockages in the event of a major earthquake. Additionally, the locations with rates of street blockage of 20% or above were distinguished from those with rates below 20% by the intensity of the marker color.

4.2 Experimental Results

The participant movements were compared. The reader can see (Fig. 10) that (a) in Experiment 1 [without information collection support], the participants tended to patrol the locations where they had been at the onset of the disaster. In contrast, (b) during Experiment 2 [with information collection support], they moved preferentially toward the regions expected to undergo the greatest damage (the regions

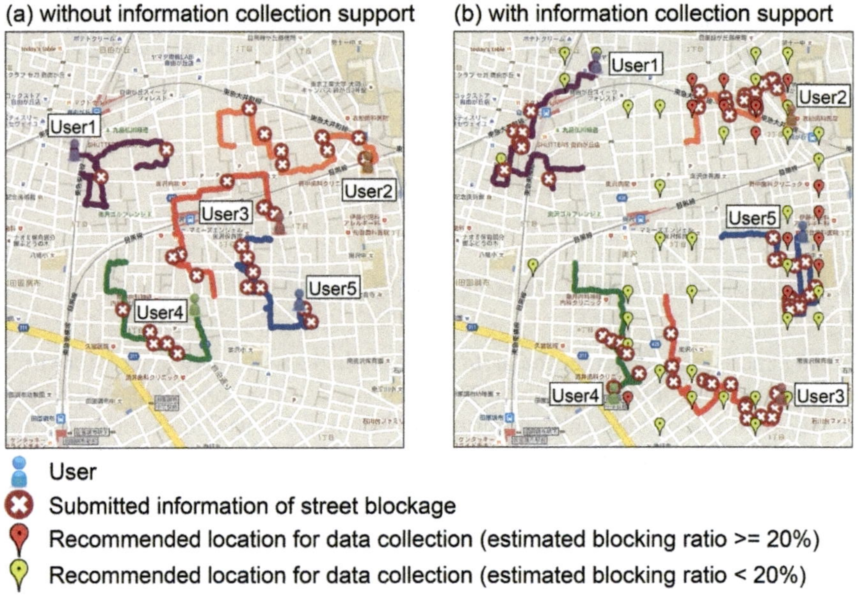

Fig. 10 Comparison of the participant movements in experiments 1 and 2

with the highest concentrations of prioritized locations) and searched those regions. As a result, when using the information collection support function, the mean number of events reported by the participants within the 15-min period rose by a factor of 2.2 (Fig. 11a) and the per-event distance traveled by the participants fell to 47% of that traveled previously (Fig. 11b). This clearly shows that the information

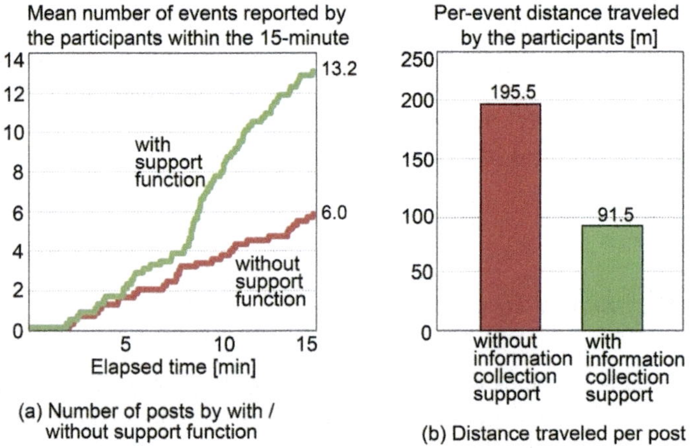

Fig. 11 The mean number of events reported by the participants within the 15-min period and the per-event distance traveled by the participants

collection support function increased the collection efficiency. It is of particular note that the number of submissions swelled significantly once each participant had arrived in an area that had been prioritized (typically, about 8 min or longer after the beginning of the experiment), thereby demonstrating the effectiveness of this information collection support.

The above results indicate that it is more effective to employ information sources such as hazard maps in order to preferentially search vulnerable or critical locations than to wander about at random to collect disaster information.

5 Experiment to Evaluate Bandwidth and User Number Limitations

Next, a simulation experiment was conducted to examine how long it would take to synchronize newly submitted disaster information (crisis mapping data) to all the users of a wide-scale version of the system described above in a network environment whose bandwidth has been crippled by a disaster. Two transmission speeds (1 Mbps [3G lines] and 15 Mbps [LTE lines]) and up to 3,000 virtual users were assumed. One hundred cycles of the simulation were performed.

The mean synchronization times in both networks were less than one second, and it was found that further increases in the number of users lengthened the delay by less than 50 ms (Fig. 12). Thus, data currency would not be degraded up to 3,000 users and could be expected to function sufficiently, even in a small-scale system environment like that used for this experiment. However, we need more precise experiments assuming an actual disaster. Also, since transmissions can become unstable when the number of users exceeds 3,000 at the present stage of development of this system, it will be necessary to investigate ways to distribute the transmission load in order to obtain more dependable real-time data sharing.

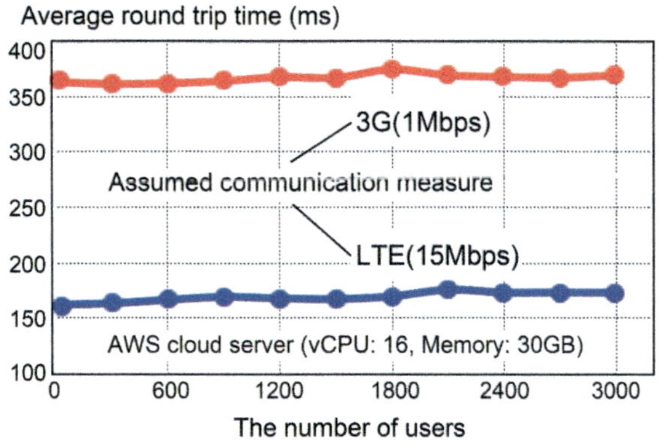

Fig. 12 The time until the disaster information submitted by one user is synchronized with all other users

6 Summary and Conclusions

A highly disaster-resistant, cloud server-based system (web application) for user-based collection, sharing, and use of real-time information about damage immediately following a disaster was developed. The system incorporates additional functions for secondary use of the submitted information about fires, a simulator that estimates the damage from spread of a fire, and an information collection support function that assigns priorities to firefighting and rescue personnel.

In the experimental stage, disaster prevention volunteers used this system in the field to collect disaster information. About a dozen volunteer participants collected data in 15-min sequences, and were observed to have collected about 0.7% of virtual disaster information distributed throughout Setagaya City. If approximately 0.2% of the population of Setagaya City were users of a system modeled on this and collected such information, they could collect over 80% of all the information about damage in the ward, and would make a significant contribution to reducing the scale of a disaster.

The field experiment also demonstrated that if an information collection support function is used, information regarding multiple damage events could be collected more efficiently, requiring much less walking distance than otherwise would be traveled by system participants. Finally, a simulation of reduced network bandwidth following a disaster was conducted and showed that even with about 3,000 simultaneous participants, the system was still able to maintain data currency while continuing to collect information.

Acknowledgements A portion of this work is supported by Cross-ministerial Strategic Innovation Promotion Program (SIP). The author wishes to express their sincere thanks to Japan Science and Technology Agency (JST).

References

Amazon.com Inc (2017) Amazon EC2 pricing. https://aws.amazon.com/jp/ec2/. Accessed 9 Nov 2017

Denis L, Palen L, Anderson J (2014) Mastering social media: an analysis of Jefferson County's communications during the 2013 Colorado floods. In: Proceedings of the 11th international conference on information systems for crisis response and management (ISCRAM), State College, Pennsylvania

Fire and Disaster Management Agency (2017) Firefighting team in your city (Tokyo Metropolitan). http://www.fdma.go.jp/syobodan/search/13.html. Accessed 9 Nov 2017

Hiltz S, Kushma J, Plotnick L (2014) Use of social media by U.S. public sector emergency managers: barriers and wish lists. In: Proceedings of the 11th international conference on information systems for crisis response and management (ISCRAM), State College, Pennsylvania

Hirokawa H, Osaragi T (2016a) Earthquake disaster simulation system: integration of models for building collapse, street blockage, and fire spread. J Disaster Res (Fuji Technology Press Ltd.) 11(2):175–187

Hirokawa N, Osaragi T (2016b) Access time of emergency vehicles under the condition of street blockages after a large earthquake. In: First international conference on smart data and smart cities, ISPRS annals of the photogrammetry, remote sensing and spatial information sciences, volume IV-4/W1:37–44
Hiruta M, Tsuruoka Y, Tada Y (2012) A proposal of a disaster-information sharing system, IEICE technical report MoMuC2012-2. Inf Process Soc Jpn 112(44):5–8
Hughes A (2014) Participatory design for the social media needs of emergency public information officers. In: Proceedings of the 11th international conference on information systems for crisis response and management (ISCRAM), State College, Pennsylvania
Japan Node.js Association (2017) Node.js. https://nodejs.jp/. Accessed 9 Nov 2017
Kimura M, Osaragi T, Oki T (2016a) Traveling method for safety confirmation after a large earthquake. In: Summaries of technical papers of annual meeting Architectural Institute of Japan, pp 963–964 (in Japanese)
Kimura M, Osaragi T, Oki T (2016b) Efficiency indices and traveling method for safety confirmation after a large earthquake. In: Papers and proceedings of the Geographic Information Systems Association (CD-ROM), C-4-4 (in Japanese)
Kubota S, Matsumura K, Yano S, Kitadani T, Kitagawa I, d Ichiuji A (2013) A proposal of disaster-information sharing system using open source GIS. In: Papers and proceedings of the Geographic Information Systems Association, 22, F-5-2 (CD-ROM)
Matthew Z, Mark G, Taylor S, Sean G (2010) Volunteered geographic information and crowd sourcing disaster relief: a case study of the Haitian Earthquake. World Med Health Policy 2 (2):7–33
McKelvey B (2017) WebSockets. https://ja.scribd.com/document/60898569/WebSockets-The-Real-Time-Web-Delivered. Accessed 9 Nov 2017
MongoDB Inc (2017) Move at the speed of your data. https://www.mongodb.com/. Accessed 9 Nov 2017
Murai K et al (2008) Necessity of structuring an effective scheme of acquiring disaster information by fire departments just after an earthquake. J Soc Saf Sci 10:89–96
Ngamassi L, Ramakrishnan T, Rahman S (2016) Use of social media for disaster management: a prescriptive framework. J Organ End User Comput 28(3)
Ngamassi L, Malik A, Zhang J, Ebert D (2017) Social media visual analytic toolkits for disaster management: a review of the literature. In: Proceedings of the 14th international conference on information systems for crisis response and management (ISCRAM), Albi, France
Osaragi T, Hirokawa N, Oki T (2015) Information collection of street blockage after a large earthquake for reducing access time of fire fighters. J Archit Plan (Trans AIJ) 80(709):465–473 (in Japanese)
Osaragi T, Hirokawa N (2017) A decision support system for fighting multiple fires in urban areas caused by large earthquakes. In: Lecture notes in geoinformation and cartography, planning support science for smarter urban futures, vol Part I. Springer, pp 77–93
Petersen L, Fallou L, Reilly PJ, Serafinelli E (2017) Public expectations of social media use by critical infrastructure operators in crisis communication. In: Proceedings of the 14th international conference on information systems for crisis response and management (ISCRAM), Albi, France
Stuart M, Lee M, Stefano M (2014) Real-time crisis mapping of natural disasters using social media. IEEE Intell Syst 29(2):9–17
Tokyo Fire Department (2001) Development and application of evaluation method for preventing ability of earthquake fire. In: The 14th report of Fire Prevention Council (in Japanese)
Tokyo Fire Department (2015) Measures to reduce human damage from fires after earthquakes. In: The 21st report of Fire Prevention Council (in Japanese)
Yates D, Paquette S (2011) Emergency knowledge management and social media technologies: a case study of the 2010 Haitian earthquake. Int J Inform Manage 31:6–13

Part III
Spatiotemporal Data Modelling and Data Mining

A Top-Down Algorithm with Free Distance Parameter for Mining Top-k Flock Patterns

Denis Evangelista Sanches, Luis O. Alvares, Vania Bogorny, Marcos R. Vieira and Daniel S. Kaster

Abstract Spatiotemporal data is becoming more and more available due to the increase in the using of location-based systems. With such data, important information can be retrieved, where co-movement patterns stand out in finding groups of moving objects moving together. However, such pattern mining algorithms are not simple and commonly require non-trivial fixed parameters as input, which are extremely dependent on the data domain and also impacted by many others context variables, being such challenging task also to domain specialists. One example of these patterns is the flock pattern that has as its most challenging parameter the distance threshold that is the size of the disks that involves the objects. Although other density-based approaches reduce the impact of the restrictions of the disk, all of them still require a distance parameter for the density connectedness. Addressing this problem, we introduce the concept of discovering of k-co-movement patterns,

D. E. Sanches (✉) · D. S. Kaster
Department of Computing, University of Londrina, Londrina, Brazil
e-mail: sanches.e.denis@gmail.com

D. S. Kaster
e-mail: dskaster@uel.br

L. O. Alvares · V. Bogorny
Department of Informatics and Statistics, Federal University of Santa Catarina, Florianópolis, Brazil
e-mail: luis.alvares@ufsc.br

V. Bogorny
e-mail: vania.bogorny@ufsc.br

M. R. Vieira
Hitachi America Ltd, R&D, Santa Clara, CA, USA
e-mail: marcos.vieira@hal.hitachi.com

A. Mansourian et al. (eds.), *Geospatial Technologies for All*,
Lecture Notes in Geoinformation and Cartography,
https://doi.org/10.1007/978-3-319-78208-9_12

which is finding the top-k patterns, according to the desired raking criterion. Especially for the flock pattern, we also define a new flock pattern query and propose a top-down algorithm with free distance parameter for the aforementioned problem.

Keywords Flock pattern · Flock discovery · Top-k flocks · Free diameter parameter · Top-down algorithm · Co-movement patterns

1 Introduction

With the increasing ubiquity of location-based systems, such as Global Positioning Systems (GPS), the amount of spatiotemporal data describing the trajectories of Moving Objects (MOs) over time is growing very quickly. As a consequence, many applications will benefit from exploring these data by finding correlations among trajectories to establish relationships between moving objects or finding typical behaviors. Trajectory analysis may be important for transportation logistics optimization, analysis and prediction of animal behavior in their natural habitat, friendship inference in social sciences, as well as applications in biology, and medicine (Zheng and Zhou 2011; Spaccapietra et al. 2008).

Over this data explosion, trajectory pattern mining is becoming more and more important in the last few years. Several types of patterns may be extracted from trajectories, including sequential patterns, periodic patterns, group patterns or co-movement patterns (Zheng 2015). Co-movement patterns are particularly interesting for detecting potential connections between MOs, such as traveling together, persecution, unusual transportation corridors, and others. Some of the most well-known co-movement patterns, which are the focus of this paper, are the flocks (Gudmundsson and van Kreveld 2006; Vieira et al. 2009; Tanaka et al. 2016), convoys (Jeung et al. 2008a, b), and swarms (Li et al. 2010a, b). These patterns share the definition of a way to *spatially connect* moving objects at a timestamp to form a group. In general, they must be inside a disk of a given diameter (disk-based approach), or each of them must not be farther than a given distance from other elements in the group (density-based approach) (Zheng and Zhou 2011; Feng and Zhu 2016).

A disk-based approach ensures that every element is distant at most the given diameter to all other elements in the group. On the other hand, a density-based pattern imposes a less rigid shape for the group as it requires that every element is not farther to a few elements in the group than the provided distance, therefore it allows that relatively distant elements may still be in the same group if there are other elements in the group that *connect* them. However, in both cases, the distance threshold is defined by the user, and the result of the patterns completely depend on this parameter. Setting a suitable distance threshold is very challenging in several situations.

Let us consider the example of flock pattern discovery in Fig. 1, extracted from the real dataset *San Francisco cabs*[1] (Piorkowski et al. 2009), which contains trajectories

[1]https://crawdad.org/epfl/mobility/20090224.

Fig. 1 Flock pattern example. Taxis in San Francisco moving towards the airport

of taxicabs moving from the center of San Francisco to its airport. The goal is to detect a flock of at least 3 taxis traveling *close enough* to each other for 3 or more consecutive timestamps. The figure (rotated 90° right to best fit) shows the positions of three taxis, identified by their colors (blue, purple and green), in four consecutive timestamps, moving from north to south. The key point is to define what means to be *close enough* in every situation. Suppose the user states that they should be distant at most 610 m to configure proximity. As it can be seen in Fig. 1, this threshold (ε) is enough to enclose the three taxis in timestamps 1, 3 and 4, as represented respectively by disks c^1, c^3 and c^4. However, the threshold fails to enclose the taxis in timestamp 2, because one of them moved faster than the others and, consequently, the pattern is lost.

This example highlights two well-known limitations of many co-movement patterns. The first one is the time consecutiveness requirement, which forces the pattern to be strictly maintained in consecutive timestamps. Some works as (Cao et al. 2016; Li et al. 2010a) proposed to relax this requirement, allowing the closeness condition among MOs in a group to be broken for a short period of time and restored afterward, maintaining the pattern. The second limitation is the difficulty to set a proper distance threshold for every situation and/or dataset. Recalling the example in the figure, it is easy to notice that the flock would be detected by stating a bit larger disk diameter. However, how large the diameter should be depends on each case. To discover a flock of taxis moving at a low speed street requires a relatively small diameter as they must be somehow close to each other to configure a co-movement. Conversely, if a group of taxis is on a highway, a much larger diameter should be considered to detect co-movement, because the safety distance between them at high speed must be larger. Another situation is when there is a traffic light on the way. Taxis moving together can become spatially dispersed due to a red traffic light if some pass through it while others do not.

In general, moving objects do not move smoothly, exactly at the same speed, such that at every time instant the distance between the points is homogeneous. Therefore, different moving objects, such as vehicles, people and animals of different families (e.g., felines, cattle, fishes or birds) require distinct distance thresholds to discover co-movement patterns. The instant location of MOs also impacts the distance threshold (low speed roads vs. high speed highways, flat land vs. mountains, air vs. land or water, etc.) as well as the application goal (e.g., traffic analysis, crowd movement and animal migration). Notice that the relaxation of the time consecutiveness does not suffice for surpassing this limitation. Furthermore, to define a suitable distance threshold is even more challenging when the dataset is not known beforehand. In these cases, exploratory data analyses would be much easier for the user if he/she could provide parameters that are simple to define.

For the aforementioned limitations, in this work we propose a new concept called k_ε-Flocks, which is an exploration mining approach based on top-k queries, and without distance threshold as input, for mining the *k*-flocks of minimum diameter in a window, so that its flock with the largest diameter has the smallest possible diameter. The concept of k_ε-Flocks relinquishes the user from defining a distance threshold, he/she should just provide the desired number *k* of flocks to be returned. We also introduce the k_ε-Flock Pattern for returning the k_ε-Flocks in all time windows of a given valid size for a dataset. Our proposals have a wide applicability in several domains as, for instance, to help domain specialists to explore a trajectory dataset in which they do not have deep knowledge of the movement behavior of the corresponding moving objects or for real-time monitoring and surveillance of the most relevant flocks, in terms of closeness, evolving in time. We also present a top-down algorithm to discover k_ε-Flocks in a given window, which is based on iteratively splitting candidate flocks into subflocks guided by border points.

The remainder of the paper is structured as follows. Section 2 presents the basic concepts and definitions, and discusses related work. Section 3 presents the proposed approach of discovering the k-co-movement patterns, focusing on the flock pattern, and defines the k_ε-Flock Pattern. Section 4 shows the proposed top-down k_ε-Flock algorithm. Section 5 presents preliminary experiments of the proposed algorithm and brings a discussion. Finally, Sect. 6 contains the conclusion and future work.

2 Preliminaries

2.1 The Flock Pattern

Given a uniquely identified moving object MO, its path can be continuously collected and represented by a sequence of location points. This entire path can be semantically segmented, in terms of its applications domains, in sub-paths called trajectories that represent movements of the MO from a start point to an end point. The following definition formalizes this concept.

Definition 1 (*Trajectory*) A trajectory T is a MO's defined record of the evolution of its position that is moving in space during a given time interval in order to achieve a given goal, perceived as a sequence of points p, where x and y are spatial

Fig. 2 Flock pattern example. Flock $\{T_1, T_2, T_3\}$ in timestamps $\{t_1, t_2, t_3\}$

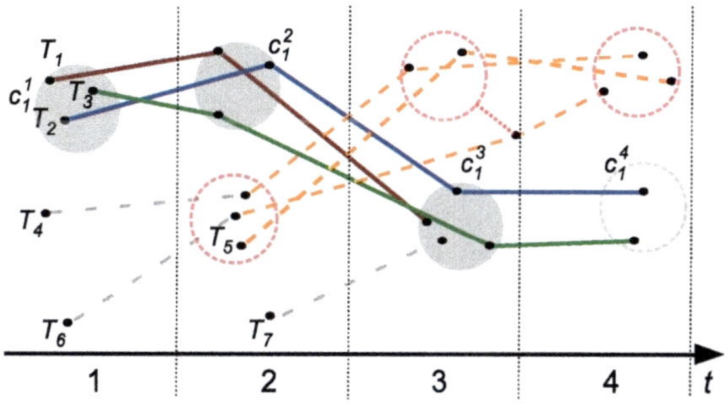

coordinates representing a location and t is the time instant when this point was collected (Bogorny et al. 2014; Spaccapietra et al. 2008).

In real applications, trajectory points usually do not follow a strict sampling rate. Therefore, the time instants of these points collection may be discretized in order to facilitate pattern mining. In this synchronization process, known as calibration (Parent et al. 2013), all points are resampled to share the same discretized sequence of regularly sampled instants.

A well-known trajectory pattern that can be mined with discretized points is the flock pattern, which, according to the definition proposed by Benkert et al. (2008), is a set of trajectories such that for every trajectory point within a time window, there is a disk of a given radius that contains all these trajectories. Adapting the notation of Vieira et al. (2009), let $\mathcal{T}$ be a set of trajectories, $\mu > 1$ be a minimum number of trajectories ($\mu \in \mathbb{N}$), $\varepsilon > 0$ be a distance threshold regarding a distance metric d ($\varepsilon \in \mathbb{R}_{>0}$), and $w = [t_{initial}, t_{final}]$ be a time window between timestamps $t_{initial}$ and t_{final}, a flock is defined as follows.

Definition 2 (*Flock*) A flock $f_w(\mu, \varepsilon)$ is a set of pieces of trajectories in $\mathcal{T}$ of at least μ moving objects in a time window w such that for every timestamp t_i in the interval of consecutive timestamps in w, $t_{initial} \leq t_i \leq t_{final}$, there is a disk c^{t_i} with diameter ε that covers all points of their trajectories at timestamp t_i.

Figure 2 illustrates this concept. It depicts a buffer of four sequential time instances (timestamps). This buffer contains location points—black points—of six different trajectories, $\mathcal{T} = \{T_1, \ldots, T_6\}$. Supposing a window of size 3 or 4, we can identify some flocks in this figure, being ε the size of the circumferences drawn, such as: $f_{[t_1,t3]}(3, \varepsilon) = \langle\{T_1, T_2, T_3\}, [t_1, t_3]\rangle$ (denoted by the gray disks), and $f_{[t_1,t4]}(2, \varepsilon) = \langle\{T_2, T_3\}, [t_1, t_4]\rangle$ (gray disks plus the dashed gray disk). Indeed, there may be many flocks with the same parameters (w, μ and ε) that must be properly identified, for instance: $f^1_{[t_2,t4]}(2, \varepsilon) = \langle\{T_2, T_3\}, [t_2, t_4]\rangle$ and $f^2_{[t_2,t4]}(2, \varepsilon) = \langle\{T_4, T_5\}, [t_2, t_4]\rangle$.

A flock is of maximal size if it contains the maximum number of MOs whose trajectories can be enclosed in diameter ε. For instance, flock $f^1_{[t_1,t3]}(2, \varepsilon) = \langle\{T_1, T_2\}, [t_1, t_3]\rangle$ in Fig. 2 is not of maximal size while $f^2_{[t_1,t3]}(2, \varepsilon) = \langle\{T_1, T_2, T_3\}, [t_1, t_3]\rangle$ is. After having defined flocks, the flock pattern is given by Definition 3.

Definition 3 (*Flock Pattern*) A flock pattern FlockPattern(μ, ε, δ) reports a set $\mathcal{F}$ containing all the flocks of maximal size $f_w(\mu, \varepsilon)$, for all time windows w with fixed length $|w| = \delta$ that are valid for the set of trajectories $\mathcal{T}$.

The two main classes of flock pattern algorithms report either (i) the longest flocks (with maximal duration), including the works of Gudmundsson and van Kreveld (2006), Arimura et al. (2014), and Geng et al. (2014), or (ii) flock patterns of fixed length, i.e., flocks are reported as soon as they complete the minimum length threshold (δ), such as the proposals of Vieira et al. (2009) and Tanaka et al. (2016). The former approaches are offline since they require to load the whole dataset to run, and the latter are online, being able to deal with data streaming using the sliding-window model (Silva et al. 2013).

The flock pattern has two main problems. The first one is the assumption that all points of different moving objects must be synchronized, what is not always the case when dealing with real movement. The second is that all works assume that the radius that contains all entities is of fixed size, defined by the user, what is not only a problem in flock detection but in several trajectory pattern mining methods and similarity measures (Furtado et al. 2018). For example, considering the *FlockPattern*($3, \varepsilon, 3$) in Fig. 2, only the flock $f_{[t_1,t3]}(3, \varepsilon) = \langle\{T_1, T_2, T_3\}, [t_1, t_3]\rangle$ is reported (gray disks). Another potential flock that could be of interest is "lost" (light pink dashed disks) because trajectories $\{T_4, T_5, T_6\}$ do not form a flock due to insufficient diameter size to enclose all three trajectories in timestamp t_3. This work aims at addressing the second problem by proposing a new class of co-movement pattern that does not require providing the distance threshold. The next section reviews related proposals.

2.2 Related Work

In several application scenarios, users start the pattern discovery process through data exploration queries. These queries usually have roughly defined conditions as users do not know a priori which may be proper conditions for the dataset according to their interest. After issuing a few queries, they can use the obtained results to gradually refine the search. With regard to trajectory pattern mining, users usually become frustrated during the exploration effort as they have to iteratively adjust a number of interrelated parameters through costly queries, which can take a long time. In the literature several patterns have been proposed, many of them addressing specifically limitations of flock pattern, but, as we can see below, none of them can effectively ease the task of estimating a suitable distance parameter.

Jeung et al. (2008a, b) were precursors in addressing what they called the "lossy-flock problem", which is a direct consequence of the difficulty of defining a proper disk size and that this circular shape may not correspond to real group patterns.

The proposed solution (Convoy Pattern) was to employ a density-based approach for initial clustering step instead of a disk-based one as in flock pattern. In this way, a convoy is based on a less restrictive proximity condition than a flock, allowing to report groups of different shapes by requiring that the moving objects remain close to a few others in the group (density-connected), instead of requiring that the moving objects remain close to all others in the group (enclosed by a disk), as it must occur to be a flock.

Other patterns focus on the time consecutiveness limitation. The Swarm Pattern (Li et al. 2010b, a), for example, fully relaxes the temporal threshold, mining even loose patterns, e.g., cars meeting in gas station sporadically. The Platoon Pattern (Li et al. 2015) also relaxes the consecutiveness in time, however, it tries to mitigate these loose patterns by requiring a local consecutiveness threshold too. These patterns form groups based on density-connectedness. Moving Clusters (Kalnis et al. 2005) can also be highlighted as co-movement patterns, nevertheless, they allow that groups/clusters start and end with completely different objects.

Regarding relaxing the size parameter, which determines the minimum number of trajectories to consider a group of MOs relevant, we can cite the Group Pattern (Wang et al. 2003, 2006). This pattern does not require this minimum size parameter, finding groups with at least two MOs. Similarly, the Group Query (Li et al. 2013) adjusts the minimum size of a group according to a score function that can the balanced to prioritize either the cardinality or the duration of the pattern found, aiming at retrieving the most significant patterns using a top-k approach.

Lastly, the works of Wachowicz et al. (2011) and Ong et al. (2013) focus on the estimation of the distance parameter of the flock pattern. Both works stand out that it is the most challenging parameter to be set for these patterns, even for domain specialists, due to the specificity of each data context. Wachowicz et al. (2011) and Ong et al. (2013) propose to estimate the distance parameter based on the pairwise point distance distribution in a timestamp or in a time window, based on the basic strategy to find the minimum distance to establish connectivity (the EPS parameter) for the DBSCAN clustering algorithm (Ester et al. 1996). The idea is to calculate the distance between each point and its k-th nearest neighbor, where k is related to the minimum group size, and then plot these values as a line graph with objects ordered in descending order, which is the so-called sorted k-distance plot. The portion of the graph such that when there is a sudden decrease suggests an upper bound for the distance parameter. However, this strategy relies on additional contextual knowledge to recommend such an estimation as well as it requires a huge amount of distance calculations, which may prevent it from being useful for a real-time (online) solution. Differently from existing proposals, in this work, we introduce the approach of discovering k-co-movement patterns and present a new pattern for retrieving flocks with minimum diameter, which does not make necessary to state the distance parameter.

3 The New Pattern k_ε-Flock

As previously mentioned, setting a suitable distance threshold as input for co-movement pattern mining algorithms requires knowing the movement behavior of the objects in the input dataset, which vary according to the nature of the object (person, vehicle, animal, etc.), to the local conditions (road network with varying speeds, traffic lights, land use, etc.), and so on. This can be fairly difficult, especially for trajectory mining tasks, in which the goal is to detect unknown but relevant patterns that evolve over time. That is, besides being hard to define an initial value for this parameter, it is not enough to provide a fixed threshold for a stream of trajectory data, as the dynamic behavior of real data movement would demand to continuously adjust this parameter.

Our proposal is to reduce this problem to the monitoring of a given *number* of patterns, identified according to a specific condition that may vary over time. The idea is to issue a kind of *top-k* query over a time window and retrieve the k answers that are the most relevant at that time regarding the desired ranking criterion. We name this approach as the ***discovery of k-co-movement patterns***. This approach has variations that refer to the mined pattern (e.g., flock, convoy, etc.) and to the parameter that defines the ranking criterion (e.g., distance, temporal length, etc.). The main advantage of such a strategy is to switch a hard parameter definition to a simpler one—the desired number k of patterns—while providing to the user an adaptative view of the patterns formed in the dataset.

In the context of this work, we are interested in making the distance threshold a free parameter for the flock pattern. Therefore, we propose the new concept **k-Flocks with Minimum Diameter** (k_ε-Flocks), which are the k flocks in a given time window and of a given minimum size such that the diameter of the flock with the largest diameter among the k_ε-Flocks is the smallest possible, i.e., for any diameter that is less than this value the number of flocks returned is less than k. The k_ε-Flock Pattern is the set of k_ε-Flocks for all time windows in the trajectory dataset. These concepts are formalized through the following definitions.

Definition 4 *(Minimum Diameter of a Flock)* The minimum diameter of a flock $f_w(\mu, \varepsilon)$ is the smallest value $\varepsilon' \in \mathbb{R}$ such that $f_w(\mu, \varepsilon') = f_w(\mu, \varepsilon)$.

Definition 5 *(Widest Flock)* Given a set of flocks F, the widest flock in this set is the flock whose minimum diameter is the largest among flocks in F. Ties are broken arbitrarily.

Definition 6 *(k_ε-Flocks)* The k_ε-Flocks regarding window w and minimum number of trajectories μ, represented as k_ε-Flocks(μ, w), is the set $\mathcal{F}_w^k$ containing k flocks such that for every flock $f_w(\mu, \varepsilon) \notin \mathcal{F}_w^k$ we have that $\varepsilon \geq \varepsilon_k$, where ε_k is the minimum diameter of the widest flock in $\mathcal{F}_w^k$. If there are not enough flocks in the window, then less than k flocks are reported.

Definition 7 *(k_ε-Flocks Pattern)* A k_ε-Flock pattern k_ε-FlockPattern(μ, δ) reports a set $\mathcal{F}^k$ containing the k_ε-Flocks(μ, w) with regard to every time window w of size δ valid for the set of trajectories $\mathcal{T}$, that is $\mathcal{F}^k = \bigcup_{\forall w} \mathcal{F}^k_w$.

The basic approach to return the k_ε-FlockPattern is to iteratively identify the k_ε-Flocks over a sliding-window buffered from an input data stream. In the next section, we present properties of the proposed concepts that allow defining an exact top-down algorithm for the identification of the k_ε-Flocks in a time window.

3.1 Properties for the Identification of the k_ε-Flocks

Given k, μ and w, the main property to identify the k_ε-Flocks(μ, w) is that a flock with more than μ trajectories can be divided into subflocks with fewer trajectories than the original flock. The minimum diameter of the subflocks will be smaller than the minimum diameter of the original flock. Thus, starting with a single flock containing all trajectories in the time window, this splitting process can be iteratively applied until reaching the answer.

To support this observation, Fig. 3 shows the number of flocks of maximal size discovered in a given time window for decreasing diameters for two well-known trajectory datasets: location points labeled as *cars* of *Geolife*[2] and *San Francisco cabs* which are detailed in Sect. 5. It can be seen that in Geolife dataset (Fig. 3a) the number of flocks augments for decreasing diameters until a saturation point and becomes smaller for diameters below this point until no more valid flocks are found. For the San Francisco cabs dataset (Fig. 3b) the behavior is similar, nonetheless, the number of flocks is not always increasing until the saturation point and not always decreasing after it. Therefore, the stop condition for the top-down approach must consider such instability.

The process consists of splitting always the widest flock in the current set of candidate flocks into subflocks. The minimum diameter of a flock in a window is the diameter of the largest disk among the disks tightly involving the points of the trajectories in the flock duration. For example, Fig. 4 (assuming both flocks *FlockPattern*$(3, \varepsilon, 3)$ of Fig. 2 are in the same window) shows the disks tightly involving the trajectories of flocks $f^1_{[t_1,t_3]}(3, \varepsilon_1) = \langle\{T_1, T_2, T_3\}, [t_1, t_3]\rangle$ and $f^2_{[t_1,t_3]}(3, \varepsilon_2) = \langle\{T_4, T_5, T_6\}, [t_1, t_3]\rangle$. Comparing Fig. 2 with Fig. 4 can be seen that in the first one all disks have the same size, whereas in the latter not because in thc latter case the disks have the smallest diameter possible. The minimum diameters of candidate flocks $f^1_{[t_1,t_3]}(3, \varepsilon_1)$ and $f^2_{[t_1,t_3]}(3, \varepsilon_2)$ are ε_1 and ε_2, which are respectively the diameters of the red solid line gray disk and red dashed line white disk. Consequently, the widest flock between these two flocks is $f^2_{[t_1,t_3]}(3, \varepsilon_2)$.

---

[2]https://www.microsoft.com/en-us/download/details.aspx?id=52367.

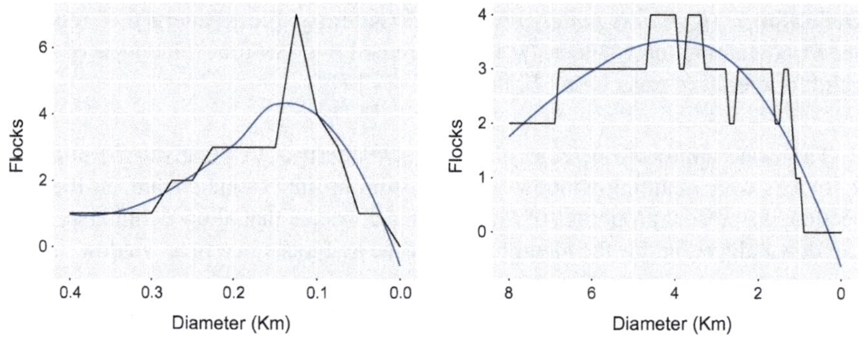

(a) Cars in Geolife (b) Cabs in San Francisco

Fig. 3 Behavior of quantity of flocks of maximal size in a time window varying ε

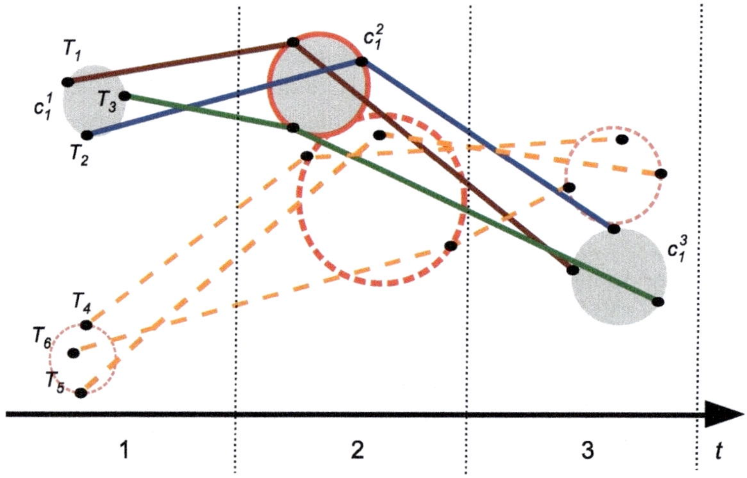

Fig. 4 Flocks' diameters in the window, with the flock with dashed disks as the widest one

After having chosen the widest flock $\overline{f_w}$ among the candidates in the window, Fig. 5 illustrates how split should happen in two possible situations: with two and with three border points, the red ones. The idea is to discover which trajectory, if removed from $\overline{f_w}$, would split it into the two widest possible subflocks of $\overline{f_w}$. It is worth mentioning that the split must guarantee that the two widest subflocks are produced to avoid missing answers in next steps of the top-down approach. To achieve this, for each border point of $\overline{f_w}$ is created a subflock with all the remainder points of $\overline{f_w}$, except this point. On the left side of the figure, $\overline{f_w}$ contains only two border points, then the two possible widest subflocks f_w^1 and f_w^2 are found. However, in a future iteration f_w^1 becomes the current $\overline{f_w}$ and have to split. As it is shown on the right side of the figure, this flock contains three border points, therefore three sub-

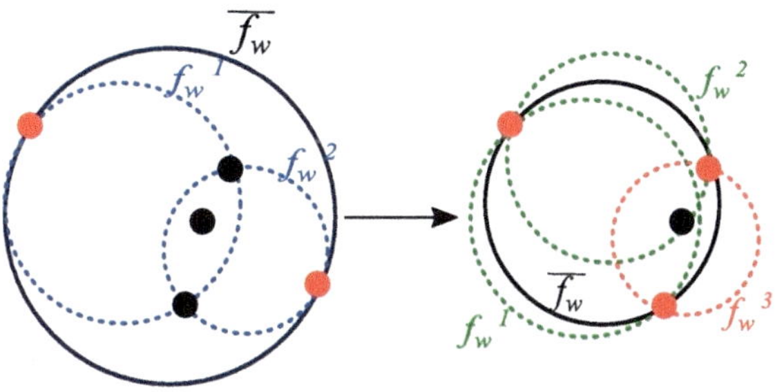

Fig. 5 Example of the split of two widest flock $\overline{f_w}$ in two different iterations. On the left side, $\overline{f_w}$ is split only into the two possible subflocks. However, on the right side, with three border points, three subflocks are found and the smaller one (f_w^3) is ignored

flocks can be generated. The algorithm chooses the two widest subflocks (f_w^1 and f_w^2 in green) and ignores the smallest one (f_w^3 in red). A fundamental property of this way of splitting is that the diameters of the subflocks are smaller than the original flock candidate, except for geometric cases that are rare in real datasets (we provide a discussion about these cases in Sect. 5.1). Therefore, this process yields an iterative reduction in the diameter that ensures to reach an exact answer.

The split process has to stop on two occasions: when the k_ε-Flocks are discovered, or when no better answer can be found within the current set of candidate flocks found so far ($currentF$). These alternative stop conditions are based on Eq. 1.

$$maxSubflocks(current\mathcal{F}) = \sum_{j=1}^{n} 2^{|f_w^j|-\mu} \begin{cases} if < k & current\mathcal{F} \not\supset k_\varepsilon\text{-Flocks} \\ if \geq k & current\mathcal{F} \text{ may contain } k_\varepsilon\text{-Flocks} \end{cases} \quad (1)$$

Equation 1 estimates the largest number of subflocks $current\mathcal{F}$ may contain. As each flock of size $s > \mu$ splits into two subflocks of size $s - 1$ each, a flock f_w^j can be split into $2^{|f_w^j|-\mu}$ subflocks. That is, for any flock of minimum size $|f_w^j| = \mu$ no more subflocks can be produced and the flock itself is counted ($2^0 = 1$); for flocks with size $|f_w^j| = \mu + 1$ one split could be done, resulting in $2^1 = 2$ subflocks; and so on. In this way, the total number of possible subflocks is the sum of the estimation on each candidate flock in $current\mathcal{F}$.

4 A Top-Down Algorithm for the Discovery of k_ε-Flocks

This section details the proposed algorithm for the retrieval of the k_ε-Flocks in a given window based on the presented top-down approach (Algorithm 1). The inputs for the algorithm are the time window w to be processed, the minimum number of trajectories μ to determine a flock, and k that is the desired quantity of flocks. The answer set $\mathcal{F}_w^k$ is initialized with a single flock, maximal in size, whose diameter is enough to wrap all points for all timestamps of w, as well as the current set of candidate flocks $current\mathcal{F}$, which contains the set of flocks found so far, being updated

Algorithm 1: Top-down k_ε-Flocks in a time window

```
Input: μ: minimum number of trajectories
w: time window regarding a set of trajectories T
k: requested number of flocks with minimum diameter
Output: F_w^k: k_ε-Flocks
/* Initialization                                                        */
1  F_w^k ← add the single flock containing the maximum number of trajectories in w
2  currentF ← F_w^k
/* Can F_w^k converge to k_ε-Flocks or at least be increased?            */
3  while (maxSubflocks(currentF) ≥ k) or (maxSubflocks(currentF) ≥ |F_w^k|) do
4      \overline{f_w} ← removeWidestFlock(currentF)
5      if size(\overline{f_w}) > μ then                        // can \overline{f_w} split?
           /* process subflocks                                          */
6          Split \overline{f_w} into the two widest subflocks f_w^1 and f_w^2 based on the border points in the
           circumference of the \overline{f_w}'s biggest disk
           /* calculate disks for the subflocks                          */
7          f_w^1.ε ← 0;  f_w^2.ε ← 0;
8          foreach timestamp t_i in window w do
9              foreach subflock f_w ∈ {f_w^1, f_w^2} do
10                 f_w[c^{t_i}] ← calculate the minimum bounding disk c^{t_i} of f_w in timestamp t_i
11                 if f_w[c^{t_i}].diameter > f_w.ε then        // Did c^{t_i} increase f_w.ε?
12                     f_w.ε ← f_w[c^{t_i}].diameter
13                 end if
14             end foreach
15         end foreach
16         currentF ← currentF ∪ {f_w^1, f_w^2}                // Add the subflocks
17     end if
18     if |currentF| ≥ |F_w^k| or |currentF| ≥ k then          // Should F_w^k be updated?
19         F_w^k ← currentF
20     end if
21 end while
22 return F_w^k, with the k_ε-Flocks, or with the largest number of answers possible (< k)
```

after each refinement iteration (lines 1–2). What differs $\mathcal{F}_w^k$ from *current*$\mathcal{F}$ is that the former is always a valid answer, while the latter may end up with an incomplete set of answers, thus requiring to have the last valid answer set saved in $\mathcal{F}_w^k$.

Lines 3–21 comprehend the answer set refinement, always regarding the widest flock at each iteration. It is performed until finding the k_ε-Flocks, if they exist, or, otherwise, the largest possible number of answers in w. This loop is repeated while Eq. 1 returns that it is possible to exist at least k flocks within the current candidate set (*maxSubflocks*(*current*$\mathcal{F}$) $\geq k$). The premise is that as long as there exists the possibility of finding the k_ε-Flocks in *current*$\mathcal{F}$, the current diameter ε is decreased by dividing the wider flocks until k flocks can be found. However, the window may not have k_ε-Flocks, so Eq. 1 gives a value less than k. In this case, the loop also continues iterating if the estimation of the number of subflocks is greater than the number of flocks in $\mathcal{F}_w^k$, because there is still the possibility to find more flocks to augment the answer set even that it will not contain k answers.

The refinement loop removes the widest flock so far in the window w ($\overline{f_w}$) to process it. If its size is greater than the minimum (μ), it can be split into two subflocks, otherwise, $\overline{f_w}$ is discarded, reducing the size of $current\mathcal{F}$ (lines 4–6). To correctly split $\overline{f_w}$, given all disks of the flock, each of which in a timestamp in the window ($\overline{f_w}[c^{t_i}], t_{initial} \leq t_i \leq t_{final}$), the algorithm selects two trajectory points in the boundary of the minimum bounding disk that is the biggest one among the flock's disks (refer to the explanation of Fig. 5 in Sect. 3.1). The minimum bounding disk can be obtained by employing any algorithm for finding the smallest-circle that encloses a set of points. It should be noted that there may be more than two boundary points, usually three. When this case occurs, the two chosen points are those boundary points that result in the two widest subflock in w among the possible subflocks of $\overline{f_w}$. Then, two subflocks f_w^1 and f_w^2 are calculated within the entire window w, each of which containing the trajectories of $\overline{f_w}$ except one of the two points. It is worth noting that this split process may produce subflocks that are subflocks of greater flocks still in $current\mathcal{F}$. In this case, this subflock also has to be discarded because it will be processed later when splitting the other candidate that contains it. Lines 7–16 calculate the minimum bounding disks for the subflocks at every timestamp in the window, set the subflocks' diameter ($f_w \cdot \varepsilon$) to the diameter of their respective largest disks, and add the subflocks to the current set of flock candidates.

After having processed the widest flock $\overline{f_w}$, an important check is made to determine whether $\mathcal{F}_w^k$ must be updated with $current\mathcal{F}$ or not (lines 18–20). At this point, there are two possibilities: either the widest flock $\overline{f_w}$ was split, increasing the size of $current\mathcal{F}$ that now may contain more candidates than $\mathcal{F}$, or it was discarded, reducing the set size. If it was split, subflocks with minor diameters were added, thus the answer was improved by having reduced the diameter. On the other hand, if it was discarded, it is necessary to verify if $current\mathcal{F}$ is a potential answer. If the size of $current\mathcal{F}$ has dropped below k it is not a potential answer and therefore cannot supersede $\mathcal{F}_w^k$ as the latter contains the last valid answer. Notice that the loop refinement continues with $current\mathcal{F}$, because although at this moment it has less than k flocks, it may have k or more potential subflocks, which can be generated in future iterations, increasing the set again. When the set of candidates does not have relevant subflocks anymore to explore, the set $\mathcal{F}_w^k$ is returned, containing the k_ε-Flocks, or, if it is not possible, less than k flocks, but with the smallest possible diameter in w.

5 Experiments

This section presents experiments to evaluate the proposed algorithm. The objective of the experiments is to depict the behavior of the algorithm regarding time elapsed with the variation of parameters. The proposed algorithm was implemented in C++ and the tests were run on a computer with an Intel(R) Core(TM) i7-3630QM

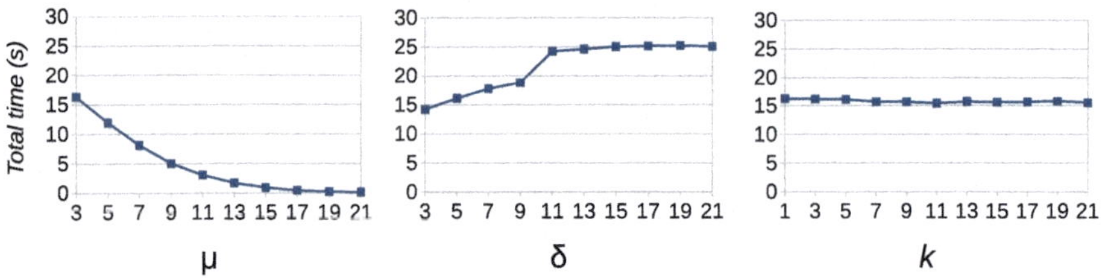

Fig. 6 Performance behavior of the proposed algorithm for varying parameters (μ, δ, k)

2.40 GHz CPU and 8 GB of RAM. The employed dataset is the well known *Geolife* (Zheng et al. 2008, 2009, 2010), specifically using the location points labeled as *cars*. In order to maximize the number of possible flock patterns, data points are assumed to be taken from the same day, with collection time instances synchronized in 2 s.

Figure 6 depicts the execution time behavior of the sliding window algorithm, varying each of its three parameters. When varying one of them, the other two are fixed with the following default values: $\mu = 3, \delta = 5$ and $k = 1$. As expected, the execution time reduces with increasing values of μ, because high values restrict the flocks that can be found, thus reducing the processing. Regarding the window length (δ), the execution time rises for small windows but stabilizes for greater ones. Finally, the execution time remains stable for all values of k as the algorithm explores a large part of the answer space, which is usually much greater than the provided k. If a single window is considered, our approach is a few times slower than a flock pattern algorithm with the distance parameter set up to retrieve a comparable number of answers. However, our algorithm automatically adjusts the distance parameter when sliding the window, while existing algorithms employ always the same distance, therefore varying considerably the number of answers along time.

5.1 Discussion: Splitting of Geometric Cases

It is worth noting a known issue that the proposed solution might face regarding what we call *geometric cases* while selecting the points to perform a split. Figure 7 depicts one of the possible theoretical scenarios. The figure shows a flock composed by four trajectories whose points are exactly equidistant (corner points of a square) for three consecutive timestamps. In examples like this, the removal of any of the four points would produce a disk with the very same diameter. Consequently, the trajectory that corresponds to the removed point cannot be removed from the flock, as its points are enclosed by the "new" disk. This flock cannot be split into two flocks of size $\mu = 3$, however it may be split into four flocks of size $\mu = 2$, for example.

The aforementioned situations are extremely rare to happen with real spatiotemporal data. Firstly, the collection of location points suffers from noises and systems

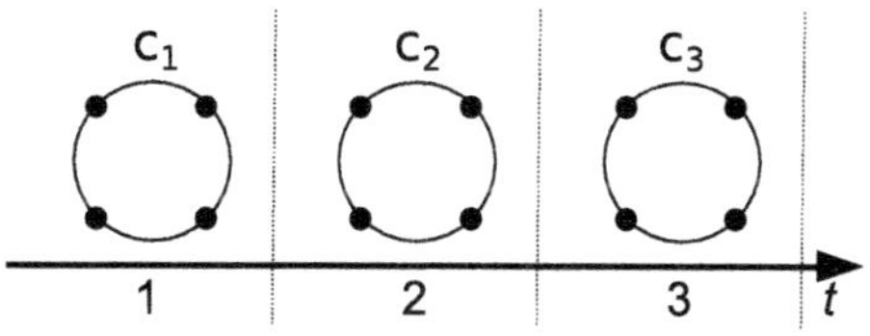

Fig. 7 Example of a *geometric case* in a flock split. A tentative of removal of any of the four points would result in the same flock

errors. Added to this, the coordinates have sufficient decimal precision that highly collaborates to the rarity of these cases. However, an additional maintenance strategy must be included in the algorithm to treat cases when the removal of one point does not result in diameter reduction.

6 Conclusion and Future Work

The ubiquity of spatiotemporal data is notorious. With such amount of data, mining co-movement patterns are of great importance. For this, several co-movement patterns have been defined in the literature, including the flock pattern addressed in this work. However, as discussed, all of them require difficult parameters as input, being a challenging task even for domains specialists.

Addressing this limitation, we propose a new concept of the discovery of k-co-movement patterns, which the main idea is an exploratory mining, releasing the user from providing a non-trivial parameter but just the number k of desired answers to be found. In this way, focusing on the flock pattern, specifically in its challenging distance parameter, we propose a new flock pattern mining and algorithm which retrieves the k flocks in a given time window w so that its largest flock has the minimum size as possible. With such approach, specialists can initially understand the data behavior and what is a good distance threshold to discovery such flocks. Finally, we presented experiments that show the general performance behavior of the proposed algorithm for varying parameters.

Our proposed algorithm is not optimal in the number of iterations. Therefore, future works include to develop a bottom-up algorithm for this problem, which we expect to be a near optimal solution, to extend our *top-k* approach to a density-based pattern, and to explore variations of the pattern that reduces the amount of overlap among the pattern answers.

Acknowledgements This work has been supported by the Brazilian Funding Agencies CAPES and CNPq.

References

Arimura H, Takagi T, Geng X, Uno T (2014) Finding all maximal duration flock patterns in high-dimensional trajectories, Hokkaido University

Benkert M, Gudmundsson J, Hübner F, Wolle T (2008) Reporting flock patterns. Comput Geom 41(3):111–125

Bogorny V, Renso C, Aquino AR, Lucca Siqueira F, Alvares LO (2014) Constant-a conceptual data model for semantic trajectories of moving objects. Trans GIS 18(1):66–88

Cao Y, Zhu J, Gao F (2016) An algorithm for mining moving flock patterns from pedestrian trajectories. In: APWeb, pp 310–321

Ester M, Kriegel H-P, Sander J, Xu X et al (1996) A density-based algorithm for discovering clusters in large spatial databases with noise. In: SIGKDD, pp 226–231

Feng Z, Zhu Y (2016) A survey on trajectory data mining: techniques and applications. IEEE Access 4:2056–2067

Furtado AS, Alvares LOC, Pelekis N, Theodoridis Y, Bogorny V (2018) Unveiling movement uncertainty for robust trajectory similarity analysis. Int J Geogr Inf Sci 32(1):140–168

Geng X, Takagi T, Arimura H, Uno T (2014) Enumeration of complete set of flock patterns in trajectories. In: SIGSPATIAL IWGS, pp 53–61

Gudmundsson J, van Kreveld M (2006) Computing longest duration flocks in trajectory data. In: GIS, pp 35–42

Jeung H, Shen HT, Zhou X (2008a) Convoy queries in spatio-temporal databases. In: ICDE, pp 1457–1459

Jeung H, Yiu ML, Zhou X, Jensen CS, Shen HT (2008b) Discovery of convoys in trajectory databases. PVLDB 1(1):1068–1080

Kalnis P, Mamoulis N, Bakiras S (2005) On discovering moving clusters in spatio-temporal data. SSTD 3633:364–381

Li X, Ceikute V, Jensen CS, Tan K-L (2013) Effective online group discovery in trajectory databases. IEEE TKDE 25(12):2752–2766

Li Y, Bailey J, Kulik L (2015) Efficient mining of platoon patterns in trajectory databases. D&KE 100:167–187

Li Z, Ding B, Han J, Kays R (2010a) Swarm: mining relaxed temporal moving object clusters. PVLDB 3(1–2):723–734

Li Z, Ji M, Lee J-G, Tang L-A, Yu Y, Han J, Kays R (2010b) Movemine: mining moving object databases. In: SIGMOD, pp 1203–1206

Ong R, Nanni M, Renso C, Wachowicz M, Pedreschi D (2013) Parameter estimation and pattern validation in flock mining. In: Int'l workshop on new frontiers in mining complex patterns, pp 3–17

Parent C, Spaccapietra S, Renso C, Andrienko G, Andrienko N, Bogorny V, Damiani ML, Gkoulalas-Divanis A, Macedo J, Pelekis N et al (2013) Semantic trajectories modeling and analysis. CSUR 45(4):42

Piorkowski M, Sarafijanovoc-Djukic N, Grossglauser M (2009) CRAWDAD dataset epfl/mobility (v. 2009-02-24). https://doi.org/10.15783/C7J010, https://crawdad.org/epfl/mobility/20090224

Silva JA, Faria ER, Barros RC, Hruschka ER, de Carvalho AC, Gama J (2013) Data stream clustering: a survey. CSUR 46(1):13

Spaccapietra S, Parent C, Damiani ML, de Macedo JA, Porto F, Vangenot C (2008) A conceptual view on trajectories. D&KE 65(1):126–146

Tanaka PS, Vieira MR, Kaster DS (2016) An improved base algorithm for online discovery of flock patterns in trajectories. JIDM 7(1):52–67

Vieira MR, Bakalov P, Tsotras VJ (2009) On-line discovery of flock patterns in spatio-temporal data. In: GIS, pp 286–295

Wachowicz M, Ong R, Renso C, Nanni M (2011) Finding moving flock patterns among pedestrians through collective coherence. IJGIS 25(11):1849–1864

Wang Y, Lim E-P, Hwang S-Y (2003) On mining group patterns of mobile users. DEXA 2736:287–296

Wang Y, Lim E-P, Hwang S-Y (2006) Efficient mining of group patterns from user movement data. D&KE 57(3):240–282

Zheng Y (2015) Trajectory data mining: an overview. ACM (TIST) 6(3):29

Zheng Y, Li Q, Chen Y, Xie X, Ma W-Y (2008) Understanding mobility based on GPS data. In: Proceedings Int'l Conference on Ubiquitous Computing, pp 312–321

Zheng Y, Xie X, Ma W-Y (2010) Geolife: a collaborative social networking service among user, location and trajectory. IEEE Data Eng Bull 33(2):32–39

Zheng Y, Zhang L, Xie X, Ma W-Y (2009) Mining interesting locations and travel sequences from GPS trajectories. In: WWW, pp 791–800

Zheng Y, Zhou X (2011) Computing with spatial trajectories, Springer Science & Business Media

Optimization and Evaluation of a High-Performance Open-Source Map-Matching Implementation

Karl Rehrl, Simon Gröchenig and Michael Wimmer

Abstract Map matching, i.e. matching a moving entity's position trajectory to an underlying transport network, is a crucial functionality of many location-based services. During the last decade, numerous map-matching algorithms have been proposed, tackling challenging aspects like sparse trajectory data or online matching. This work describes GraphiumMM, an open-source map-matching implementation combining and optimizing geometrical and topological matching concepts from previous works. The implementation aims at highly accurate and performant map matching in online and offline mode taking trajectories with average sampling intervals between 1 and 120 s as input. For evaluating its runtime performance and matching quality, results are compared to results from the open-source map-matcher Barefoot. Results indicate better matching quality and runtime performance especially for sampling intervals from 1 to 15 s in offline and online mode.

Keywords Map matching · Trajectories · Evaluation · Open source

1 Introduction

Map matching denotes the matching of a moving entity's position trajectory (most likely Global Navigation Satellite System (GNSS) measurements) to an underlying map, typically a transport network (Newson and Krumm 2009). In the context of moving object analysis, map matching tackles the question of determining an

K. Rehrl (✉) · S. Gröchenig · M. Wimmer
Salzburg Research Forschungsgesellschaft mbH, Jakob-Haringer-Straße 5, 5020 Salzburg, Austria
e-mail: karl.rehrl@salzburgresearch.at

S. Gröchenig
e-mail: simon.groechenig@salzburgresearch.at

M. Wimmer
e-mail: michael.wimmer@salzburgresearch.at

A. Mansourian et al. (eds.), *Geospatial Technologies for All*, Lecture Notes in Geoinformation and Cartography,
https://doi.org/10.1007/978-3-319-78208-9_13

entity's most likely path out of possible paths. In the domain of Intelligent Transport Systems (ITS), map matching is used for locating a navigating vehicle on the road network or for mapping time-referenced positions of a moving vehicle (so-called Floating Car Data or FCD) to segments of a road network. With respect to the individual use case, map matching uses completed ("offline map-matching") or iteratively expanding trajectories being processed while the entity is still moving ("online map-matching").

Due to the relevance for many domains, the map-matching problem has been tackled by the transport and GIS research community for years (i.e. Greenfeld 2002; Newson and Krumm 2009; Liu and Li 2017). Handling varying spatial accuracies of position measurements (e.g. due to GNSS errors) as well as diverse sampling intervals (the temporal intervals at which the position of a movement entity is recorded) makes map matching an ambitious research question. While the problem has been tackled by numerous approaches in the past, most of the algorithms focus on the offline case, which is easier to handle due to complete input trajectories (see Table 1). Online map matching has to cope with additional challenges related to incomplete trajectories and an iterative matching process. Determining continuously the most accurate path (also in case of lacking information) is one of the biggest map-matching challenges. Especially in the case of real-time traffic applications where trajectories of thousands of moving entities have to be processed simultaneously, the runtime performance of the map-matching algorithm plays an important role. Until now, most of the proposed map-matching algorithms have been designed and evaluated with respect to matching quality while disregarding runtime performance (cf. Sect. 2). Evaluating map-matchers against both requirements, map-matching quality and runtime performance, is still an open issue.

The current work introduces GraphiumMM, an open-source Java map-matching implementation combining the best of breed geometrical and topological matching concepts from previous works. The implementation introduces several algorithmic optimizations for pursuing the goal of high runtime performance for sampling intervals from 1 to 30 s (which are typical sampling intervals of fleet tracking systems; see Fig. 1) while ensuring high map-matching quality. For evaluating GraphiumMM, results based on well-defined quality and performance metrics are compared to results from the open-source Java map-matching library Barefoot. Results show that GraphiumMM outperforms Barefoot in terms of runtime performance (processed track points and kilometers per minute) for sampling intervals between 1 and 15 s, while providing better map-matching quality.

The remaining of the work is structured as follows: Sect. 2 discusses previous map-matching approaches. Section 3 introduces the map-matching implementation "GraphiumMM". Section 4 introduces the evaluation methodology and compares map-matching performance and quality of GraphiumMM and Barefoot. Section 5 concludes the work and provides an outlook.

Table 1 Summary of the main characteristics of previously proposed map-matching algorithms

Work	Sampling interval (s)	Online	Offline	Peculiarities
White et al. (2000)	1		x	Geometric approach with similarities outperforms approach which includes connectivity information
Greenfeld (2002)	1		x	Weight-based, topological approach
Brakatsoulas et al. (2005)	30		x	Geometric approach with Fréchet distance; incremental approach is faster, global algorithm produces better quality
Hummel (2006)	1		x	Probabilistic approach based on Bayesian classification and HMM; good results even in city scenarios
Quddus et al. (2006)	1		x	Fuzzy logic approach optimized for high density road networks; 99.2% matched correctly in rural environment
Newson and Krumm (2009)	1–600		x	Probabilistic approach based on HMM; perfect results for sampling intervals up to 30 s
Lou et at. (2009)	30–360		x	Optimized for high sampling intervals
Velaga et al. (2009)	1		x	Weight-based, topological approach considering connectivity and turn restrictions
Yuan et al. (2010)	30–630		x	Voting-based for high sampling intervals; outperforms (Lou et al. 2009) especially for sampling intervals over 2 min
Goh et al. (2012)	3–300	x		HMM and variable sliding window for online processing
Tang et al. (2012)	1–30		x	Probabilistic approach using the track point and segment geometries
Song et al. (2012)	1–30		x	Uses extensive multi-threading
Liu et al. (2012)	1–30		x	Uses simplified road network
Sauerwein (2013)	1–120	x	x	Weight-based, topological approach combining geometrical and topological measures
Mattheis et al. (2014)	Not specified	x	x	Probabilistic approach based on HMM, open-source is the best future option
Liu and Li (2017)	90		x	Probabilistic approach for high sampling intervals

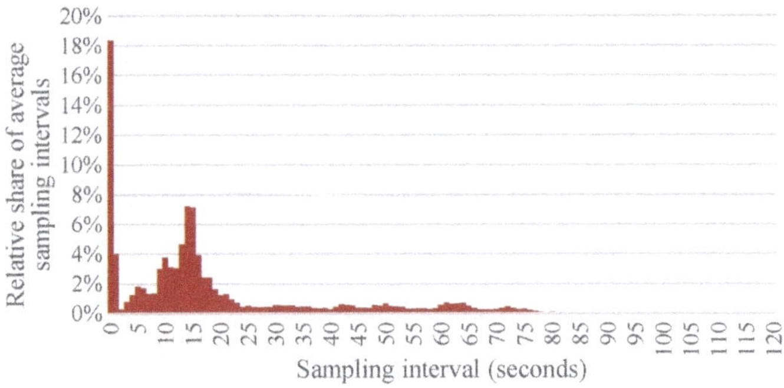

Fig. 1 Distribution of average sampling intervals of trajectories from typical fleet tracking systems (*Source* Austria's national floating car data platform)

2 Related Work

Early map-matching approaches investigate the geometry of the road segments (Kim et al. 1996; White et al. 2000). Brakatsoulas et al. (2005) proposed a geometrical approach mapping parts or the whole trajectory onto the road network using the Fréchet distance. Purely geometrical approaches are error prone due to measurement noise and the influence of sampling intervals (Newson and Krumm 2009). Topological approaches improve the matching quality by taking connections between road segments into account. Greenfeld (2002) published one of the first topological map-matching algorithms that also works with low quality GNSS data. A more advanced approach has been introduced by Hummel (2006) who applied a Bayesian classification and a Hidden Markov Model (HMM). This probabilistic approach has been adopted and refined by Newson and Krumm (2009) who also performed a quality evaluation using an artificially increasing sampling interval and a varying geometrical accuracy. Route mismatch analyses indicate best results for sampling intervals up to 30 s. Lou et al. (2009) proposed another algorithm for low sampling rates considering geometrical and topological structures as well as temporal constraints. Matching quality and runtime performance are evaluated better compared to the incremental algorithm by Greenfeld (2002) and Fréchet distance algorithm by Brakatsoulas et al. (2005). Goh et al. (2012) extended previous works (Lou et al. 2009; Newson and Krumm 2009) by adding an online mode for real-time applications. To improve the performance, track points are removed artificially in order to reduce the amount of data. Additionally, the "Viterbi"-algorithm (Goh et al. 2012) identifies a variable sliding window to only take relevant track points into account. Sauerwein (2013) combines geometrical and topological approaches for low and high sampling intervals and implemented several path validations and consistency checks in order to detect wrong matching results. Mattheis et al. (2014) also rely on the HMM approach for offline and online modes. They have been the first to publish their algorithm as open-source software called "Barefoot". Liu and Li (2017) developed a probabilistic approach for low sampling rates taking the topological structure of the road network and speed information into account. Other probabilistic approaches apply Fuzzy

Logic (Quddus et al. 2006), Extended Kalman Filter or Belief Theory. Although these algorithms typically improve map-matching quality especially for lower sampling rates, they suffer from high runtime complexity that makes them slow and not the first choice for online map matching and higher sampling rates. For the 2012 ACM SIGSPATIAL GIS Cup, Ali et al. (2012) organized a map-matching competition. From 31 submissions, the organizers ranked the top-5 submissions by runtime, correctness and overall score. One of the best submissions applied a probabilistic approach based on the correlation between consecutive track points and the shape of road segments (Tang et al. 2012). Another submission used a multi-threading infrastructure and improved the HMM by taking the maximum speed into account to avoid matching on minor parallel roads (Song et al. 2012). A third approach evaluated a drastically simplified road network (Liu et al. 2012).

An important characteristic of map-matching algorithms is concerned with the handling of different sampling intervals, which typically requires different map-matching strategies. Another characteristic is concerned with offline/online map matching. Most of the proposed map-matching algorithms (Greenfeld 2002; Newson and Krumm 2009; Lou et al. 2009; Hummel 2006) are designed for offline map-matching only, while some recent approaches provide online map-matching functionality as well (Goh et al. 2012; Sauerwein 2013). Table 1 summarizes the main characteristics of previously proposed map-matching algorithms.

For this work the geometrical and topological map-matching algorithm originally proposed by Sauerwein (2013) is used and further optimized for offline and online processing of thousands of trajectories in parallel with varying GNSS accuracies and sampling intervals. The focus of this work is on runtime optimization (especially for sampling intervals between 1 and 15 s) while ensuring equal or better map matching quality. For the first time, a comparison between two map-matchers (GraphiumMM and Barefoot, two open-source implementations) based on well-defined metrics has been conducted.

3 "GraphiumMM" Map-Matcher Implementation

In this work, the term "map-matching" is defined as the process of matching each point of a time-ordered sequence of georeferenced locations to a road network for determining the most valid path of a moving entity (object or person) with respect to a road network. Although map matching is a basic functionality of many transport-related applications (e.g. navigation, real-time traffic state estimation and traffic data analysis), the availability of open-source map-matching implementations or services is rare. The GraphiumMM project has been started in 2012, due to the lack of a scalable, high-quality online map-matching implementation. Similar to Barefoot (which was published in 2015 only), GraphiumMM pursued the goal of providing an open-source online as well as offline map-matching implementation being optimized for low sampling intervals and providing high runtime performance for thousands of parallel map-matching processes.

3.1 Map-Matching Algorithm

Based on the work by Sauerwein (2013), the GraphiumMM algorithm uses several geometrical and topological measures for selecting and evaluating path candidates from a digital road network for a given trajectory of time-ordered track points, where each track point is represented as WGS84 coordinate linked to a timestamp. The digital road network is modelled as a graph of geographically referenced road segments, being topologically connected in a way that these connections represent possible movement options within the network. This requires the modelling of routing restrictions like one-way rules as well as some information about the maximum feasible speed of movement (e.g. speed limits or at least road classes). The output of the map-matching algorithm is a path (an ordered sequence of road segments containing the direction of movement for each segment) that has been most likely followed by the entity. The algorithm works in offline-mode (with completed trajectories) as well as in online-mode (as incremental map-matcher matching only new track points of a trajectory since the last map-matching run). During map matching, it tries to match as many track points onto the road network graph as possible. Due to the use of the open-source project Graphium[1] for managing the transport graph, arbitrary road graphs can be used. Moreover, the algorithm is capable of determining the most likely position of each track point as offset on a directed road segment, which is a clear difference to other map-matching algorithms only determining the path. The overall map-matching procedure works as follows:

As a first step, the map-matcher tries to find an appropriate entry point to the road network. This is accomplished by identifying road segments that are within a defined search radius of the first track point of the trajectory. The initial search radius is defined as a variable. Numerous tests with varying search radii revealed 150 m as recommended start value for vehicle data taking accumulations of GNSS errors at the beginning of trajectories into account (Montenbruck et al. 2006). A fixed radius instead of variable search radii is preferred as a trade-off between runtime performance and matching quality. If the initial search with the first track point returns no road segment, the algorithm steps from track point to track point until a road segment is located within the search radius. In case that the search returns more than one road segment, all segments are handled as potential entry segments. The successive track points are used to identify next segments and to create initial directed path candidates.

[1]https://github.com/graphium-project

For iteratively identifying the next segments, an incremental workflow is applied for each path candidate. The first step of the workflow considers road segments as matching candidates that (i) are within the search radius around the track point (from experimental tests it is recommended to set the radius to 30 m for the rest of the workflow), (ii) are topologically connected to the last segment in the path (see Fig. 2) and (iii) are not constrained by turn restrictions. If such a segment candidate is identified, it is added to the respective path candidate. So far, the implemented selection of path candidates is based on geometrical as well as topological constraints. A major difference to previous works concerns the handling of track points being located within the search radius to the last segment in a path. For performance reasons, these track points are linked to the last segment without searching for other potential mapping candidates. The first track point that is not within the search radius to this segment determines the next segment candidates and the track points on the previous segment are matched again in order to obtain the correct matching segment. Another performance improvement is achieved by identifying and skipping short segments, having a length of less than the sum of the search radius and the median distance between the last five track points. In such cases, the segments being topologically connected to the short segment are considered for the path candidates.

If the selection of a connected road segment candidate within the search radius of the current track point fails, the algorithm switches from topological matching to a routing approach. This situation typically occurs if (i) track points are missing (e.g. in tunnels), if (ii) the trajectory follows roads that are not present in the road network or if (iii) the next track point cannot be matched on a connected road segment because of a high sampling interval. The algorithm tries to fill these gaps with a series of route queries (Fig. 3). The last segment of each path candidate is selected as start segment for the routing. Road segments within the search radius of the next track point are selected as route targets. Afterwards, shortest paths are queried between all start segments and all target segments. Apart from topological connectivity, the routing algorithm also considers access rules (e.g. one-ways and turn restrictions).

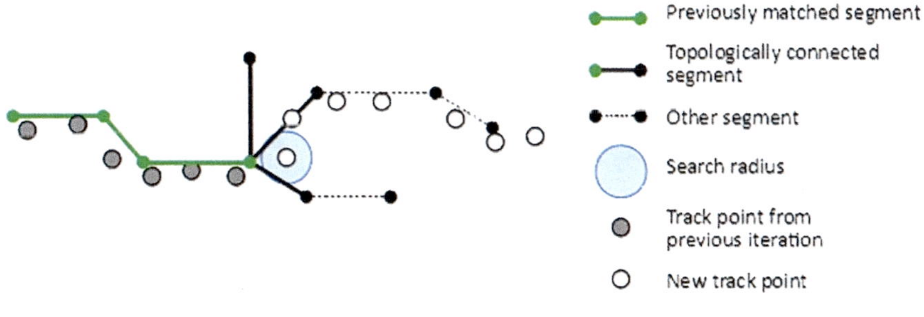

Fig. 2 Identification of the topologically connected segments. Black, thick segments are connected to the last matched segment, however, only two of them are within the matching radius

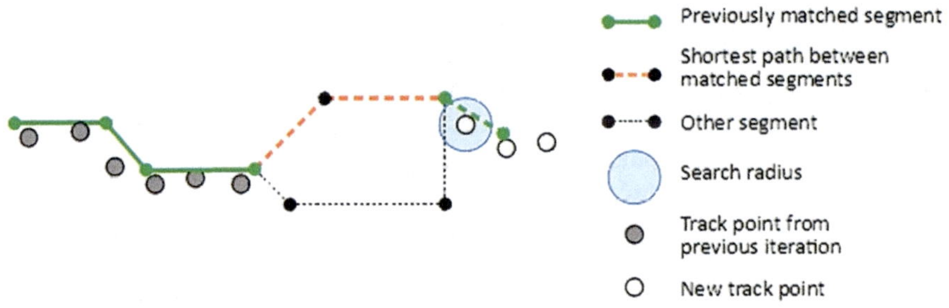

Fig. 3 Estimating possible paths between track points based on shortest path routing

The shortest path search might result in paths containing detours and/or short cuts due to missing or erroneous road segments or a high sampling interval (Fig. 10a). To identify unreasonable paths, the algorithm evaluates the calculated driving speed of the entity whether it exceeds the maximum feasible speed for the route. The calculation is based on the route length and the measured travel time between track points. While a previous approach (Newson and Krumm 2009) uses a speed restriction of three times the speed limit or a maximum of 180 km/h, GraphiumMM classifies the road segments by road type. By default, the maximum feasible speed is set to 150 km/h for motorway segments and 120 km/h for all other road segments. If the calculated speed between two track points exceeds the speed thresholds, the route is discarded.

If both approaches fail, this part of the trajectory is skipped. However, for each consecutive track point, the algorithm repeats the search for road segments being located within the search radius of 30 m. In case a new road segment has been identified, the algorithm starts new paths and tries to expand them following the previously described topological as well as routing approaches (Fig. 4). In order to accelerate the search for the next matchable track point, a bounding box with a height and width of 200 m (recommended value from empirical tests taking runtime performance and matching quality into account) is laid around a track point that is not near a road segment. All track points within the bounding box are skipped and the algorithm proceeds with the first one being outside the box.

Fig. 4 A part of a trajectory cannot be matched to the road network

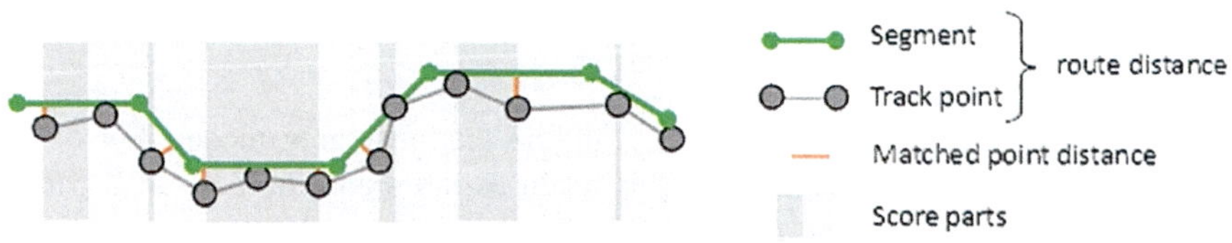

Fig. 5 Segment-based (green line) and track point-based (grey line) route distances as well as matched point distance (orange) for path rating; grey background areas indicate score parts

In order to select the most suitable path according to the trajectory, each calculated path candidate is rated according to two distance measures *matched point distance* and *route distance* (Fig. 5). These measures are calculated for each part of a path. A part ranges across all track points matched to the same segment or between two track points matched to different segments. The matched-point distance (f_{Point}) calculates the average distance between each track point and the road segment candidate (Eq. 1). It is set to zero for road segments without matched track points. The route distance (f_{Route}) determines a relative score between the length of the route being calculated using the length of the road segment and the length of the straight lines between the track points (Eq. 2).

$$f_{Point} = avg\left(distance_{track\,Point\,To\,Segment}\right) \tag{1}$$

$$f_{Route} = \left| \frac{distance_{segments} - distance_{track\,Points}}{distance_{track\,Points}} \right| \tag{2}$$

The matching score for a part is calculated according to Eq. (3). Both factors are normalized by the maximum value and weighted. The recommended weights ($W_{Point} = 0.4$ and $W_{Route} = 0.6$) have been empirically derived emphasizing route distance over point distance (Sauerwein 2013). To determine a score for the whole path, the scores of each part are summed up and normalized by the number of track points. The end segments from all path candidates are retrieved and only the best-rated path per end segment is retained. At maximum of seven best-rated paths are followed for the next track point iteration.

$$score = w_{Point} * \frac{f_{Point}}{max_{Point}} + w_{Route} * \frac{f_{Route}}{max_{Route}} \tag{3}$$

After all track points have been processed, the best-rated path is selected. The path consists of a list of matched road segments along with driving directions and metadata describing path properties like the matched length, a matching quality parameter (being derived from matched point distance and route distance), and the number of routed parts.

3.2 Online Mode

GraphiumMM supports both offline and online map-matching modes. For offline map matching, a completed trajectory is taken as input and the best-rated path is returned. The idea of online map matching is that parts of the trajectory are processed in certain time intervals while the entity (vehicle/person/object) is still moving. Online map matching is applied in an iterative way if the current location of a navigating entity on the road network should be continuously calculated or traffic information (e.g. traffic state) should be derived in near real-time. As the online map-matcher has to cope with incomplete trajectories, erroneous path assumptions for certain calculation intervals may arise (Fig. 10b).

For preparing GraphiumMM for online map matching, the offline map-matching algorithm was extended with a state handler, which manages all previous paths and track points. "Certain" paths are previously identified paths, which have been evaluated at high matching quality and therefore may not be realigned in further processing steps. The last road segment of such paths ends at the convergence point and is together with the matched track points the starting point for the next iteration. Without the certain path evaluation, the consecutive map-matching process could start from an incorrect road segment (e.g. due to GNSS inaccuracies). The last "certain" road segment is identified in every map-matching iteration using the "variable sliding window"-algorithm (Goh et al. 2012). Usually, the last certain segment is not the last segment in the path. In some cases, e.g. at the beginning of a trajectory, there exists no certain segment at all which leads to the case that all track points have to be re-matched. Figure 6 illustrates the identification of the last certain segment and the convergence point as well as the track points to be sent in the next iteration.

For starting a new map-matching iteration, the state handler sends the last certain path segment as well as all uncertain segments along with associated track points and the new unmatched track points to the map-matcher. The matching process itself follows the same rules as for the offline mode.

3.3 Implementation

GraphiumMM has been implemented as a Neo4J plugin for the Graphium framework.[2] Graphium is an open-source framework for distributed storage, management and versioning of transport graphs from different sources, which can be accessed via a REST-API. Besides deploying Graphium as a stand-alone server, it can also be deployed as a plugin of the Neo4J graph database (version 3.2 has been used). Such a deployment has to be chosen for running the GraphiumMM map-matcher, which uses several graph processing and routing features of Neo4J (e.g. graph

[2]https://github.com/graphium-project/

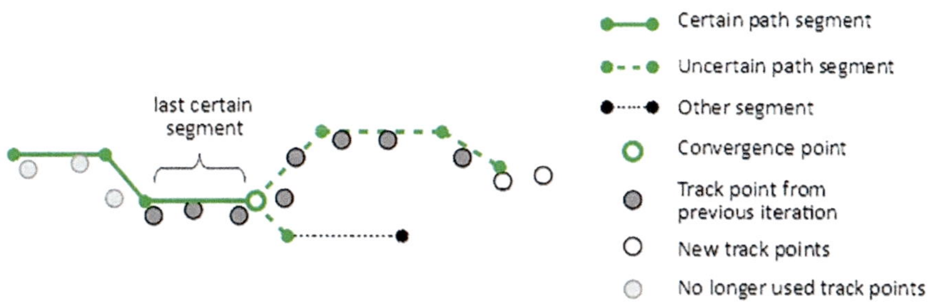

Fig. 6 Set-up of the next map-matching iteration with certain and uncertain segments as well as matched and unmatched track points

traversals, Dijkstra and A* algorithms). Using a graph database has the benefit of fast graph traversal and routing performance. All the road segment connections are represented as native relationships in the graph database, which allows for direct traversals. For the spatial search of appropriate start and target road segments for shortest path routings, the STR-tree from JTS[3] is applied. This is a R-tree using the sort-tile-recursive algorithm (Rigaux et al. 2002). The routing itself is done via a shortest path search using the A* algorithm. In analogy to Graphium, GraphiumMM provides a HTTP API for request handling. JSON has been selected as request and response message format.

4 Evaluation

This section presents evaluation results with respect to runtime performance and map-matching quality and compares the results of GraphiumMM to the open-source map-matcher Barefoot[4] (Mattheis et al. 2014). Barefoot implements a HMM map-matching algorithm supporting online and offline map matching. It can be deployed as a standalone server ("Matcher Server" for offline and "Tracker Server" for online map matching) and it is designed for use in parallel and distributed systems.

For comparisons, exactly the same road network as well as trajectory dataset were used. Furthermore, all test runs were executed on the same VMware Virtual Platform running on a server with Intel® Xeon® E3-1275 @ 3.40 GHz, Quad Core CPU with Hyper-Threading Technology (three cores are used by the virtual platform) and 20 GB RAM. Both tests were executed consecutively in order to avoid possible performance losses in case of simultaneous execution.

[3]https://sourceforge.net/projects/jts-topo-suite/

[4]https://github.com/bmwcarit/barefoot/wiki

4.1 Trajectory and Road Network Data

As evaluation dataset, 32 different trajectories with GNSS-measured vehicle locations were selected. The dataset includes 79,479 track points covering a total length of 1,904 km. All selected trajectories are originally sampled at an interval of 1 s. To evaluate the map-matching algorithm with higher sampling intervals, all trajectories were re-sampled to intervals of 2, 5, 10, 20, 30, 45, 60, 90 and 120 s. Figure 7 illustrates the trajectories and the coverage on the corresponding OSM road network. In order to be representative, the trajectory dataset includes tracks on all road categories (from motorways to residential roads). It should be noted that some parts of the trajectories are intentionally not covered by the road dataset since this is considered a realistic test scenario.

Since the extent of the road network graph is considered an important factor for runtime performance and matching quality, the map-matchers have been evaluated using two different road network configurations with different levels of detail. The high-level configuration (OSM_AT_HL) contains roads tagged as *motorway*, *trunk*, *primary*, *secondary* and *tertiary*. The low-level configuration (OSM_AT_LL) additionally contains roads tagged as *residential*, *service* and *unclassified*. Beside the road class, which is used to derive the maximum feasible speed for shortest path evaluations, additionally required road attributes are access restrictions like one-ways and segment connections that consider topological relationships as well as traffic rules like turn restrictions. GraphiumMM offers the Osm2Graphium module (available online in GraphiumMM repository) to (i) handle the segmentation of road segments by intersections and to (ii) add turn restrictions. Barefoot performs the road network preprocessing internally. To foster external validation of results, the trajectory dataset and the test cases have been made available as open data.[5]

4.2 Method

The evaluation was conducted for offline and online mode. While in offline mode, each complete trajectory was sent to the map-matcher, in online-mode an iterative map-matching process with consecutive track points or batches of 30 s was applied. In order to compare the matching quality of both map-matchers, two quality metrics are introduced: The metric **'matched track points'** indicates the relative share of track points that could be matched by the map-matcher. The metric **'matching error'** (Newson and Krumm 2009) indicates the quality of a matched path by taking detours and missing parts into account. For the test data, the expected path has been manually determined based on ground truth data (the real paths of the vehicles are known). As shown in Eq. (4), the matching error is calculated by

[5]https://github.com/graphium-project/graphium-neo4j/tree/master/data

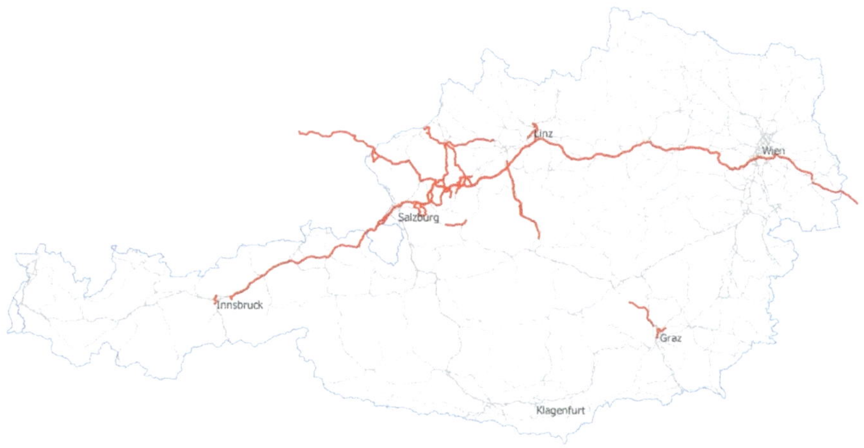

Fig. 7 32 GNSS trajectories (red lines) as sample data visualized on top of the Austrian OSM dataset (grey lines)

dividing the distance of all detours plus the distance of all missing track parts by the total distance of the correct path. A disadvantage of this method is that the correct order of matched segments cannot be verified. Furthermore, multiple matched segments will be ignored because they are eliminated after creating the geometry objects.

$$matching\_error = \frac{d_{detours} + d_{missing}}{d_{correct}} \tag{4}$$

In addition to the quality metrics, the metric **'runtime'** measures the duration of a single map-matching run for a trajectory. To obtain comparable results, GraphiumMM and Barefoot were both executed as single threads. For evaluating the map-matching performance, the throughput is measured in **'kilometers per minute'** and **'track points per minute'**.

For online matching, Barefoot comes with a Python script, which sends each track point sequentially. The Barefoot implementation is not able to receive a batch of track points. For GraphiumMM, the trajectories were processed in track point batches, each covering a time interval of 30 s. The decision to process batches rather than single points represents the most likely behavior of connected vehicles sending their GPS data in certain time intervals to a map-matching server. Consequently, tracks with sampling intervals of 30 s and more are always processed as single-point batches. After a trajectory was processed, the matching quality of the resulting path geometries was validated as described above. As performance indicator, **'runtime per batch'** in milliseconds was added for online mode. It represents the average time needed to process a batch of track points for a 30 s time interval with GraphiumMM or each a single track point with Barefoot.

4.3 Results

Quality and performance indicators were calculated for (i) GraphiumMM and Barefoot, (ii) online and offline mode, for (iii) the OSM_AT_HL and OSM_AT_LL datasets and for (iv) ten different sampling intervals. All metrics except runtime metrics were calculated using the weighted average over all trajectories. The number of track points or the length per trajectory is used as weight. The runtime is the summed runtime for all tracks.

Table 2 summarizes the results for offline mode. A general observation is that matching results for the high-level map dataset are less accurate because the trajectories run on some residential roads, which are missing in the high-level road dataset. However, as the results clearly indicate, map-matching with the high-level dataset leads to better performances with both map-matchers, as routings are faster.

Table 2 Evaluation of the offline algorithms for the OSM_AT_HL and OSM_AT_LL dataset and ten sampling intervals. Quality indicators include the matching error, the number of matched points and the runtime

Sampling interval	Matched points (%)		Matching error × 100		Runtime (s)	
	Graphium MM	Barefoot	Graphium MM	Barefoot	Graphium MM	Barefoot
OSM_AT_HL (s)						
1	87.9	90.2	0.0	1.6	6.7	135.0
2	88.3	90.6	0.6	2.9	4.7	71.2
5	88.2	90.5	0.1	3.3	2.9	28.6
10	88.0	90.6	0.3	3.2	3.7	14.7
20	87.8	90.6	0.0	3.7	3.9	7.7
30	88.2	90.8	0.1	4.2	4.2	5.3
45	87.3	90.6	0.5	7.0	4.4	3.8
60	87.7	90.7	1.1	4.3	4.5	3.0
90	86.0	90.8	2.5	4.0	3.6	2.2
120	86.2	89.8	2.2	6.4	4.4	1.8
OSM_AT_LL (s)						
1	92.7	93.0	0.2	0.3	13.7	418.6
2	95.2	93.4	0.4	0.3	11.8	218.1
5	92.8	93.3	0.3	0.6	10.1	86.7
10	92.9	93.4	0.8	0.9	31.1	44.1
20	93.4	93.4	0.9	1.2	34.7	27.9
30	92.9	93.4	0.9	1.6	41.6	23.7
45	92.9	93.3	1.7	1.8	46.6	22.1
60	93.0	93.5	2.1	1.8	49.3	22.1
90	92.7	93.5	3.8	2.0	46.1	24.2
120	92.8	93.4	5.5	2.3	58.5	23.2

Contrarily, using the low-level road dataset leads to better matching results while the runtime is higher since routing takes more time.

Taking a closer look at matching error and runtime (cf. Table 2 and Fig. 8), results for the OSM_AT_HL network indicate lower matching errors of GraphiumMM for all sampling intervals, while runtime is significantly lower from 1 to 5 s and a little lower from 10 to 30 s. For higher sampling intervals, the runtimes of Barefoot is a little lower compared to GraphiumMM. For the OSM_AT_LL network, matching errors of GraphiumMM are lower from 1 to 45 s while runtime is significantly lower from 1 to 5 s and a little lower at 10 s. For higher sampling intervals, the runtime of Barefoot is lower compared to GraphiumMM. These results clearly indicate that the proposed optimizations of GraphiumMM paid off. Concerning track points, Barefoot matches a higher share of track points compared

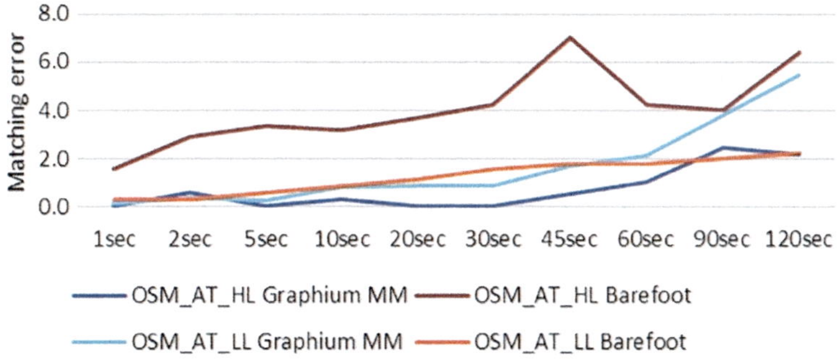

Fig. 8 Matching error for GraphiumMM and Barefoot for both datasets

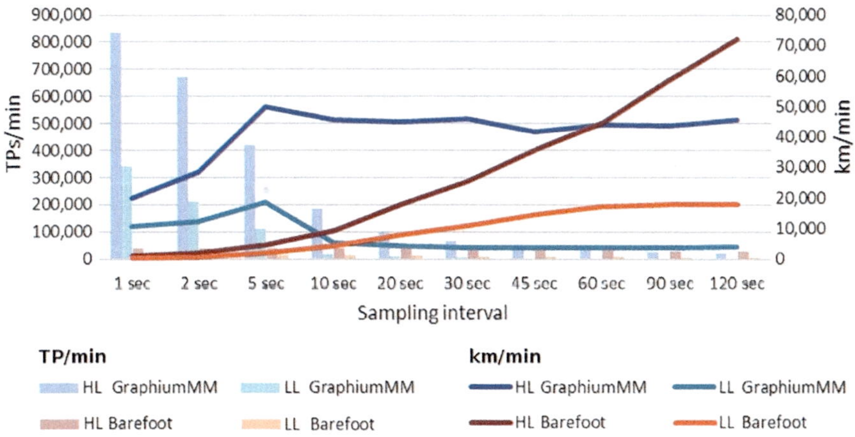

Fig. 9 Performance indicators track points per minute (bars) and kilometers per minute (lines) for GraphiumMM (blue, turquoise) and Barefoot (red, orange) as well as OSM_AT_HL (blue, red) and OSM_AT_LL (turquoise, orange) for offline map matching

to GraphiumMM. Likely reasons are that Barefoot also accepts detours (see Fig. 10c) and uses a search radius of 200 m by default while GraphiumMM uses a radius of only 30 m (except for the first track points). The lower search radius of GraphiumMM may account for lower runtimes and a lower number of matching errors recorded in the evaluation.

Taking a closer look at performance metrics (Fig. 9), results confirm that GraphiumMM is optimized for fast processing of a high number of track points occurring at sampling intervals fewer than 15 s, while Barefoot processes a nearly constant number of track points per second. Concerning the metric "kilometers per minute" reveals higher results for GraphiumMM (from 1 to 10 s for the OSM_AT_LL dataset and from 1 to 45 s for the OSM_AT_HL dataset) while Barefoot shows an increasing performance for higher sampling intervals. The rather constant performance values of GraphiumMM may be explained with a limited

Table 3 Evaluation of the online algorithms for the OSM_AT_HL (HL) and OSM_AT_LL (LL) dataset and ten sampling intervals. Quality indicators include the matching error, the amount of matched points and the runtime

Sampling interval	Matched points (%)		Matching error × 100		Runtime per batch (ms)	
	Graphium MM	Barefoot	Graphium MM	Barefoot	Graphium MM	Barefoot
OSM_AT_HL (s)						
1	87.6	90.2	0.2	2.2	11.8	10785.5
2	88.3	90.6	0.5	2.2	7.4	3783.8
5	87.9	90.6	0.2	2.3	5.8	1040.4
10	88.2	90.6	0.3	1.8	7.6	486.1
20	87.5	90.7	0.5	2.3	9.4	344.9
30	86.9	90.8	0.4	2.4	10.5	247.5
45	87.3	90.6	0.7	2.1	13.5	250.4
60	87.4	90.8	1.4	2.5	16.6	232.8
90	87.2	90.8	2.4	2.5	17.4	239.7
120	86.6	90.4	3.2	1.9	26.8	243.1
OSM_AT_LL (s)						
1	92.5	90.9	0.2	1.6	17.4	21278.6
2	92.8	93.4	0.3	0.7	15.1	7244.4
5	92.8	93.3	0.6	1.1	14.6	1810.3
10	92.8	93.4	0.9	1.3	53.2	781.9
20	93.1	93.4	1.0	1.6	76.5	521.7
30	92.3	93.4	1.1	2.1	92.4	344.0
45	93.1	93.3	1.7	2.1	130.4	334.9
60	92.1	93.5	2.3	2.7	172.2	338.6
90	92.6	93.5	3.8	2.9	204.0	324.9
120	92.6	93.4	5.4	2.8	304.6	323.2

routing performance appearing as a clear advantage of Barefoot. The obvious peaks in both figures (Barefoot's matching quality at 45 s, GraphiumMM's km/min at 5 s) can be explained by changing path validation strategies at these sampling intervals.

Comparing the metric "matching error" between offline and online mode (cf. Tables 2 and 3) reveals differences between both algorithms. While Barefoot performs better in online mode, GraphiumMM reveals a slightly better matching quality in offline mode. A possible reason is the less restrictive path validation in order to avoid losing potential track parts. The comparison of runtime performance reveals significantly lower values of Barefoot in online mode being predominately a result of the inefficient processing of track point batches. It should be noted that further tuning of the Barefoot server, which was not in the focus of this work, might improve evaluation results.

5 Conclusions

In this work, for the first time, a comprehensive comparison of two open-source map-matching implementations using a representative dataset of 32 vehicle trajectories, offline and online mode, two road network configurations and sampling intervals from 1 to 120 s has been conducted. The evaluation not only uses map-matching quality metrics, but also, for the first time, runtime performance metrics. Results clearly indicate that although both implementations use similar map-matching concepts and algorithms, dedicated optimizations may lead to better map-matching performance while keeping or even increasing map-matching quality. Especially the exact parametrization of the map-matching algorithms is a crucial question that can only be solved by taking the exact requirements into account (e.g., road network configuration, offline/online mode, mode of data collection, spatial accuracy of track points, expected sampling intervals). Knowing these requirements can be used for specific optimizations leading to performance and quality improvements. Conclusions may be summarized as follows:

- Scaling down the underlying road network to road classes that should be definitely matched, leads to better runtime performance and map-matching quality in offline and online mode.
- Using a bounding box to ignore track points being distant to the road network proved successful to improve runtime performance (e.g., GraphiumMM processes some batches in less than 1 ms while Barefoot needs equal processing time for all track points, even for the distant ones). Further performance improvements can be achieved by fast processing of multiple track points near the same road segment and by skipping very short road segments.
- Map-matching quality may be significantly improved by using multiple path validations (e.g. accepting all routing results may lead to paths differing significantly from the original trajectory as indicated in Fig. 10c).

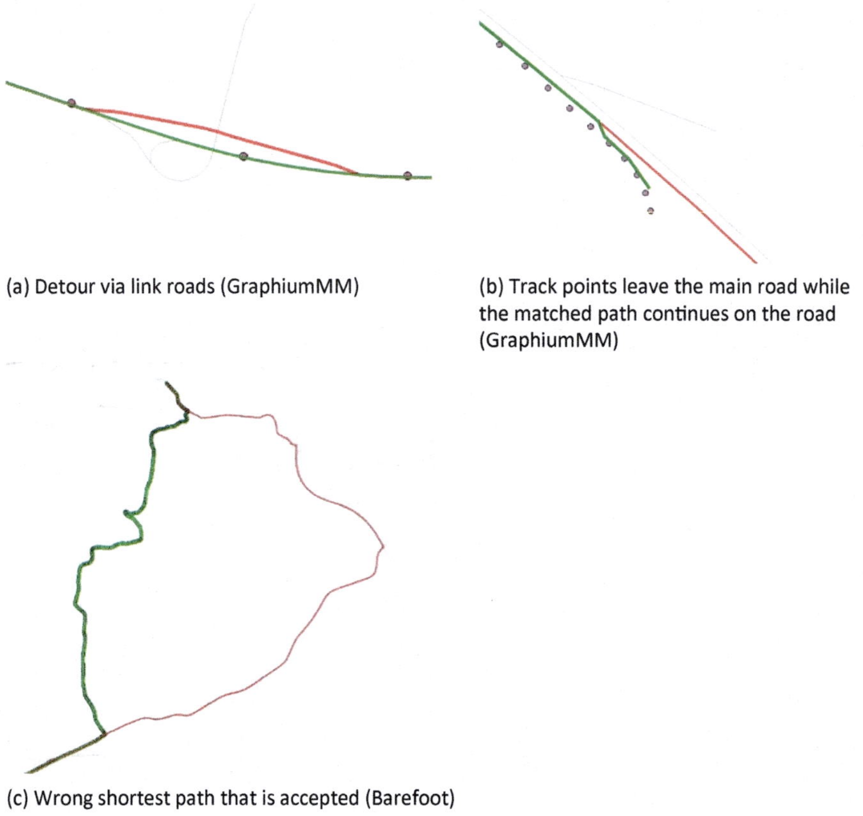

Fig. 10 Examples of identified map-matching problems; green lines indicate correctly matched paths, red lines reveal wrong paths

- In online mode, map-matching performance greatly benefits from the introduction of a varying sliding window (as proposed by Goh et al. (2012)) as well as the processing of track point batches instead of single points.
- Evaluation results indicate that for vehicle data, sampling intervals between 5 and 10 s may be considered as best deal with respect to matching quality and runtime performance.
- Since both map-matchers aim at high scalability for productive environments, the support of multi-threaded execution of requests for horizontal scaling (simultaneous processing of multiply tracks) is beneficial.

In the future, additional performance and map-matching quality improvements of both map-matchers based on these results are expected. Furthermore, it is expected that opening of test data fosters additional evaluations of map-matching implementations.

Acknowledgements This work has been partly funded by the Austrian Ministry for Transportation, Innovation and Technology (bmvit).

References

Ali M et al (2012) ACM SIGSPATIAL GIS Cup 2012. In: Proceedings of the 20th international conference on advances in geographic information systems—SIGSPATIAL'12, p 597. http://dl.acm.org/citation.cfm?doid=2424321.2424426

Brakatsoulas S et al (2005) On map-matching vehicle tracking data. In: Proceedings of the 31st VLDB conference, Trondheim, Norway, pp 853–864

Goh CY et al (2012) Online map-matching based on hidden Markov model for real-time traffic sensing applications. In: The 15th international IEEE conference on intelligent transportation systems, 117543, pp 776–781. http://ieeexplore.ieee.org/xpls/abs_all.jsp?arnumber=6338627

Greenfeld JS (2002) Matching GPS observations to locations on a digital map. Trans Res Board 3:13

Hummel B (2006) Map matching for vehicle guidance. In: Dynamic and mobile GIS: investigating space and time

Kim JS et al (1996) Node based map-matching algorithm for car navigation system. In: Proceedings of the 29th ISATA symposium, Florence, pp 121–126

Liu K et al (2012) Effective map-matching on the most simplified road network. In: Proceedings of the 20th international conference on advances in geographic information systems—SIGSPATIAL'12, p 609. http://dl.acm.org/citation.cfm?doid=2424321.2424429

Liu Y, Li Z (2017) A novel algorithm of low sampling rate GPS trajectories on map-matching. EURASIP J Wirel Commun Netw 2017(1):30

Lou Y et al (2009) Map-matching for low-sampling-rate GPS trajectories. In: Proceedings of the 17th ACM SIGSPATIAL international conference on advances in geographic information systems—GIS'09, p 352. http://portal.acm.org/citation.cfm?doid=1653771.1653820

Mattheis S et al (2014) Putting the car on the map: a scalable map matching system for the open source community. In: Lecture notes in informatics (LNI), Proceedings—Series of the Gesellschaft fur Informatik (GI), P-232, pp 2109–2119

Montenbruck O, Garcia-Fernandez M, Williams J (2006) Performance comparison of semicodeless GPS receivers for LEO satellites. GPS Solutions 10(4):249–261

Newson P, Krumm J (2009) Hidden Markov map matching through noise and sparseness. In: Proceedings of the 17th ACM SIGSPATIAL international conference on advances in geographic information systems—GIS'09, pp 336–343. http://portal.acm.org/citation.cfm?doid=1653771.1653818

Quddus M, Noland R, Ochieng WY (2006) A high accuracy fuzzy logic based map matching algorithm for road transport. J Intell Trans Syst Technol Plann Oper 10(3):103–115

Rigaux P, Scholl M, Voisard A (2002) Spatial databases: with application to GIS

Sauerwein T (2013) Optimization and evaluation of an online map-matching algorithm for mid-range sampling rates. Universität Marburg

Song R et al (2012) Quick map matching using multi-core CPUs. In: Proceedings of the 20th international conference on advances in geographic information systems—SIGSPATIAL'12, pp 605–608. http://dl.acm.org/citation.cfm?doid=2424321.2424428

Tang Y, Zhu AD, Xiao X (2012) An efficient algorithm for mapping vehicle trajectories onto road networks. In: Proceedings of the 20th international conference on advances in geographic information systems—SIGSPATIAL'12. New York, NY, USA, ACM, pp 601–604. http://doi.acm.org/10.1145/2424321.2424427

Velaga NR, Quddus MA, Bristow AL (2009) Developing an enhanced weight-based topological map-matching algorithm for intelligent transport systems. Trans Res Part C Emerg Technol 17 (6):672–683

White CE, Bernstein D, Kornhauser AL (2000) Some map matching algorithms for personal navigation assistants. Trans Res Part C Emerg Technol 8(1–6):91–108

Yuan J et al (2010) An interactive-voting based map matching algorithm. In: Proceedings—IEEE international conference on mobile data management, pp 43–52

A Personalized Location-Based and Serendipity-Oriented Point of Interest Recommender Assistant Based on Behavioral Patterns

Samira Khoshahval, Mahdi Farnaghi, Mohammad Taleai and Ali Mansourian

Abstract The technological evolutions have promoted mobile devices from rudimentary communication facilities to advanced personal assistants. According to the huge amount of accessible data, developing a time-saving and cost-effective method for location-based recommendations in mobile devices has been considered a challenging issue. This paper contributes a state-of-the-art solution for a personalized recommender assistant which suggests both accurate and unexpected point of interests (POIs) to users in each part of the day of the week based on their previously monitored, daily behavioral patterns. The presented approach consists of two steps of extracting the behavioral patterns from users' trajectories and location-based recommendation based on the discovered patterns and user's ratings. The behavioral pattern of the user includes their activity types in different parts of the day of the week, which is monitored via a combination of a stay point detection algorithm and an association rule mining (ARM) method. Having the behavioral patterns, the system exploits two recommendation procedures based on conventional collaborative filtering and K-furthest neighborhood model to recommend typical and serendipitous POIs to the users. The suggested POI list contains not only relevant and precise POIs but also unpredictable and surprising items to the

---

S. Khoshahval · M. Farnaghi (✉) · M. Taleai · A. Mansourian
Faculty of Geodesy and Geomatics Engineering,
K. N. Toosi University of Technology, Tehran, Iran
e-mail: mahdi.farnaghi@nateko.lu.se

S. Khoshahval
e-mail: s.khoshahval@email.kntu.ac.ir

M. Taleai
e-mail: taleai@kntu.ac.ir

A. Mansourian
e-mail: ali.mansourian@nateko.lu.se

M. Farnaghi
Department of Physical Geography and Ecosystem Science, GIS Center,
Lund University, Lund, Sweden

A. Mansourian
Center for Middle Eastern Studies, Lund University, Lund, Sweden

A. Mansourian et al. (eds.), *Geospatial Technologies for All*,
Lecture Notes in Geoinformation and Cartography,
https://doi.org/10.1007/978-3-319-78208-9_14

users. To evaluate the system, the values of RMSE of each procedure were computed and compared. Conducted experiments proved the feasibility of the proposed solution.

Keywords Personalized recommender assistant • Point of interest (POI) Association rule mining • Behavioral pattern • Serendipity • K-furthest neighborhood

1 Introduction

The increasing amount of registered point of interests (POIs), generated by users in the complex urban environments, has caused difficulties in search and discovery of POIs that are favorable to users. While finding POIs that match user preferences are an arduous task, location-based services made this issue possible via recommender assistants on mobile devices. Recommender systems are designed as essential tools to provide users with suitable suggestions based on their desires and preferences (Ricci et al. 2015). In other words, recommender systems suggest items which users may not have been able to find and with specific characteristics that match the user's preferences.

Knowing the user's desires and preferences is the most vital factor in constructing a satisfactory recommender system. In fact, extracting user's desires and preferences with appropriate accuracy can significantly ascend the quality of suggestions of the recommender systems. In case of POI recommender systems, getting familiar with the user's taste and activity type in each part of the day of the week can be a key point for offering the most suitable POIs to the user. Knowing about user's daily activities can provide recommender system with a rich source of information about the frequent activities done by the user during a day. Awareness of the user's activity type enables recommender system to compare the user's current location and time to their previously monitored, frequently happening activities and generate interesting suggestions for the user.

User's activity type is valuable, hidden information in GPS trajectory data. Trajectory analysis methods (Dodge et al. 2008; Gudmundsson et al. 2011; Zheng and Zhou 2011; Etienne et al. 2012; Choong et al. 2016) provide appropriate means to extract different places visited by the user in their daily activities. Having these visited points for each user, the association rule mining (ARM) method can be used to extract most frequently visited places for the user and therefore their activity type in each part of the day of the week. The ARM is a well-known data mining method that extracts frequent structures and relationships among item-sets in a huge transactional database (Agrawal et al. 1993). The ARM divides the problem of finding the most frequent item-set into two subsections of discovering the items with measures of more than a predefined threshold and generating rules based on the discovered candidate item-sets.

This paper provides a new perspective on the generation, usage, and combination of the behavioral pattern of the user, extracted from their trajectory data, in a personalized recommender system to present POI suggestions. The proposed method includes two stages of behavioral pattern extraction and POI recommendation based on collaborative filtering and serendipity-oriented recommendation techniques. To extract the behavioral patterns, the outputs of trajectory analysis are fed to a well-known ARM algorithm, called Apriori (Agrawal et al. 1993) to extract the most frequent activity types in each part of the day of the week. Having the user's activity type, two different recommendation procedures are used to provide users with personalized recommendations. The two recommendation procedures are developed based on conventional collaborative filtering and K-furthest neighborhood model (Said et al. 2013). While the former recommendation procedure based on conventional collaborative filtering leads to specialized suggestion based on user's previously ranked POIs, the later procedure focuses on diversity and tries to propose serendipitous, unexpected suggestions that can change the user's taste and expose them to new experiences. A serendipitous POI recommendation must be surprising to the user but, at the same time, it should be accurate to be useful; therefore, both of the procedures follow the user's behavioral pattern to provide relevant suggestions at the proper time. To evaluate the system, the values of RMSE of each procedure were computed and compared.

The remainder of the paper is organized as follows. First, the related works and literature are explains in Sect. 2. Section 3 thoroughly explains the proposed solution, consisting of the behavioral pattern extraction mechanism and POI recommendations procedures. Section 4 is devoted to implementation and experimental evaluation, where we compare and discuss the results. Eventually, conclusions are drawn in the final section including the future work.

2 Related Work

The recent decade has witnessed a growing trend in the process of monitoring user's activities through trajectory analysis (Zheng and Zhou 2011; Gudmundsson et al. 2011; Ying et al. 2014). Stay point detection is one of the trajectory analysis techniques which utilize different algorithms to discover user's stay point from trajectory data. A stay point is a specific place where the user's movement continues in that location for a specific time duration to do a specific task. Various researchers, including Cao et al. (2007), Li et al. (2008) and Li et al. (2010) proposed various stay point detection algorithms, mostly work based on distance and time interval between points. Another technique is clustering which focuses on finding groups of points with similar features to locate the stay points (Palma et al. 2008; Schreck et al. 2009; Rocha et al. 2010; Fu et al. 2005).

Several authors have proposed different solutions to add recommendation capabilities to location-based services and applications (Liu and Xiong 2013; Gao et al. 2015; Ye et al. 2011; Yuan et al. 2013; Liu et al. 2013). These studies tried to recommend POIs to users using collaborative filtering, content-based filtering, or hybrid recommendation methods. Meanwhile, a few studies have focused on incorporating the information that has been extracted from the user's trajectories into their POI recommendation solutions. Ying et al. (2014) categorized mobile users according to the semantic similarity of their trajectories and then suggested the next location to them. In another study, sparse data of location-based services has been used to model the user's activities using tensor factorization (Yang et al. 2015). A more recent research demonstrated a hybrid context-aware approach which recommends items based on implicit ratings using user's movement (Celdrán et al. 2016).

To the best of the authors' knowledge, none of the previous studies in the geospatial community has addressed the problem of recommendation diversity in their location-based recommender solutions. Recommendation diversity is a practical way to provide attractive suggestion and overcome the problem of overspecialization (Ziegler et al. 2005). Overspecialization is a problem, caused by conventional recommendation techniques, which propose already-known items with similar attributes to the items that user saw them before (Iaquinta et al. 2008). Recently, researchers have implemented a measure of diversity called serendipity to tackle the overspecialization problem. Serendipity is defined as a method which reveals attractive, surprising and unexpected items which are not ascertainable for user (Herlocker et al. 2004; McNee et al. 2006; Shani and Gunawardana 2011; Ge et al. 2010; Murakami et al. 2007; de Gemmis et al. 2015; Yamaba et al. 2013). This problem has led researchers such as Iaquinta et al. (2008) to use a hybrid solution which combines conventional filtering models with serendipitous suggestions. Yamaba et al. (2013) proposed a recommender system in which the user's impression of an item is extracted from folksonomy and used as an indicator to recommend unexpected suggestions to the user. Another study proposed a graph-based recommendation mechanism based on extracting the hidden correlations among items and using similarity measures to find surprising items for recommendation (de Gemmis et al. 2015). In another research, level of curiosity was considered as a factor of serendipitous recommendation for tourism (Menk et al. 2017).

The previous works on recommender systems mostly relied on user's rankings to discover user's tastes and preferences. In this sense, our proposed solution takes benefit of the ARM method to form user's behavioral pattern and provides serendipitous POIs recommendations based on user's behavioral pattern as well as their rankings.

3 Method

With the purpose of providing users with appropriate POI recommendations, a recommender assistant was designed in this study. As shown in Fig. 1, the system is composed of server-side and client-side components. The client-side components

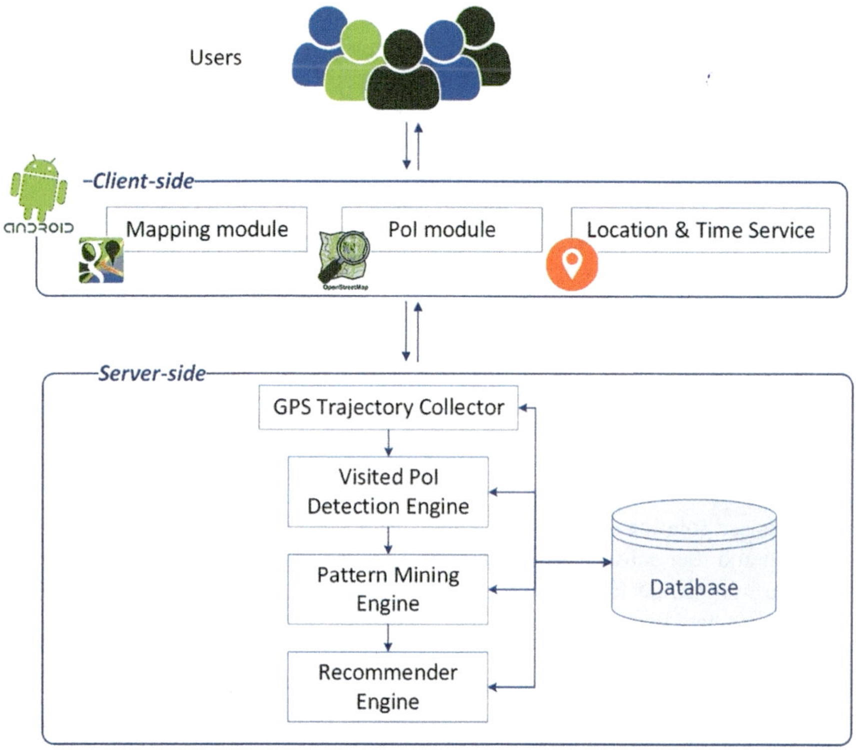

Fig. 1 Proposed recommender assistant structure

including, a *mapping module*, a *POI module*, and a *location and time service* are implemented as a mobile application on the handheld devices of the users. The mobile application is responsible for interaction with end users. It provides users with a map on which they can visualize and browse different data layers, including POIs. The POIs were downloaded from OSM (Open Street Map) website and saved in the database. When the user selects a specific POI, more information about the POI is retrieved and visualized on the screen. They can also rank each POI with a number between 1 and 5. These rankings are sent to the server and saved in the database. The rankings form the user-item rating matrix, $R_{m \times n}$, where, m is the number of users and n is the number of POIs. This matrix is then used for recommendation purposes. The mobile application is also aware of user's movement and time using the location and time service and sends those movements as trajectories to the server to be saved in the database and used as input data for pattern mining. The mobile application also visualizes the recommended POIs to users.

On the other hand, server-side components are dealing with the trajectory analysis to extract the behavioral pattern of the users and provide them with POI recommendation. A *GPS trajectory collector* component keeps track of the users' daily activities continuously and saves the related data in the database.

The collected trajectory data is sent to a *visited POI detection engine*. This engine applies a stay point detection algorithm to extract the visited POIs of the user (see Sect. 3.1.1). The extracted visited POIs are then inserted into a *pattern mining engine* to discover the users' behavioral pattern through a rule generation and interpretation mechanism based on the ARM method. The extracted behavioral patterns include activity type, the day of the week and the part of the day when that activity frequently happens. Having the users' behavioral pattern along with users' rankings, a *recommender engine* recommends POIs to the user in different parts of the day and different days of the week. The system offers accurate and unexpected suggestions in two procedures using conventional collaborative filtering (CF) and serendipity-oriented (SO) models.

3.1 Behavioral Pattern Extraction

The proposed solution works in two consecutive steps of spatiotemporal visited POI detection and user activity pattern mining (Fig. 2) to extract the behavioral patterns from the user's trajectories. Here, the system receives user's trajectories as input data and extracts their behavioral pattern as output.

3.1.1 Spatiotemporal Visited POI Detection

The system keeps track of user's movements by saving their trajectories. To extract the stay points out of numerous points in a user's trajectory, we implemented a hybrid time-based and distance-based stay point detection algorithm. The general concept of our spatiotemporal stay point detection algorithm bears a close resemblance to algorithms proposed by Mamoulis et al. (2004) and Li et al. (2010).

First, the algorithm receives the trajectory data in the form of $\langle UserTraj = (lng_1, lat_1, t_1), \ldots, (lng_n, lat_n, t_n) \rangle$, containing a sequential order of triples, where lng_j is

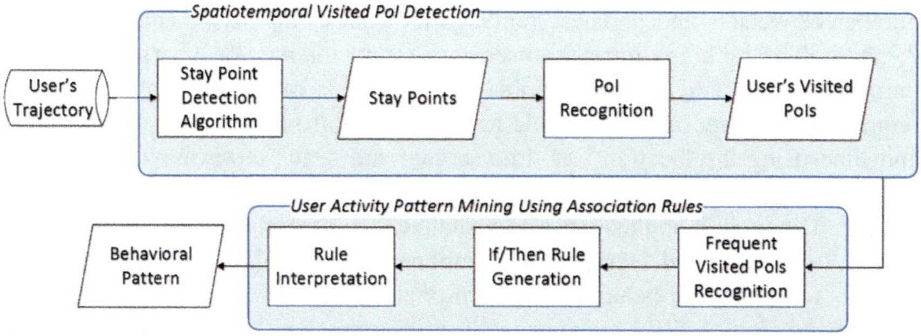

Fig. 2 Behavioral pattern extraction workflow

longitude, lat_j is latitude and t_j is timestamp $(\forall 0 \leq j < n,\ t_j < t_{j+1})$. The stay point detection algorithm proceeds through time and measures the distance between GPS points. If the distance is less than the predefined threshold $(\delta_d < \delta_{d-threshold})$, the point is added to a series of candidate stay point. As the algorithm goes on, the other GPS points which pass the distance criteria are aggregated with the set. If the distance is higher than the threshold, the algorithm checks the time criteria. If the time interval is more than the predefined time threshold $(\delta_t > \delta_{t-threshold})$ the centroid of the set is considered as a stay point, the set gets emptied and the algorithm searches for the next stay point. Eventually, all the stay points are extracted in the form of $(lng, lat, t_{enter}, t_{exit})$.

For each stay point, the distances between the point and the POIs are computed, and the stay point is assigned to the POI with the minimum distance. In this way, the extracted stay points are merged by the POIs to form the sets of visited POIs of each user. The output of spatiotemporal visited POI detection is saved as tuples of Eq. 1 for each user.

$$(PoI\ ID,\ PoI\ Type,\ Day\ of\ Week,\ Part\ of\ Day) \tag{1}$$

where *PoI ID* is the unique identifier of the POI, *PoI Type* is the activity type of the POI. The activity type is defined as a nominal variable which can take different activity names such as educational, religious, shopping, work, entertainment, health and etc. *Day of Week* is an ordinal variable that takes a value of 1 to 7, pertaining to each week day, starting from Saturday. Moreover, *Part of Day* is defined as other ordinal variables that can take eight values of early morning, late morning, noon, early afternoon, late afternoon, early evening, late evening and night.

3.1.2 User Activity Pattern Mining Using Association Rules

The behavioral pattern of the user is derived from their frequently visited POIs using the ARM method. In a database with a set of n items $I = \{i_1, i_2, \ldots, i_n\}$, an association rule is defined as $X \Rightarrow Y$, where $X, Y \subseteq I$ and $X \cap Y = \varnothing$. For each user, the output tuples of the spatiotemporal visited POI detection, excluding their *PoI ID*, are fed as input items to the Apriori algorithm. The Apriori algorithm finds the most interesting rules among the possible rules using interestingness measures of minimum support and minimum confidence

The support of a rule, $X \Rightarrow Y$, is the number of all transactions in the database with both X and Y occurrence to the number of all transactions (Eq. 2). The confidence of that rule equals to the occurrence probability of the transactions with both X and Y to the number of transactions having X (Eq. 3).

$$Support = Pr(X \cup Y) = \frac{\#both\, X\, \&\, Y}{\#total} \tag{2}$$

$$Confidence = Pr(Y|X) = Pr(X \cup Y)/Pr(X) = \frac{\#both\,X\,\&\,Y}{\#X} \tag{3}$$

Apriori, first, searches the whole database and calculates the measure of support for all items. Next, items that have support measure of higher than minimum support are selected as candidate item-sets. Then, candidate item-sets are joined together to form datasets of the next pass of the database. The items with the support of lower than a defined minimum support are deleted in each pass. The searching and pruning continues until the most frequent dataset is found. The values of minimum support and minimum confidence in our research were defined experimentally.

By generating and interpreting these rules for each part of the day and each day of the week, the behavioral pattern is generated as a list of time-activity sequences as shown in Eq. 4.

$$(PoI\ Type, Day\ of\ Week,\ Part\ of\ Day) \tag{4}$$

3.2 POI Recommender Assistant

Having the behavioral patterns of the users in the form of Eq. 4, the system knows what each user usually does in each part of the day of the week. Therefore, the system tries to provide the user with particular recommendations based on their previously monitored patterns. With the aim of recommending both accurate and unexpected POIs at the right time, the system utilizes two recommendation procedures. The first procedure, called CF (collaborative filtering) POI recommendation, exploits the conventional CF method using user-based and item-based similarity measures. Following that, the second procedure, called SO (serendipity-oriented) POI recommendation, provides suggestions using the K-furthest neighborhood model.

Figure 3 shows the workflow of the proposed approach where user-item rating matrix, $R_{m \times n}$, and the extracted behavioral patterns of the users are entered as inputs. The outputs are generated recommendations as a list of top-N highly ranked POIs and a list of surprising POIs for each user at each part of the day of the week.

3.2.1 Collaborative Filtering POI Recommendation

The first procedure calculates the similarity measures between users and between POIs based on their previously rated POIs to create two lists of user-based and item-based POI recommendations to users, respectively. We used the Pearson coefficient (Adomavicius and Tuzhilin 2005) as the similarity measure between each pair of users. Pearson correlation measure calculates the user-based similarity

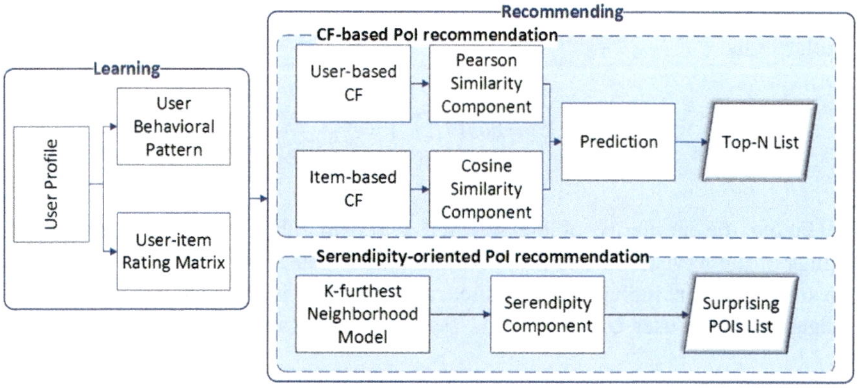

Fig. 3 Recommender assistant procedures

value from the user-item matrix $R_{m\times n}$. The similarity between the active user U_a and an arbitrary user U_x is calculated using Eq. 5.

$$sim(a,x) = \frac{\sum_{i\in I_{a,x}}(r_{a,i}-\overline{r_a})(r_{x,i}-\overline{r_x})}{\sqrt{\sum_{i\in I_{a,x}}(r_{a,i}-\overline{r_a})^2}\sqrt{\sum_{i\in I_{a,x}}(r_{x,i}-\overline{r_x})^2}} \tag{5}$$

where $I_{a,x} = \{i\in I | r_{a,i} \neq \varnothing \,\&\, r_{x,i} \neq \varnothing\}$ is the set of all POIs rated by both the user U_a and the user U_x, $r_{a,i}$ and $r_{x,i}$ represent the ratings of the user U_a and the user U_x to POI $i\in I_{a,x}$, respectively. $\overline{r_a}$ is the average ratings of the user U_a and $\overline{r_x}$ is the average rating of the user U_x to all rated POIs. To find the similar users to the user U_a, the ratings of unvisited POIs is calculated using the prediction measure of Eq. 6.

$$pred(a,i) = \overline{r_a} + \frac{\sum_{b\in N} sim(a,x) * (r_{x,i}-\overline{r_x})}{\sum_{b\in N} sim(a,x)} \tag{6}$$

where, $\overline{r_a}$ is the average ratings of the user U_a, $sim(a,x)$ is the similarity measure calculated between the user U_a and the user U_x, and $r_{x,i}$ is the ratings of the similar user U_x to the POIs. Having the rankings of unvisited POIs, the first list of top-N POI recommendation based on the similar user's tastes, including only the unvisited POIs matching the user activity type at the current part of the day of the week is generated and recommended to the user U_a.

When the user U_a selects one of the POIs, the system recommends another list of POIs to the user based on the item-based CF similarity measure. The major purpose of the item-based CF similarity measure is to suggest unvisited POIs based on the similarity between POIs, not the similarity between users (Sarwar et al. 2001, Wang et al. 2006). We adopted Cosine similarity measure (Eq. 7) to calculate the similarity between the selected POI and other POIs in order to discover the most similar POIs.

$$sim(I_x, I_y) = cos\left(\vec{I}_x, \vec{I}_y\right) = \frac{\vec{I}_x \cdot \vec{I}_y}{\left\|\vec{I}_x\right\|_2 \times \left\|\vec{I}_y\right\|_2} \quad (7)$$

Having the similarity of the selected POI and other POIs, we predicted the ratings of the unvisited POIs that have the highest similarities using Eq. 8. Finally, the top-N list, including the unvisited similar POIs with the highest ratings are suggested to the user U_a.

$$pred(a, i) = \frac{\sum_{j \in rated\ Items(a)} sim(j, i) * r_{a,j}}{\sum_{j \in rated\ Items(I_x)} sim(j, i)} \quad (8)$$

3.2.2 Serendipity-Oriented POI Recommendation

In the second procedure, to provide users with a POI recommendation list containing not only precise and accurate POIs but also surprising and unexpected items, another recommendation procedure is developed based on a dissimilarity neighborhood model. To find diverse and less common but still precise and relevant neighborhoods according to the behavioral pattern of the user, dissimilar users to the user U_a are discovered using the K-furthest neighborhood model (Said et al. 2013). The K-furthest neighborhood model is known as a serendipity-oriented method which detects less common neighborhoods to create disparate suggestions from the detected uncommon neighborhoods. A dissimilar user is an arbitrary user, U_x, who have not rated the POIs that the user U_a have already rated and the dissimilarity measure is computed by Eq. 9. In fact, the discovered users are not completely the opposite of the active user, U_a, but they have different tastes which are useful to reveal unexpected neighborhoods.

$$Dissim(a, x) = sim(a, x') \quad (9)$$

where dissimilarity measure equals to similarity measure between the user U_a and the virtual user $U_{x'}$. The virtual user is the one whose ratings are the inverted ratings of the arbitrary user U_x. To compute the dissimilarity measure, a modified Pearson similarity measure is computed (Eq. 10).

$$sim(a, x') = \frac{\sum_{i \in I_{a,x'}} (r_{a,i} - \overline{r_a})(r'_{x,i} - \overline{r'_x})}{\sqrt{\sum_{i \in I_{a,x'}} (r_{a,i} - \overline{r_a})^2} \sqrt{\sum_{i \in I_{a,x'}} (r'_{x,i} - \overline{r'_x})^2}} \quad (10)$$

where $I_{a,x'}$ is a list of POIs rated by both users U_a and $U_{x'}$, $r_{a,i}$ is the ratings of the user U_a to all POIs $i \in I_{a,x'}$ that are rated by both the user U_a and the user $U_{x'}$. $r'_{x,i}$ is the inverted ratings of $r_{x,i}$ and equals to the subtraction of $r_{x,i}$ from the sum of maximum and minimum amounts of all of the ratings. Discovering the dissimilar user, the POIs that are previously rated by the dissimilar users are expected to be surprising and also favorable to the user U_a. Therefore, the unexpected list of POIs, which are related to the current user's activity type is suggested to the user U_a.

4 Implementation and Experimental Evaluation

4.1 Implementation

The proposed system was developed using Java programming language. The client-side components were implemented as an Android application, where Google Map API was used to provide the mapping functionality. The server-side components where implemented as a Java Web Application using Java Servlet and Apache Mahout library was used to support recommendation functionalities. The system was implemented and tested in Tehran, Iran for 2 months.

Figure 4 shows the main page of the mobile application of the developed recommender assistant. The mobile application keeps track of the users' movement. The system extracts visited POIs of each user from their trajectory data using the spatiotemporal visited POI detection method (see Sect. 3.1.1). Figure 5 depicts the visited POIs of a user on Saturday, which is a weekday in Iran and Fig. 6 shows the visited POIs of the same user on Friday (a weekend day in Iran). Each one of the timescales and locations is assigned to one part of the day and one activity type, respectively.

Having the visited POIs of each user, the Apriori algorithm was used to obtain user activity patterns in the form of if/then rules. Table 1 shows a sample of the resulted if/then rules. The first rule in Table 1, with the support of 1.2% and the confidence of 0.98%, says that the user activity type is educational in the early morning of Saturday. The system was able to detect various activity types of the users in various parts of the day and days of the week. Therefore, the recommender assistant was aware of the user's tastes and preferences and used this information to provide users with personalized suggestions.

Recommendations based on the two procedures are available through "See what other users like" and "Surprise me!" buttons for CF POI recommendation and SO POI recommendation, respectively (Fig. 3). By clicking the "See what other users like" button, the system first compares the current time and day to the behavioral pattern of the user in the database and send the activity type as input to the user-based CF POI recommendation procedure (Sect. 3.2.1). The user-based CF POI recommendation procedure provides users with personalized suggestions. As an example (Fig. 7), the user's activity type in the late afternoon of Thursdays

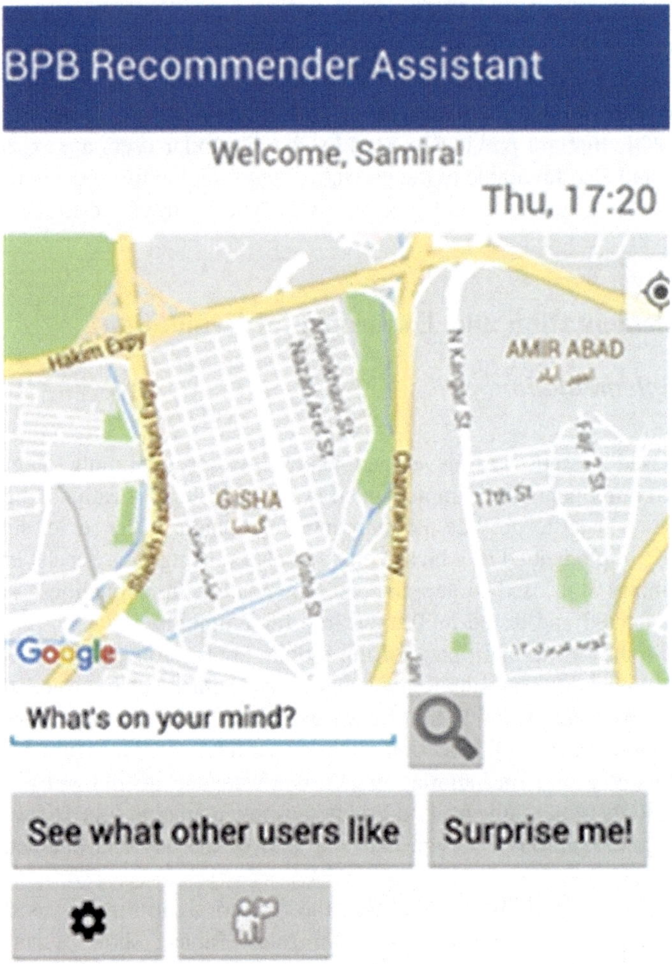

Fig. 4 Recommender assistant main menu

was entertainment, and the user frequently visits the POIs of cinema, theatre, swimming pool and park. Therefore, the user-based CF recommendation procedure presented related POIs to the user (Fig. 7).

Following that, when the user selects one of those POIs that are recommended by the user-based recommendation procedure, the system executes the item-based CF recommendation procedure (Sect. 3.2.1), and the upcoming page offers the most similar items to the selected item to the user. For example, as shown Fig. 8, the user selected Olive pool, and the system recommended her the similar POIs, Sadaf and Velenjak pools, as the most similar pools.

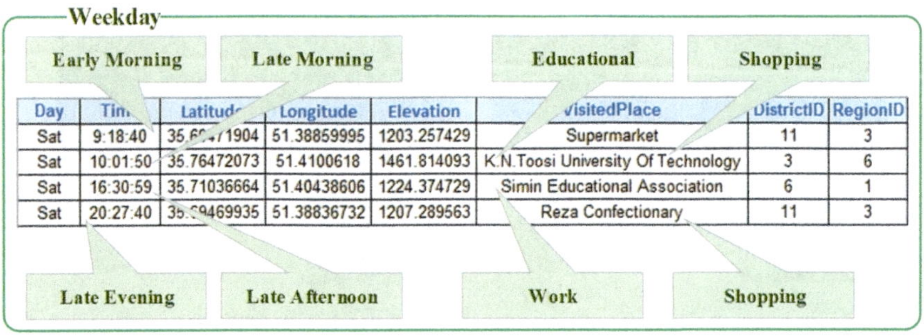

Fig. 5 User's activity types on a weekday

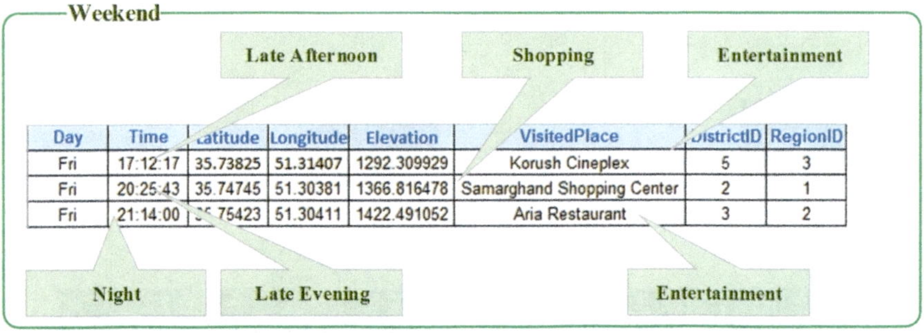

Fig. 6 User's activity types on a weekend

Table 1 Resulted if/then rules

Rules	
If	Then
Day of week = "*Saturday*", part of day = "*Early Morning*"	*Educational*
part of day = "*Early Afternoon*"	*Work*
part of day = "*Late Evening*", place = "*Work*"	*Home*

On the contrary, when the user touches the "Surprise me!" button, the system employs the SO recommendation procedure. Based on the current activity type of the user, the K-furthest neighborhood model and dissimilarity measure are used to create a surprising suggestion list (Sect. 3.2.2). As an example, in Fig. 9 the user activity type in the late afternoon of Thursdays is entertainment, and the user frequently selects one of the four entertainment activities of cinema, theatre, park and swimming pool. The system provided the user with a list of other entertainment POIs in Tehran such as bowling, sled, Tennis, Café, aquarium, etc. as shown (Fig. 9).

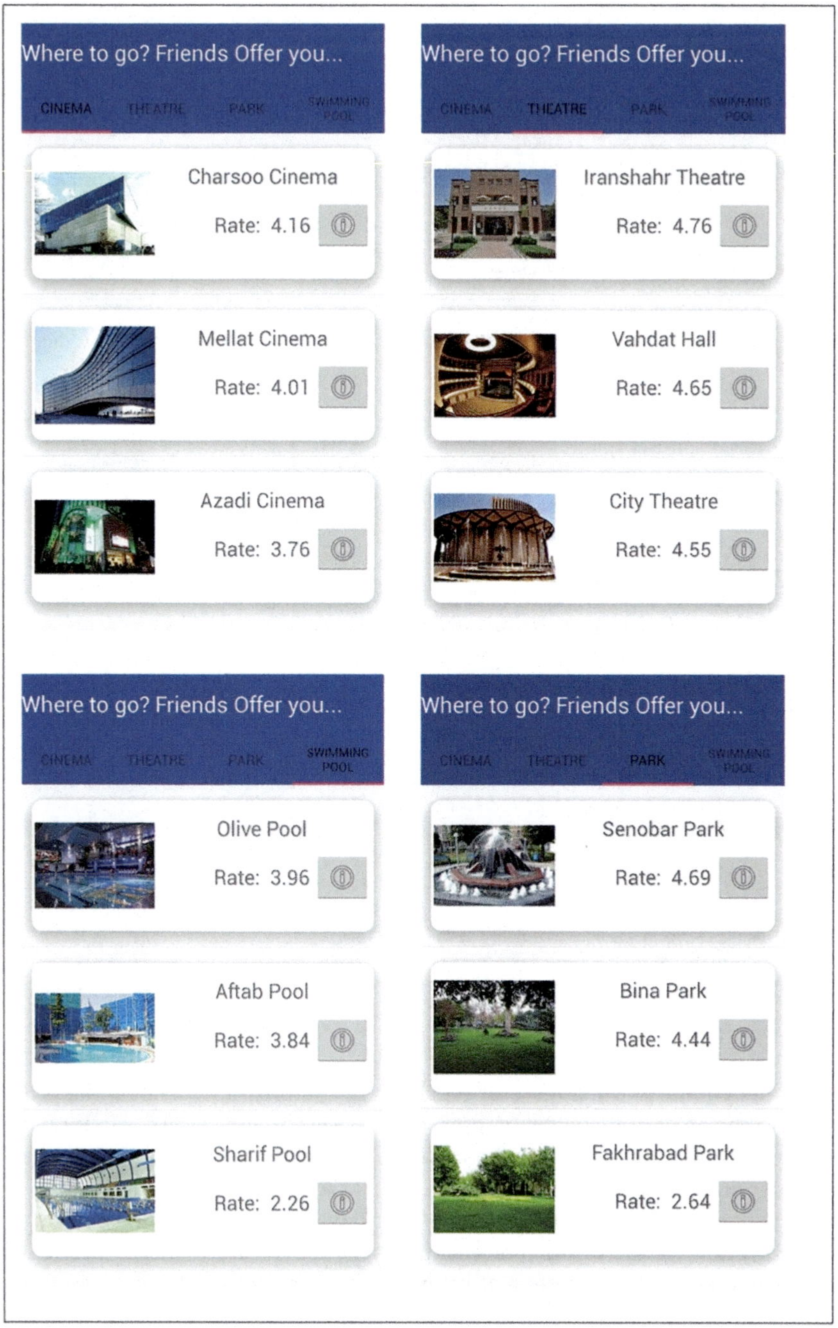

Fig. 7 User-based CF recommendation

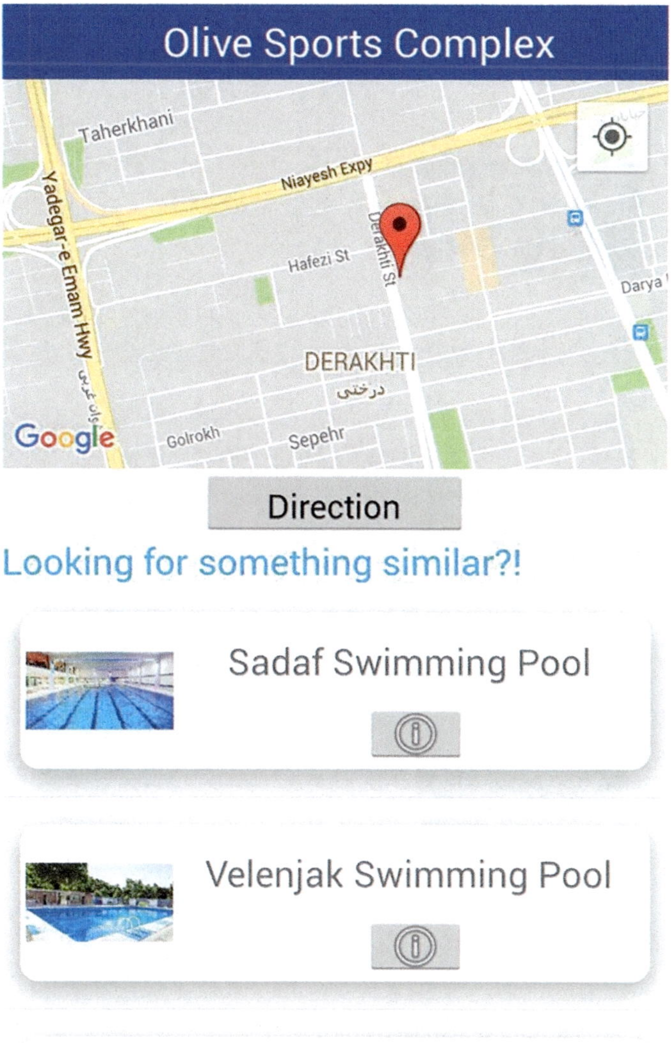

Fig. 8 Item-based CF recommendation

4.2 Experimental Evaluation

To evaluate the feasibility of the proposed recommendation method, an offline evaluation approach (Herlocker et al. 2004) were considered in which ten university students and academic staff were asked to use the system for 2 months. During this time, the mobile application saved users' trajectory data during their daily activities. The users were also asked to keep a log file and save the time, location and type of their activities in that file. After this 2-month period, they used the system as a

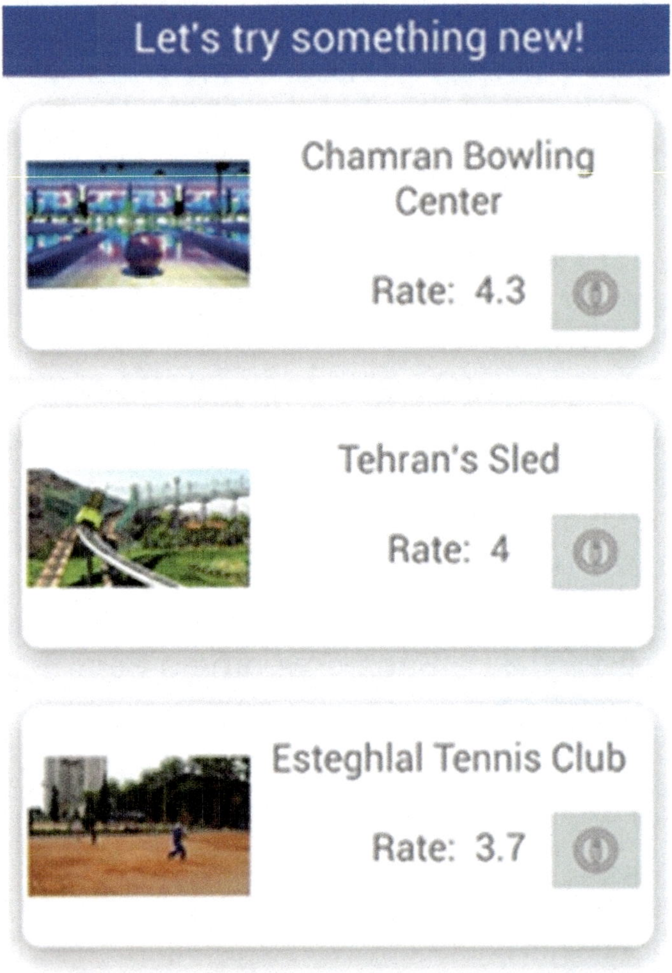

Fig. 9 Serendipity-oriented POI recommendation

recommender assistant and ranked some POIs. The system provided them with recommendations based on each recommendation procedure, and they graded their satisfaction of the recommended results via a five-point rating scale for each procedure. Eventually, the users' ratings and the computed ratings of the system were used to calculate the RMSE measure for each procedure using Eq. 11.

$$RMSE = \sqrt{\frac{\sum_{i,j}\left(r_{i,j} - \hat{r}_{i,j}\right)^2}{N}} \quad (11)$$

where N is the number of ratings, $r_{i,j}$ is the users' ratings, and $\hat{r}_{i,j}$ is the systems' calculated ratings. For the first POI recommendation procedure, the computed RMSE measure resulted 0.32 and 0.28 for the user-based recommendation and item-based methods, respectively. The resulted RMSE measure for the item-based CF recommendation method confirms the ability of this method in providing more accurate results in comparison to the user-based CF recommendation. The dynamic nature of users and the changes in user's taste through time increases data sparsity which results in weaker accuracy of the user-based POI recommendation method.

For the serendipitous POI recommendation procedure, the RMSE was 0.30. Although serendipitous POI recommendations are interesting and not easy-to-predict for the user, it is still necessary that these recommendations be relevant and useful. Since there should always be a proper trade-off between accuracy and serendipity, user's activity pattern was used in this study to inhibit the accuracy loss. By using the user's activity pattern and filtering the outputs of the SO recommendation based on the current activity type of the user, the developed system guarantees the usefulness of the SO recommendation. Therefore, the RMSE of the SO recommendation was relatively better than user-based CF recommendation but still lower than the item-based CF recommendation. In another word, there is a tradeoff between the accuracy and diversity of the SO recommendation, where losing accuracy may cause the serendipitous recommendations to be disappointing to the user and vice versa. Our system was able to impede the accuracy loss through utilization of the behavioral pattern as a filter on the results of the SO recommendation. Receiving an RMSE within an acceptable range, in comparison to the item-based and user-based recommendation, for the serendipity-oriented POI recommendation ensures the feasibility of the presented solution.

5 Conclusion and Future Work

In this paper, we presented a personalized location-based and serendipity-oriented recommender system to suggest POIs to users, based on their previously monitored behavioral patterns. The proposed system follows two recommendation procedures which give accurate and surprising suggestions. By exploiting the ARM methods, the system extracts the behavioral patterns of the users from their trajectory data and provides recommendations in specific parts of the day using collaborative filtering method. Furthermore, unexpected suggestions are created using a K-furthest neighborhood model which finds dissimilar users and suggests unexpected items to them. Our research has highlighted the importance of extracting user activity types and involving them in the learning process of a recommender assistant.

To further our research, we intend to extend the behavioral pattern extraction method to include user location along with time in the output patterns. This way the rules will include the three components of where, when and what for each user and therefore the recommendation based on those rules would be much more personalized and accurate. Utilization of other dissimilarity measures to provide a wider range of diversity and discover interesting relationships between users and POIs is another research direction that may be pursued afterward. Moreover, the specification of other behavioral patterns, like rare and sporadic activities, in POI recommendation or utilization of sequential pattern extraction methods may be considered in the future studies.

References

Adomavicius G, Tuzhilin A (2005) Toward the next generation of recommender systems: a survey of the state-of-the-art and possible extensions. IEEE Trans Knowl Data Eng 17:734–749

Agrawal R, Imieliński T, Swami A (1993) Mining association rules between sets of items in large databases. In: ACM sigmod record, Washington, DC, USA. ACM, pp 207–216

Cao H, Mamoulis N, Cheung DW (2007) Discovery of periodic patterns in spatiotemporal sequences. IEEE Trans Knowl Data Eng 19:453–467

Celdrán AH, Pérez MG, Clemente FJG, Pérez, GM (2016) Design of a recommender system based on users' behavior and collaborative location and tracking. J Comput Sci 12

Choong MY, Chin RKY, Yeo KB, Tze Kin Teo K (2016) Trajectory pattern mining via clustering based on similarity function for transportation surveillance. Int J Simul Syst Sci Technol 17

de Gemmis M, Lops P, Semeraro G, Musto C (2015) An investigation on the serendipity problem in recommender systems. Inf Process Manage 51:695–717

Dodge S, Weibel R, Lautenschütz A-K (2008) Towards a taxonomy of movement patterns. Inf Visual 7:240–252

Etienne L, Devogele T, Bouju A (2012) Spatio-temporal trajectory analysis of mobile objects following the same itinerary. Adv Geo-Spatial Inf Sci 10:47–57

Fu Z, Hu W, Tan T (2005) Similarity based vehicle trajectory clustering and anomaly detection. In: IEEE international conference on image processing ICIP 2005. IEEE, pp II-602

Gao H, Tang J, Hu X, Liu H (2015) Content-aware point of interest recommendation on location-based social networks. In: AAAI, pp 1721–1727

Ge M, Delgado-Battenfeld C, Jannach D (2010) Beyond accuracy: evaluating recommender systems by coverage and serendipity. In: Proceedings of the fourth ACM conference on recommender systems, 2010 Barcelona, Spain. ACM, pp 257–260

Gudmundsson J, Laube P, Wolle T (2011) Computational movement analysis. In: Springer handbook of geographic information. Springer, Berlin

Herlocker JL, Konstan JA, Terveen LG, Riedl JT (2004) Evaluating collaborative filtering recommender systems. ACM Trans Inf Syst (TOIS) 22:5–53

Iaquinta L, De Gemmis M, Lops P, Semeraro G, Filannino M, Molino P (2008) Introducing serendipity in a content-based recommender system. In: 2008 Eighth international conference on hybrid intelligent systems HIS'08, Barcelona, Spain. IEEE, pp 168–173

Li Q, Zheng Y, Xie X, Chen Y, Liu W, Ma W-Y (2008) Mining user similarity based on location history. In: 2008 Proceedings of the 16th ACM SIGSPATIAL international conference on advances in geographic information systems, Irvine, California. ACM, p 34

Li Z, Ding B, Han J, Kays R, Nye P (2010) Mining periodic behaviors for moving objects. In: 2010 Proceedings of the 16th ACM SIGKDD international conference on knowledge discovery and data mining, Washington, DC, USA. ACM, pp 1099–1108

Liu B, Fu Y, Yao Z, Xiong H (2013) Learning geographical preferences for point-of-interest recommendation. In: 2013 Proceedings of the 19th ACM SIGKDD international conference on knowledge discovery and data mining, Chicago, Illinois, USA. ACM, pp 1043–1051

Liu B, Xiong H (2013) Point-of-interest recommendation in location based social networks with topic and location awareness. In: Proceedings of the 2013 SIAM international conference on data mining, Austin, Texas, USA. SIAM, pp 396–404

Mamoulis N, Cao H, Kollios G, Hadjieleftheriou M, Tao Y, Cheung DW (2004) Mining, indexing, and querying historical spatiotemporal data. In: 2004 Proceedings of the tenth ACM SIGKDD international conference on knowledge discovery and data mining, Seattle, WA, USA. ACM, pp 236–245

Mcnee SM, Riedl J, Konstan JA (2006) Being accurate is not enough: how accuracy metrics have hurt recommender systems. In: 2006 CHI'06 extended abstracts on human factors in computing systems, Montréal, Québec, Canada. ACM, pp 1097–1101

Menk A, Sebastia L, Ferreira R (2017) Curumim: a serendipitous recommender system based on human curiosity. Procedia Comput Sci 112:484–493

Murakami T, Mori K, Orihara R (2007) Metrics for evaluating the serendipity of recommendation lists. In: Annual conference of the Japanese society for artificial intelligence, 2007. Springer, Berlin, pp 40–46

Palma AT, Bogorny V, Kuijpers B, Alvares LO (2008) A clustering-based approach for discovering interesting places in trajectories. In: Proceedings of the 2008 ACM symposium on applied computing, Fortaleza, Ceara, Brazil. ACM, pp 863–868

Ricci F, Rokach L, Shapira B, Kantor PB (2015) Recommender systems handbook. Springer

Rocha JAM, Times VC, Oliveira G, Alvares LO, Bogorny V (2015) DB-SMoT: a direction-based spatio-temporal clustering method. In: 2010 5th IEEE international conference intelligent systems (IS), London, UK. IEEE, pp 114–119

Said A, Fields B, Jain BJ, Albayrak S (2013) User-centric evaluation of a K-furthest neighbor collaborative filtering recommender algorithm. In: Proceedings of the 2013 conference on computer supported cooperative work, San Antonio, Texas, USA. ACM, pp 1399–1408

Sarwar B, Karypis G, Konstan J, Riedl J (2001) Item-based collaborative filtering recommendation algorithms. In: 2001 Proceedings of the 10th international conference on World Wide Web, Hong Kong, Hong Kong. ACM, pp 285–295

Schreck T, Bernard J, von Landesberger T, Kohlhammer J (2009) Visual cluster analysis of trajectory data with interactive Kohonen maps. Inf Visual 8:14–29

Shani G, Gunawardana A (2011) Evaluating recommendation systems. In: Recommender systems handbook, pp 257–297

Wang J, De Vries AP, Reinders MJ (2006) Unifying user-based and item-based collaborative filtering approaches by similarity fusion. In: 2006 Proceedings of the 29th annual international ACM SIGIR conference on research and development in information retrieval, Seattle, Washington, USA. ACM, pp 501–508

Yamaba H, Tanoue M, Takatsuka K, Okazaki N, Tomita S (2013) On a serendipity-oriented recommender system based on Folksonomy and its evaluation. Procedia Comput Sci 22:276–284

Yang D, Zhang D, Zheng VW, Yu Z (2015) Modeling user activity preference by leveraging user spatial temporal characteristics in LBSNs. IEEE Trans Syst Man Cybern Syst 45:129–142

Ye M, Yin P, Lee W-C, Lee D-L (2011) Exploiting geographical influence for collaborative point-of-interest recommendation. In: 2011 Proceedings of the 34th international ACM SIGIR conference on research and development in information retrieval, Beijing, China. ACM, pp 325–334

Ying J-C, Chen H-S, Lin KW, Lu EH-C, Tseng VS, Tsai H-W, Cheng KH, Lin S-C (2014) Semantic trajectory-based high utility item recommendation system. Expert Syst Appl 41:4762–4776

Yuan Q, Cong G, Ma Z, Sun A, Thalmann NM (2013) Time-aware point-of-interest recommendation. In: 2013 Proceedings of the 36th international ACM SIGIR conference on research and development in information retrieval, Dublin, Ireland. ACM, pp 363–372

Zheng Y, Zhou X (2011) Computing with spatial trajectories. Springer Science & Business Media

Ziegler C-N, Mcnee SM, Konstan JA, Lausen G (2005) Improving recommendation lists through topic diversification. In: 2005 Proceedings of the 14th international conference on World Wide Web, Chiba, Japan. ACM, pp 22–32

Identifying and Interpreting Clusters of Persons with Similar Mobility Behaviour Change Processes

David Jonietz, Dominik Bucher, Henry Martin and Martin Raubal

Abstract With the emergence of new mobility options and various initiatives to increase the sustainability of our travel behaviour, it is desirable to gain a deeper understanding of our behavioural reactions to such stimuli. Although it is now possible to use GPS-tracking to record people's movement behaviour over a longer period, there is still a lack of computational methods which allow to detect and evaluate such behaviour change processes in the resulting datasets. In this study, we propose a data mining method for describing individual persons' mobility behaviour change processes based on their movement trajectories and clustering participants based on the similarity of these behavioural adaptations. We further propose to use a decision tree classifier to semantically explain the derived clusters in a human-interpretable form. We apply our method to a real, longitudinal movement dataset.

1 Introduction

Recent developments, such as electric and autonomous vehicles, the sharing economy, the wide availability of modern smartphones, and fast and affordable IT infrastructure are profoundly changing the ways we travel and perceive our everyday mobility (Gossling 2017). Today's GPS-enabled devices and vehicles constantly generate large data streams which not only contain their time-stamped locations, but also allow the inference of rich semantic information, such as transport mode choices or trip purposes. These data provide exciting new opportunities for understanding the general rules guiding our mobility behaviour as well as the optimization of travel networks and business plans (Yuan and Raubal 2016). This is particularly interesting in the context of energy consumption and CO_2 production of the passenger transport sector, and the potential for an increase in sustainability provided by new forms of mobility and location-aware information and communication technology (Weiser et al. 2016). In fact, there are numerous studies which use movement

D. Jonietz (✉) · D. Bucher · H. Martin · M. Raubal
Institute of Cartography and Geoinformation, ETH Zurich,
Stefano-Franscini-Platz 5, 8093 Zürich, Switzerland
e-mail: jonietzd@ethz.ch

A. Mansourian et al. (eds.), *Geospatial Technologies for All*,
Lecture Notes in Geoinformation and Cartography,
https://doi.org/10.1007/978-3-319-78208-9_15

trajectory data for investigating human mobility behaviour and its potential changes towards a higher sustainability, for instance by tracking the mobility behaviour of a small sample of the population, and exposing them to a behavioural stimulus in the form of detailed feedback on their personal ecological footprint (e.g. Montini et al. 2015; Bucher et al. 2016), or other voluntary travel behaviour change initiatives (e.g. Stopher et al. 2013).

These and also other studies which aim to assess people's mobility behaviour changes through analysing longitudinal GPS-trajectory datasets face the common challenge of detecting and quantitatively evaluating such change of travel habits. At present, this involves manual checking of the data or simple pre-post comparisons of computed index values such as average daily distances travelled by car, or the produced amount of CO_2 (e.g. Pendyala et al. 2001; Stopher et al. 2013). Other approaches have compared the structure of temporal travel sequences (e.g. Langlois et al. 2016), e.g. the stability of weekly patterns with regards to travel distances, but have not examined the behaviour change process per se.

In contrast to previous studies, however, it would be desirable to be able to quantify people's behaviour change (e.g. as a reaction to a behavioural stimulus) both in an automated fashion, and at a much greater level of detail than simple pre-post-comparisons. For instance, this would allow to detect certain 'types' of persons with similar mobility behaviour change processes, such as early adopters of new technologies, or rather sceptical people with a high stability of existing travel habits. Automated methods which search individual persons' longitudinal trajectory datasets for behaviour change indicators, and provide detailed information with regards to the semantic (*in what way?*) and the temporal dimension (*how fast?*) of the detected mobility behaviour change process, would be valuable to answer questions such as: How do people react to the availability of new mobility options (e.g. an electric car)? Are there certain types of people in terms of how quickly they adopt this new technology and how profoundly it changes their established mobility habits? In this context, the large-scale pilot study *SBB Green Class*[1] is currently conducted in Switzerland by the *Swiss Federal Railways (SBB CFF FFS)*, in which over 100 participants are offered a comprehensive Mobility as a Service (MaaS) package, containing (among others) a general public transport pass valid in the entire country and an electric car. During the entire project duration, the participants are tracked with a smart phone application. Here, it is particularly interesting how the study participants react to this behavioural stimulus (i.e. the availability of novel behavioural options), and whether certain 'types' of users with similar behavioural adaptations emerge.

In this study, we propose a data mining (DM) method for describing behaviour change processes based on movement data and building clusters of likewise participants based on the similarity of their behavioural adaptations. Apart from their mere identification, our concept additionally allows to semantically interpret the derived clusters, i.e. to produce human-interpretable explanations of the found classes, which can then be presented to a transportation analyst or other stakeholders. For this, we use a two-level abstraction process in which, first, a set of selected descriptive

[1] www.sbb-greenclass.ch.

mobility features is computed from a person's trajectory data (see e.g. Pendyala et al. 2001; Jonietz and Bucher 2018). These provide the means to abstract further until we derive a set of features which, instead of describing people's mobility behaviour, explicitly describe selected characteristics of their dynamic behaviour change processes. This set of highly abstracted features is used for clustering participants with 'similar' behaviour change processes. Finally, for their interpretation, the results are used to train a decision tree classifier. We test our concept on an exemplary dataset.

This paper is structured as follows: In Sect. 2 we present relevant prior work, including approaches for detecting and describing mobility behaviour change, the clustering of temporal processes, and the use of decision trees for cluster interpretation. Then, in Sect. 3, our method is described in detail, before it is applied to our real-world dataset in Sect. 4. After a discussion of the results, Sect. 5 concludes this paper.

2 Related Work

2.1 Examining Behavioural Variations in Travel Surveys

Nowadays, travels surveys often rely on movement data recorded by modern GPS-equipped smart phones (Shen and Stopher 2017). The vast majority of these surveys, however, merely assess the status quo of mobility behaviour, thereby neglecting its dynamic dimension (i.e. behavioural change processes). According to (Schlich and Axhausen 2003), this is due to the general assumption of our mobility routines being highly habitual and static—as also demonstrated by numerous empirical studies (e.g. Song et al. 2010)—while feasibility restrictions leading to generally short tracking periods are certainly another contributing factor (Schlich and Axhausen 2003). There are, however, also studies which explicitly focus on the dynamics of travel behaviour, which, according to Scheiner and Holz-Rau (2013), can be grouped into two lines of research: first, there are studies which focus on behavioural changes related to interventions with the explicit aim of triggering such effects with behavioural stimuli (e.g. Montini et al. 2015; Bucher et al. 2016). Secondly, there are studies which aim to describe the stability and regularity of travel habits (e.g. Schlich and Axhausen 2003; Gonzalez et al. 2008; Song et al. 2010), or the effect of life course-related transitions of travel behaviour (e.g. Van der Waerden 2003; Lanzendorf 2003).

In the context of this study, we are particularly interested in approaches to quantitatively describe mobility behaviour change processes. A frequently used strategy for this purpose is to calculate descriptive measures which aggregate selected aspects of the recorded mobility behaviour of each individual person or the entire study sample for discrete time intervals, and compare the derived values at different points in time (e.g. Stopher et al. 2013; Jonietz and Bucher 2018). Other studies calculate a range of indexes to measure the temporal variability of travel behaviour, such as Schönfelder and Axhausen (2001), who count and compare the visited locations per week, Pendyala et al. (2001) who examine variations of the number of trips per day, Kim et al. (2017), who describe people's tendency to travel on the same public transport route, or Heinen and Ogilvie (2016) who apply common measures for market concentration and entropy to modal splits. Other approaches have used DM techniques, in particular clustering algorithms, to group persons with similarly structured travel sequences, such as stable working day travel patterns with reduced travel on the weekends (e.g. Langlois et al. 2016; Ma et al. 2013; Bhaskar et al. 2015). None of the available methods, however, focus on extracting detailed characteristics of the behaviour change process itself, such as how fast people changed certain aspects of their mobility behaviour (e.g. used certain transport modes more or less frequently), or how stable their travel habits were before or after the change occurred.

2.2 Clustering Dynamic Processes Based on Their Similarity

In general, despite the fact that time is an essential dimension for modelling space as well as spatial behaviour (Claramunt and Thériault 1995; Golledge 1997), its integration into analytical processes is not trivial. Two questions are of particular interest when analysing dynamic processes: How can we automatically detect when abrupt changes in a process occur, and how can we quantify and compare changes over a longer period of time?

The detection of sudden changes in a process is closely related to the general task of finding outliers, which is commonly required to detect faulty sensors or other system failures (Venkatasubramanian et al. 2003; Isermann 2005). The problem of outlier detection has been studied extensively (Rousseeuw and Van Zomeren 1990; Gupta et al. 2014; Aggarwal and Yu 2001) and is often performed by using a clustering algorithm, where outliers are 'noisy' points that do not lie in a cluster (Ester et al. 1996; Hartigan and Wong 1979; Yuan and Raubal 2012). One possibility to use outlier detection for the identification of clusters of similar dynamic processes is to engineer a set of suitable features based on some chosen time measure (e.g. on a weekly or a daily base). Such features often simply encode the change in value of some original features, i.e. they are computed as a derivative of the original data. For example, the weekly change of the share of a certain mode of transport can be used to detect sudden changes in the mobility behaviour of a person (Jonietz and Bucher 2018), and testing which features exhibit similar change patterns yields different (weekly) clusters.

Singer and Willett (2003) give an overview of longitudinal data analysis, and focus particularly on how to detect and quantify change in processes with values changing systematically over time—both within-individual as well as to explain inter-individual differences in change. They present a two-level model, which first models how some features change on an individual level, and then considers the change of features in inter-individual comparison. In contrast to our research, this form of longitudinal analysis aims to explain certain characteristics in the data, but has a set of predictors previously available which implicitly form the clusters of interest. Functional cluster analysis is similar in that it first fits curves to longitudinal data (usually spline approximations), which are then used in a k-means, hierarchical, or density-based clustering algorithm (Garcia-Escudero and Gordaliza (2005), Chiou and Li (2007), cf. Xu and Wunsch (2005) for a summary of different clustering algorithms).

Finally, it is possible to view dynamic processes as trajectories in high-dimensional spaces. Trajectory clustering is usually based on a distance measure, such as dynamic time warping or the Fréchet distance, and many approaches have been proposed to cluster trajectories efficiently (Lee et al. 2007; Chen et al. 2005; Yuan and Raubal 2012). Often all the above approaches are combined with a dimension reduction technique, such as principal component analysis and singular value decomposition (Jolliffe 2016), as the data become increasingly sparse in high dimensions.

2.3 Interpreting Cluster Results with Decision Trees

Decision trees refer to a popular approach for supervised classification, in which a feature space is split into two or more sub-spaces at each internal node of a directed, rooted tree. This incremental divide occurs on the basis of certain discrete uni- or multivariate functions of the input values, and stops when a certain stopping criterion is reached. The resulting leaf nodes then describe distinct classes of target labels, which can be explained by inspecting the sequence of rules which led to its allocation. Decision trees are typically optimized by minimizing the generalization error (Rokach and Maimon 2005) and rely on heuristic methods such as classification and regression trees (CART) (Breiman et al. 1984).

Apart from classification, decision trees can also be used for interpreting the results of a clustering process. Whereas there are numerous methods for assessing the quality and stability of tendency of computed clusters (Zaki et al. 2014), explaining the exact reasons for cluster assignments is a non-trivial task (Parisot et al. 2014). For this reason, Parisot et al. (2014) have proposed to use the feature values with the derived cluster labels to train a decision tree which then forms and visualizes a set of rules which best describe the derived cluster memberships. A particular goal of the authors is to obtain a simple but at the same time accurate decision tree. In fact, for decision tree-based classification, identifying a suitable trade-off between producing simplified, under-fitted and complex over-fitted decision trees is particularly challenging. A potential solution is represented by pruning, i.e., deriving over-fitted trees as a first step and then removing unnecessary sub-branches (Rokach and Maimon 2005).

3 Method

In this section, we present our method for clustering people based on the similarity of their mobility behaviour change processes, which includes the following sub-steps:

- **Feature extraction level I (L1)**. Computation of descriptive measures of mobility behaviour at discrete time intervals.
- **Feature extraction level II (L2)**. Computation of descriptive measures of the mobility behaviour change process based on the L1 features.
- **Clustering**. Grouping of users with similar mobility behaviour change process, i.e., based on their level II features.
- **Interpretation of the derived clusters**. Training of a decision tree with the level II features and the cluster labels, visualization of the detected rules for class label assignment.

3.1 Feature Extraction Level I—Mobility Behavior

As data source, we assume a set of movement trajectories which are map matched to the transportation network and segmented into the following units (c.f. Axhausen and Frick 2005; Jonietz and Bucher 2018):

- *Track points*: position fixes with x,y coordinate pairs and a time stamp
- *Trip legs*: aggregated track points based on the same used transport mode
- *Trip*: sequence of one or more trip legs between two activities
- *Stay point*: a location where someone was stationary for longer than 5 min (a duration of more than 5 min was considered as necessary for someone to purposefully stay somewhere)
- *Activity*: a stay point which represents an actual destination of travel (e.g., work, home or a shop).

Furthermore, the data should contain a set of attributes including a user id (for all units), the mode of transport (for track points and trip legs), and the purpose for stay points (in our exemplary dataset, these purposes are *home*, *work*, *errand*, *leisure*, and *wait*). From the latter, we can indirectly infer the purpose of trips and their underlying trip legs by assigning the purposes of a trip's start and end point to its respective trip legs (a trip can therefore be allocated two purposes, e.g., *work* and *errand*). Further details about the dataset used in our case study is provided in Sect. 4.

In the first step of the analysis, we aim to compute and extract a set of descriptive measures of selected aspects of a user's mobility behaviour (L1), which will provide the basis for a downstream extraction of L2 features. Such descriptive measures are for example the total distance someone travels in a given time interval, or the modal split during the same period (i.e. how often someone takes the car in comparison to taking public transport or walking somewhere by foot). According to Claramunt

Table 1 Mobility features

Feature description	Number of resulting features
Duration-based modal split by purpose	$n_m \cdot n_p$
Distance-based modal split by purpose	$n_m \cdot n_p$
Duration per stay point by purpose	n_p
Total number of trips	1
Total distance travelled	1
Total number of trip legs	1
Duration of trip legs, sum over all purposes	1
Duration of stay points, sum over all purposes	1
Total duration of triplegs and stay points	1
CO_2 emissions	1

and Thériault (1995), a critical step for such spatio-temporal analyses is to define the chronon, i.e., the shortest duration or discrete interval of time considered in the analysis. At this stage, we choose a one-day interval (i.e., 24 h) for the calculation of mobility features. To smooth out sudden and steep feature value variations, however, as could be caused for instance by a one-time business trip, we do not work directly with the daily values, but rather calculate a moving average over a sliding window (for instance with a size of 1 week and a step size of 1 day).

On this basis, we can calculate a range of descriptive features aggregating selected aspects of a user's mobility behaviour within the respective time span, such as the modal split in total and for each trip purpose, based on the distance and the duration of travelling. In the latter case, a total of $2 \cdot n_m \cdot n_p$ features are produced, where n_m is the number of modes and n_p is the number of purposes available in the dataset, as we have to encode each combination of travel mode and purpose in its own feature.

Due to this study being set in the context of sustainable mobility, our list of features also includes total CO_2 emissions per user and day, on the basis of mode-specific average values as provided by various sources (e.g., *mobitool*[2]). In addition, we propose to extract numerous general features, including durations at stay points, trip counts, and distances, to come up with the list shown in Table 1, and a total number of features $N_{L1} = 2 \cdot n_m \cdot n_p + n_p + 7$.

These L1 features capture behaviour at certain points in time, but cannot directly be used to assess and classify the change of behaviour itself. For example, knowing that someone travelled far on one day does not tell us anything about a changing behaviour. If we see an increase in travelled distance over several days, on the other hand, we can conclude that some change is taking place.

[2]See www.mobitool.ch.

3.2 Feature Extraction Level II—Mobility Behaviour Change

In a second step, we thus process the L1 features with the intent of quantifying changes in their values, i.e., a user's behaviour change. As it is common that certain L1 features lack data on some days (a user might have turned off GPS tracking or ran out of battery), we interpolate these missing feature values linearly between the closest previous and next available value. For example, if a certain user lacks data in week 2 of a 3-week tracking period, the feature values in week 2 would simply be the mean of the values in week 1 and 3.

As we ultimately want to cluster people according to the degree with which they *similarly* change their behaviour, we found the following descriptors suitable for further analysis. These descriptors are computed for every user and for each L1 feature individually:

- A first order approximation of the L1 feature values (i.e., a polynom of the form $v_t = a \cdot t + b$, where v_t is the value of the L1 feature at time t). This yields the L2 features *trend line intercept* (b) and *trend line slope* (a), which capture how a person's initial behaviour is, and how she generally changes this behaviour. For instance, if someone increasingly takes the bicycle for commuting instead of the car, the trend line slope (a) of the L1 feature CO_2 emissions would be negative, as bicycles produce less CO_2 than cars.
- The overall *volatility* of the feature, resp. of the underlying process. As we look at the volatility over the whole sampling period, this simply corresponds to the *variance* of all feature values. This feature captures the steadiness of a certain L1 feature over the whole study period, i.e. if someone often changes a certain behaviour or exhibits very regular patterns.
- The *min* and *max deviation* of the feature. To get this value, we first compute the differences between all consecutive samples. From these we take the 5th and 95th percentile as the min resp. max deviation. These deviations capture the extremes in behaviour change from one day to another.
- The *number of anomalies*, which are computed using a sliding window (of size 16) over all samples. For each window, an anomaly can appear at the rightmost sample, i.e. at position 16. We first recenter this value by subtracting the mean of all samples in the window from it. If the absolute value of this re-centered sample is larger than $\lambda \cdot \sigma$ (where σ is the standard deviation of all values in the window, and λ is a tuning parameter, set to 3), the sample is counted as an anomaly. If a person shows many anomalies (over a duration larger than 16 days, otherwise there would only be one window), it indicates that there are frequent large changes in behaviour that happen after steady periods.
- A first order approximation (similar to the first L2 feature) of a sliding window (of size 16) computation of the *variance*. From this trend line, we again take the *slope* and *intercept*. These features capture the change in variance, e.g. when someone starts to use a certain mode of transport very unsteadily.

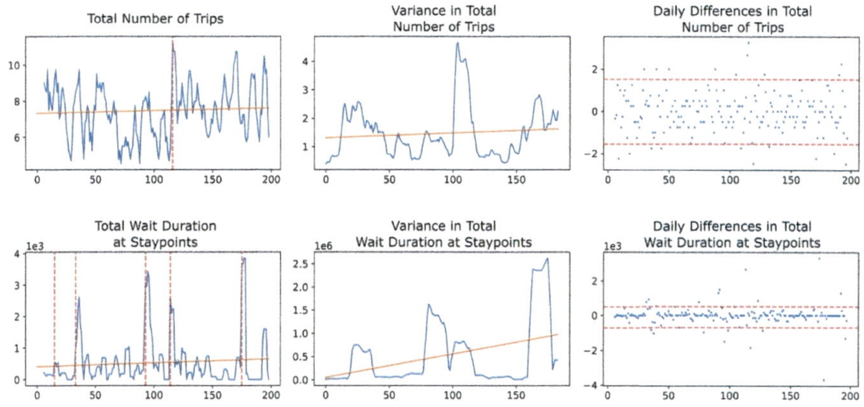

Fig. 1 Two L1 features and the extracted L2 features for an exemplary user

In total, this results in a 2-dimensional matrix of the form $n \times N_{L2}$, where n denotes the number of users and $N_{L2} = 8 \cdot N_{L1}$ the number of L2 features. For better understanding, Fig. 1 shows a feature excerpt for an exemplary user from the dataset used in Sect. 4. The first row depicts the *total number of trips* the user takes on a given day (smoothed using the moving average described in Sect. 3.1). The first plot shows the raw L1 feature values in blue, the (single) anomaly in red, and the trend line in orange. The second and third plot show the sliding window variance resp. the differences between consecutive samples in blue, and the thus derived L2 features in orange (trend line of variance) and red (5th and 95th percentile of daily changes). The second row treats the L1 feature *total wait duration at staypoints*, i.e. how long someone waits every day at staypoints classified as "wait" (for public transport). Notable differences include the increased number of anomalies in the second feature, a more steep trend line of the variance, and different feature ranges. Before further processing and clustering, the L2 features are normalized again by performing a rank transformation, i.e., replacing each feature value with its rank divided by the total number of ranks.

3.3 Cluster Computation and Interpretation

Before clustering the data derived by the feature extraction (L2), we need to cope with their high dimensionality. In order to preserve the explanatory value of each separate feature dimension, we choose a feature selection approach rather than feature extraction. Suitable feature subsets should have a dispersed distribution to be of discriminate value, and not be redundant (i.e. correlate with each other) (Hall 2000). Therefore we first compute the interquartile ratio, a robust measure of relative dispersion (Abernethy 1986), for each feature dimension:

$$v_q = \frac{X_{0.75} - X_{0.25}}{X_{0.25} + X_{0.75}}$$

where $X_{0.25}$ is the first and $X_{0.75}$ is the third quartile.

Based on this score, we select those with an above average interquartile ratio v_q for all further analyses. As a second step, we eliminate redundant features, assuming high linear correlation as a simplified measure for redundancy. We therefore compute the pair-wise linear correlation, and eliminate the feature with the lower v_q score in case of an absolute Pearson correlation coefficient $|r| > 0.8$ if the statistical significance $p \leq 0.05$.

At this stage, the features are ready for clustering. We choose the DBSCAN algorithm (Sander et al. 1998) for its ability to independently identify the number of clusters and for being robust to noise. In order to identify suitable parameter settings, we test various values for the maximum neighbourhood size ε, the minimum number of points *minPts* within distance ε of a cluster core point and the used distance metric (euclidean and manhattan). Furthermore, we vary the number of selected features that form our matrix $A \in \mathbb{R}^{n \times d}$ where n denotes the constant number of users and d the number of selected features. We vary epsilon within the 10th and the 90th quantile of all k-nearest neighbours distances calculated between all features d of all users n, with $k = minPts$. As stepsize d we choose one tenth of the interquartile range, $d = 0.1 \times (X_{0.75} - X_{0.25})$. We vary *minPts* within the interval of $[3, 7]$, meaning that we initially choose the three features with the highest v_q (as this is the most discriminating feature) and after testing all parameter combinations, we add the feature with the next highest v_q until we use all eligible features in the final iteration.

The results of each iteration are assessed based on the number of detected clusters, as well as their silhouette score, a measure of cluster separation proposed by Rousseeuw (1987). On this basis, it is possible to identify suitable parameter settings (i.e., which produce clusters with acceptable silhouette scores and take into account enough feature dimensions), and proceed with the analysis.

Having derived cluster labels, we intend to extract meaningful, human-interpretable information about the characteristics of the mobility behaviour changes of similar users. For this, we apply the approach proposed by Parisot et al. (2014), and use the feature values together with the received labels to train a decision tree classifier which we expand until all leaves are pure, i.e., contain only samples of one class. In order to acknowledge the possibility of strongly varying class sizes, we balance the tree by applying class weights which are inversely proportional to the class frequency. By visualizing the resulting tree and inspecting the detected classification rules, it is then possible to explain the differences in mobility behaviour among the clusters, and receive insights about similarities and dissimilarities in users' mobility behaviour change processes.

4 Case Study

We tested our method on a movement trajectory dataset which was recorded from 107 users from November 2016 to September 2017 with a smart phone application,[3] and represents a sub-set of the *SBB Green Class project*[4] dataset. All processing steps were implemented in Python 3.5, using scipy and sklearn for data preprocessing and mining, and PostgreSQL 9.6 with PostGIS 2.4 for all spatial operations.

4.1 Data and Procedure

The data consist of GPS-trajectories with additional semantic information about the mode of transport and the purpose of stay points, which have both been validated by the users themselves in a prompted response survey.[5] Please note that for this study, we exclude airplane trips since our focus is on changes in everyday mobility behaviour.

After basic data preprocessing (e.g. transforming the data into a format suitable for a relational database), we calculate the level I features as described in Sect. 3.1, which includes segmenting the data into trip legs, trips, stay points and activities, and calculating the 122 features for all 107 users and 283 complete days (averaged over a sliding window of 5 working days). In order to test our method on a consistent dataset, at this stage we focus only on weekdays, and therefore exclude Saturdays and Sundays. This also means that a large share of the analysed mobility patterns stem from commuting, which is particularly interesting for behaviour change as commutes can easily be planned in advance and adjustments to commuting patterns lead to long-lasting effects. As described in Sect. 3.2, we then compute the level II features, thereby arriving at a total of 804 feature dimensions, which are, however, reduced to 36 after filtering for v_q and redundancy (see Sect. 3.3). After testing a wide range of possible parameter settings as described, we identified the values listed in Table 2 as suitable.

4.2 Results

Table 3 shows the results of the clustering with the parameters listed in Table 2. For example, cluster 0 contains 45 users, i.e. these users somehow changed their behaviour in a similar way. For a visualisation of the features that led to this clustering, consider Fig. 2 which shows the five identified clusters plotted in parallel

[3]SBB DailyTracks, developed by MotionTag GmbH.

[4]See www.sbb-greenclass.ch.

[5]The following modes of transport were available to the study participants for validation: *Airplane, Bicycle, Boat, Bus, Car, Coach, Ebicycle, Ecar, Train, Tram, Walk.* With regards to the stay points, the following purposes could be allocated by the participants: *Home, Work, Errand, Leisure, Wait.*

Table 2 Parameter settings

ε	*minPts*	*d*	Distance metric
0.688	4	5	*manhattan*

Table 3 Results of the clustering. The DBSCAN algorithm identified five different clusters (0–4), and was not able to assign 29 users to any cluster

Cluster index	Number of users in cluster
−1 (outliers)	29
0	45
1	6
2	14
3	7
4	4

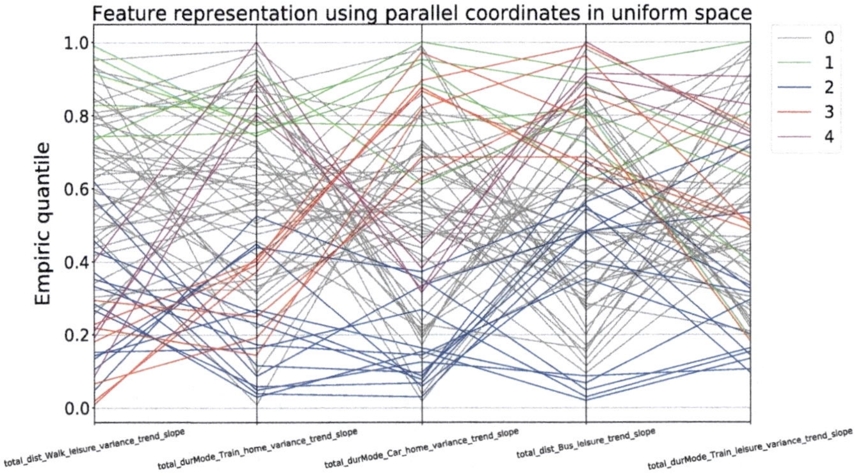

Fig. 2 Parallel plot of clustering results (excluding outliers). Each vertical line represents the domain of an individual feature and every observation is represented as a continuous line that crosses each vertical line at its corresponding feature value

coordinates (Inselberg and Dimsdale 1987). As one can see, the observation lines of a cluster show strongly correlated values across the 5 feature dimensions, and were therefore correctly allocated the same cluster label.

One can recognize that people in cluster 4 (pink) show an increasing tendency to take the bus for leisure purposes (in comparison with all other users), while people in cluster 1 (green) show an increasing unsteadiness in their use of trains to get from and to home. Since interpreting results in such a visual manner is rather time-consuming and inaccurate, we use a decision tree to estimate the importance of various features in building the clusters.

In Fig. 3, the result of training a decision tree with the derived labels and feature values is shown with the rule sequence for two exemplary classes emphasised (the

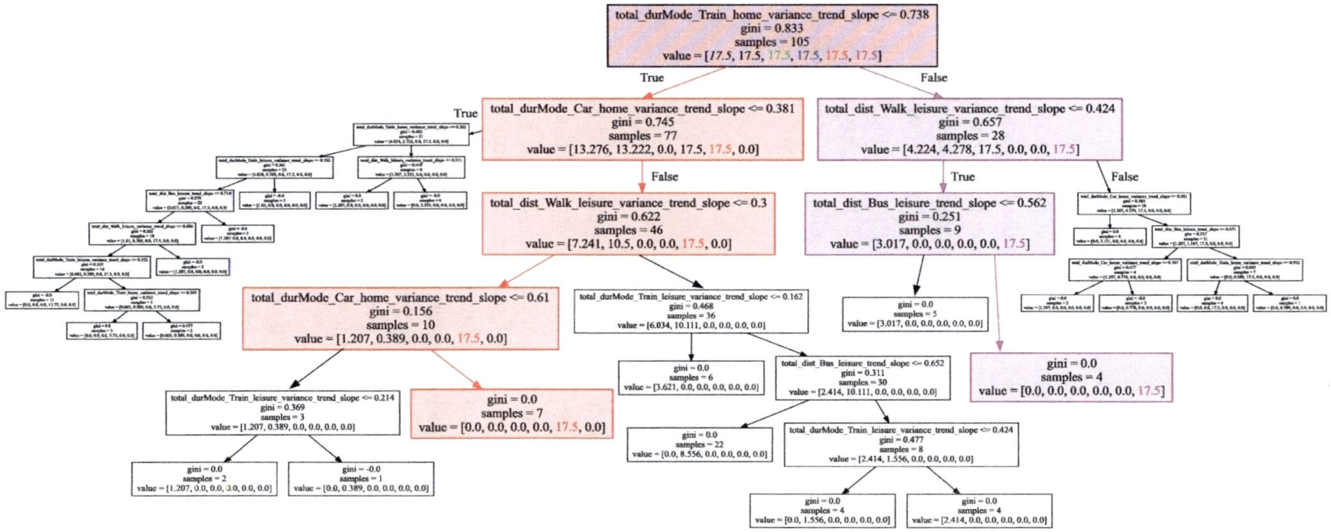

Fig. 3 Decision tree explaining the clusters of the case study (cluster 3: red; cluster 4: pink). Each node consists of a rule (first line) which is either true (left arrow) or false (right arrow), a *gini* impurity index to denote the quality of the split, the number or remaining samples, and a list of values showing for each category the relative number of samples which fall into that category at that node

other clusters are represented in the tree as well—the effect of different rules on them can be seen by looking at the *value* list in each box, which is sorted according to cluster number). At the top position, one can identify the feature found to be most discriminative by the decision tree is the gradient of a trend line over the variance found for the total duration spent on the train to or from one's own home. This means that over the whole study period, members of cluster 4 show a higher increase in inconsistent usage of the train for this purpose compared to e.g. persons in cluster 3 (this could indicate a break of the habit of using the train for cluster 4 members).

If we follow the rule sequence further, we find that cluster 4 members can be further characterized by a lower increase in inconsistency with regards to walking for leisure purposes (this could indicate either a stable walking habit or the persisting lack of such), and a higher increase in using the bus for the same purpose. Persons of cluster 3, in contrast, are marked by a medium increase in inconsistency with regards to using the car to or from their home location, and low increases in instability with regards to walking for leisure activities. These results provide indications to guide deeper analyses of the movement patterns and the formation of behavioural hypotheses for the respective cluster members. In our exemplary dataset, this could indicate that a Mobility as a Service (MaaS) offer primarily leads to changes in how people use mobility for *home* and *leisure* related activities.

5 Discussion and Conclusions

In this study, we proposed a method which analyses longitudinal movement trajectory datasets to find clusters of people that exhibit similar mobility behaviour changes. We focused on the semantic (*in what way?*) and temporal (*how fast?*) dimension of people's adaptation of their travel routines as a reaction to stimuli, such as the availability of novel mobility options. Other than previous approaches, we extracted a list of features for clustering via a two-step abstraction process, which ultimately describes not the mobility behaviour of a person, but selected characteristics of her behavioural change process itself. As demonstrated, it was possible to cluster the users based on these L2 features, and explain the derived results with a decision tree classifier.

On an exemplary dataset, we could demonstrate a potential application scenario of our method. As discussed, we could derive meaningful clusters of study participants who showed similar behavioural adaptations. It should be noted, however, that our method focuses on the nature of these processes, but does not allow to explain their exact causes, which in our case study, could include the MaaS offer itself, but also seasonal effects (e.g. weather, holidays). Nevertheless, the proposed method provides highly detailed information about the behaviour change processes, and can assist stakeholders in gaining a deeper understanding of people's reactions to novel mobility options, and ultimately support decision making. Apart from mobility behaviour, we expect our general framework to also be applicable to other related domains, such as ecological studies of animal movements in differing environmental circumstances.

There are still some shortcomings to our method. For instance, as we have described, a lot of processing needs to be done on both levels, including e.g. missing value interpolation or transformation processes. These steps introduce uncertainty to the data and might even result in a loss of information (e.g. in the case of rank transformation). When interpreting the results, therefore, there must be awareness of these measures. Furthermore, our method produces a high-dimensional L2 feature space, which makes feature selection necessary but increases the risk of accidentally filtering meaningful dimensions. Our method also requires trajectories to be annotated with transport modes and stay point purposes, data which is not available for most public datasets. Finally, as seen in our case study, the resulting explanations of cluster groups are highly detailed, but can also be very abstract depending on the list of L2 features. These should be selected with care and in relation to the exact study purpose.

For future work, we plan to use the method to examine differences in adopting electric cars in detail, and apply the method to other problem domains.

Acknowledgements This research was supported by SBB CFF FFS within the SBB Green Class Project, the Swiss National Science Foundation (SNF) within NRP 71 "Managing energy consumption", and by the Commission for Technology and Innovation (CTI) within the Swiss Competence Center for Energy Research (SCCER) Mobility.

References

Abernethy CL (1986) Performance measurement in canal water management: a discussion. Overseas Development Institute (ODI)

Aggarwal CC, Yu PS (2001) Outlier detection for high dimensional data. In: ACM Sigmod Record, vol 30, pp 37–46. ACM

Axhausen KW, Frick M (2005) *Nutzungen—Strukturen—Verkehr*, pp 61–79. Springer, Berlin, Heidelberg. ISBN 978-3-540-27010-2

Bhaskar A, Chung E et al (2015) Passenger segmentation using smart card data. IEEE Trans Intel Transp Syst 16(3):1537–1548

Breiman L, Friedman J, Stone CJ, Olshen RA (1984) Classification and regression trees, CRC press

Bucher D, Cellina F, Mangili F, Raubal M, Rudel R, Rizzoli AE, Elabed O (2016) Exploiting fitness apps for sustainable mobility-challenges deploying the goeco! app. In: ICT for sustainability (ICT4S)

Chen L, Özsu MT, Oria V (2005) Robust and fast similarity search for moving object trajectories. In: Proceedings of the 2005 ACM SIGMOD international conference on management of data, pp 491–502. ACM

Chiou J-M, Li P-L (2007) Functional clustering and identifying substructures of longitudinal data. J Royal Statist Soc Series B (Statistical Methodology) 69(4):679–699

Claramunt C, Thériault M (1995) Managing time in GIS: an event-oriented approach. Recent Adv Temp Databases, 23–42

Ester M, Kriegel H-P, Sander J, Xu X et al (1996) A density-based algorithm for discovering clusters in large spatial databases with noise. KDD 96:226–231

Garcia-Escudero LA, Gordaliza A (2005) A proposal for robust curve clustering. J Classif 22(2):185–201
Golledge RG (1997) Spatial behavior: a geographic perspective, Guilford Press
Gonzalez MC, Hidalgo CA, Barabasi A-L (2008) Understanding individual human mobility patterns. Nature 453(7196):779–782
Gossling S (2017) Ict and transport behaviour: a conceptual review. Int J Sustain Transp (Just-accepted)
Gupta M, Gao J, Aggarwal CC, Han J (2014) Outlier detection for temporal data: a survey. IEEE Trans Knowl Data Eng 26(9):2250–2267
Hall MA (2000) Correlation-based feature selection of discrete and numeric class machine learning
Hartigan JA, Wong MA (1979) Algorithm as 136: a k-means clustering algorithm. J Royal Statis Soc. Series C (Applied Statistics) 28(1):100–108
Heinen E, Ogilvie D (2016) Variability in baseline travel behaviour as a predictor of changes in commuting by active travel, car and public transport: a natural experimental study. J Transp Health 3(1):77–85
Inselberg A, Dimsdale B (1987) Parallel coordinates for visualizing multi-dimensional geometry. In: Computer graphics 1987, pp 25–44. Springer
Isermann R (2005) Model-based fault-detection and diagnosis-status and applications. Annual Rev Control 29(1):71–85
Jolliffe P (2016) Introduction. In: Learning, migration and intergenerational relations, pp 1–33. Springer
Jonietz D, Bucher D (2018) Continuous trajectory pattern mining for mobility behaviour change detection. In: Progress in location-based services 2018. Springer
Kim J, Corcoran J, Papamanolis M (2017) Route choice stickiness of public transport passengers: Measuring habitual bus ridership behaviour using smart card data. Transp Res Part C: Emerg Technol 83:146–164
Langlois GG, Koutsopoulos HN, Zhao J (2016) Inferring patterns in the multi-week activity sequences of public transport users. Transp Res Part C: Emerg Technol 64:1–16
Lanzendorf M (2003) Mobility biographies. A new perspective for understanding travel behaviour. In: 10th international conference on travel behaviour research, vol 10, p 15
Lee J-G, Han J, Whang K-Y (2007) Trajectory clustering: a partition-and-group framework. In: Proceedings of the 2007 ACM SIGMOD international conference on management of data, pp 593–604. ACM
Ma X, Wu Y-J, Wang Y, Chen F, Liu J (2013) Mining smart card data for transit riders travel patterns. Transp Res Part C: Emerg Technol 36:1–12
Montini L, Prost S, Schrammel J, Rieser-Schüssler N, Axhausen KW (2015) Comparison of travel diaries generated from smartphone data and dedicated GPS devices. Transp Res Procedia 11:227–241
Parisot O, Ghoniem M, Otjacques B (2014) Decision trees and data preprocessing to help clustering interpretation. In: DATA, pp 48–55
Pendyala R, Parashar A, Muthyalagari G (2001) Measuring day-to day variability in travel characteristics using GPS data. In: 79th annual meeting of the transportation research board
Rokach L, Maimon O (2005) Decision trees. Data mining and knowledge discovery handbook, pp 165–192
Rousseeuw PJ (1987) Silhouettes: a graphical aid to the interpretation and validation of cluster analysis. J Comput Appl Math 20:53–65
Rousseeuw PJ, Van Zomeren BC (1990) Unmasking multivariate outliers and leverage points. J Amer Statist Assoc 85(411):633–639
Sander J, Ester M, Kriegel H-P, Xu X (1998) Density-based clustering in spatial databases: the algorithm GDB scan and its applications. Data Mining Knowl Discov 2(2):169–194
Scheiner J, Holz-Rau C (2013) Changes in travel mode use after residential relocation: a contribution to mobility biographies. Transportation 40(2):431–458

Schlich R, Axhausen KW (2003) Habitual travel behaviour: evidence from a six-week travel diary. Transportation 30(1):13–36
Schönfelder S, Axhausen KW (2001) Mobidrive-längsschnitterhebungen zum individuellen verkehrsverhalten: Perspektiven für raum-zeitliche analysen
Shen L, Stopher PR (2017) Review of GPS travel survey and GPS data-processing methods. Transport reviews, 0(0):1–19. ISSN 0144-1647. https://doi.org/10.1080/01441647.2014.903530
Singer JD, Willett JB (2003) Applied longitudinal data analysis: modeling change and event occurrence, Oxford university press
Song C, Qu Z, Blumm N, Barabási A-L (2010) Limits of predictability in human mobility. Science 327(5968):1018–1021
Stopher PR, Moutou CJ, Liu W (2013) Sustainability of voluntary travel behaviour change initiatives: a 5-year study
Van der Waerden P (2003) The influence of key events and critical incidents on transport mode choice switching behaviour: a descriptive analysis. In: Proceedings of 10th international conference on travel behaviour research, 2003
Venkatasubramanian V, Rengaswamy R, Yin K, Kavuri SN (2003) A review of process fault detection and diagnosis: part i: quantitative model-based methods. Comput Chem Eng 27(3):293–311
Weiser P, Scheider S, Bucher D, Kiefer P, Raubal M (2016) Towards sustainable mobility behavior: research challenges for location-aware information and communication technology. GeoInformatica 20(2):213–239
Xu R, Wunsch D (2005) Survey of clustering algorithms. IEEE Trans Neural Netw 16(3):645–678
Yuan Y, Raubal M (2012) Extracting dynamic urban mobility patterns from mobile phone data. In: GIScience, vol 7478, pp 354–367. Springer
Yuan Y, Raubal M (2016) Analyzing the distribution of human activity space from mobile phone usage: an individual and urban-oriented study. Int J Geogr Inf Sci 30(8):1594–1621
Zaki MJ, Meira W Jr, Meira W (2014) Data mining and analysis: fundamental concepts and algorithms. Cambridge University Press

Mixed Traffic Trajectory Prediction Using LSTM–Based Models in Shared Space

Hao Cheng and Monika Sester

Abstract Real–world behaviors of human road users in a non-regulated space (shared space) are complex. Firstly, there is no explicit regulation in such an area. Users self-organize to share the space. They are more likely to use as little energy as possible to reach their destinations in the shortest possible way, and try to avoid any potential collision. Secondly, different types of users (pedestrians, cyclists, and vehicles) behave differently. For example, pedestrians are more flexible to change their speed and trajectory, while cyclists and vehicles are more or less limited by their travel device—abrupt changes might lead to danger. While there are established models to describe the behavior of individual humans (e.g. Social Force model), due to the heterogeneity of transport modes and diversity of environments, hand-crafted models have difficulties in handling complicated interactions in mixed traffic. To this end, this paper proposes using a Long Short–Term Memory (LSTM) recurrent neural networks based deep learning approach to model user behaviors. It encodes user position coordinates, sight of view, and interactions between different types of neighboring users as spatio–temporal features to predict future trajectories with collision avoidance. The real–world data–driven method can be trained with pre-defined neural networks to circumvent complex manual design and calibration. The results show that ViewType-LSTM, which mimics how a human sees and reacts to different transport modes can well predict mixed traffic trajectories in a shared space at least in the next 3 s, and is also robust in complicated situations.

Keywords Shared space · Mixed traffic · Trajectory prediction · Long short–term memory

H. Cheng (✉) · M. Sester
Institute of Cartography and Geoinformatics, Leibniz University, Hannover, Germany
e-mail: hao.cheng@ikg.uni-hannover.de

M. Sester
e-mail: monika.sester@ikg.uni-hannover.de

A. Mansourian et al. (eds.), *Geospatial Technologies for All*,
Lecture Notes in Geoinformation and Cartography,
https://doi.org/10.1007/978-3-319-78208-9_16

1 Introduction

In distinction to classic traffic designs which, in general, separately dedicate road resources to road users by time or space division, an alternative solution—shared space—has been proposed by traffic engineers. This concept was first introduced by the Dutch traffic engineer Hans Monderman in the 1970s (Clarke 2006). It was later formally defined by Reid as "a street or place designed to improve pedestrian movement and comfort by reducing the dominance of motor vehicles and enabling all users to share the space rather than follow the clearly defined rules implied by more conventional designs" (Reid 2009). This design allows mixed types of users (pedestrians, cyclists, and vehicles) to interact with each other and negotiate to take or give their right-of-way.

It is relatively easier and cheaper to construct less or non-regulated spaces than the classic traffic designs and more feasible for urban and crowded places (Karndacharuk et al. 2014). Nevertheless, efficiency and safety in shared spaces need to be fully investigated. At a micro level, understanding how road users behave and how we can foresee their behaviors after a very short observation time (e.g. 3 s) are crucial to traffic planning and autonomous driving in such areas. However, this is not a trivial task. Mixed traffic movement data, especially in shared spaces, contain various spatio–temporal features. The involved geographical space, objects and their associated multidimensional attributes change over time (Andrienko et al. 2011). A simple approach may be sufficient for simple situations, such as the Social Force model for pedestrian dynamics (Helbing and Molnar 1995). Robust approaches are required to handle complex situations when mixed traffic is present.

Human behaviors are affected by lots of factors which are very person dependent (e.g. age, gender, time pressure Kaparias et al. 2012). For this reason, modeling their decision–making process about where and when to go next in the interactions with others is a great challenge. These hidden characteristics of personality, however, will eventually be reflected by the change of their positions, orientations, speeds, acceleration, and deceleration. This phenomenon inspires us to build models which can directly leverage hidden characteristic features and mimic how a human sees and reacts based on his or her explicit motion sequences in the past together with the expected behavior of other traffic participants, and then predict his or her trajectories in the future.

There are models that take movement data as input for trajectory prediction for mixed traffic in shared spaces, but many of them still require domain experts and manual fine–tuning efforts (Schönauer et al. 2012; Rinke et al. 2017; Pascucci et al. 2017). On the other hand, data–driven models, for example, deep learning neural networks, especially recurrent neural networks have achieved massive success for sequence prediction in domains like handwriting and speech recognition (Graves 2013; Graves and Jaitly 2014). In this paper, Long Short–Term Memory (LSTM) recurrent neural network models are proposed for mixed traffic prediction in shared spaces, which circumvent manual model building and calibration procedures.

They are trained by feeding with users' motion sequences in the past along with user type and sight of view using a real–world dataset.

Outline of the paper. In this paper, we first summarize the works that have been published for mixed traffic modeling and prediction in shared spaces and the state–of–the–art data–driven approaches in related domains. Then we introduce our approach motivated by a work for pedestrian modeling in Sect. 3. A real–world dataset and evaluation metrics for our approach are described in Sect. 4. We report our experiments and results in the following sections. In the end, we conclude our paper with some interesting problems that we would like to investigate in future work.

2 Related Work

Mixed traffic in shared spaces The schemes of shared spaces have been a heated topic for an alternative traffic design. However, to the best knowledge of the authors, only a few studies have dealt with shared space modeling: simulating mixed traffic in shared spaces based on game theory (Schönauer et al. 2012) and mixed traffic modeling and prediction using an extended Social Force model with collision avoidance (Pascucci et al. 2015; Schiermeyer et al. 2016; Rinke et al. 2017; Pascucci et al. 2017). Nevertheless, the game proposed by Schönauer et al. for conflict handling is heavily hand-crafted and lacks flexibility—"the type of game, the number of players, the number of games repeated, and whether the game allows cooperation must be specified". On the other hand, in the studies of the extended Social Force model, mixed traffic is analyzed in a categorical fashion regarding involved transport modes, e.g. pedestrian versus pedestrian or pedestrian versus car. Their model does not provide a mechanism that can deal with arbitrary collisions regardless of user types.

Data–driven approaches in trajectory prediction In recent years, with the increased availability of computational power and large-scale datasets, data–driven approaches have been largely used for learning movement data. Long and Nelson summarized possible methods to learn trajectory–related movements (Long and Nelson 2013). Unsupervised learning, for example clustering, and segmentation are applied to recognize similar trajectory patterns (Morris and Trivedi 2009; Pelekis et al. 2011). Due to the divergence of mixed road users and their interdependence in shared spaces, these methods are not reliable when the involved objects and contexts change quickly.

Deep learning approaches in trajectory prediction There are deep learning approaches for behavior modeling, e.g. a conventional neural network based model for pedestrian behavior (Yi et al. 2016) and a recurrent neural network based model for car–following (Wang et al. 2017). But both of these networks are limited to homogeneous user types. Nevertheless, these works shed light on using deep learning approaches for trajectory modeling.

Initially, Long Short–Term Memory recurrent neural networks proved to be powerful for complex sequence generation, e.g. text and handwriting (Graves 2013) and speech recognition (Graves and Jaitly 2014). "They can be trained by processing real data sequences one step at a time and predict what comes next". This process is comparable to trajectory prediction—observing initial steps of movement and trying to forecast the future motion. In comparison to an isolated sequence in a text, a single trajectory cannot be predicted as an independent motion of a road user since there are other road users and factors in the vicinity impacting his or her behavior, the so-called repulsive and attractive effects (Helbing and Molnar 1995). In order to capture these social effects, a centralized bounding grid was introduced in (Alahi et al. 2016) to process the interactions with neighboring users when using LSTM for trajectory prediction (Social-LSTM). Experiments on five open datasets (Lerner et al. 2007; Pellegrini et al. 2009) showed Social-LSTM outperforming the classic model–based approaches, such as Social Force (Yamaguchi et al. 2011) and Iterative Gaussian Process (Trautman et al. 2013), for pedestrian trajectory prediction. In addition, the data–driven approach circumvents complex manual setups needed for fine–tuning these classic models.

However, Social-LSTM was only tested on pedestrians. There are distinctive patterns regarding transport modes. For example, the involved transport modes, environment, and density will impact the intensity of pedestrian reactions to conflicts; cyclists have limited flexibility to deal with collisions due to their bicycles; vehicles may behave prudently to avoid collisions with more vulnerable road users (Rinke et al. 2017). Moreover, equipped with rear mirrors and multiple sensors, vehicles have a larger sight of view compared with pedestrians and cyclists. Lacking a mechanism to handle different transport modes, Social-LSTM could not, up to now, be directly applied to mixed traffic trajectory prediction in shared spaces.

3 Methodology

In order to differentiate transport modes and apply Social-LSTM (Alahi et al. 2016) for mixed traffic trajectory prediction in shared spaces, we introduce a bounding grid which incorporates both user type and sight of view based on Social-LSTM. In the interactions, the regarding user is addressed as an ego-user, which is the same denomination used in (Rinke et al. 2017), and the other users in his or her vicinity are addressed as neighboring users.

Every user is trained as a single LSTM, whereas the interactions with neighboring users are filtered by the bounding grid mentioned above. The basic network structure for our models is derived from Social-LSTM (see Fig. 1). For the input layer, it has a spatial input part to store the user's x and y coordinates and a tensor input part to capture the neighboring users within a predefined bounding grid for each ego-user (see Fig. 3a). Instead of simply pooling a binary indicator to tell the ego-user about the existence of other users in a uniformly sized grid as in the Social-LSTM, the tensor input here also customizes the grid according to the ego-user's sight of

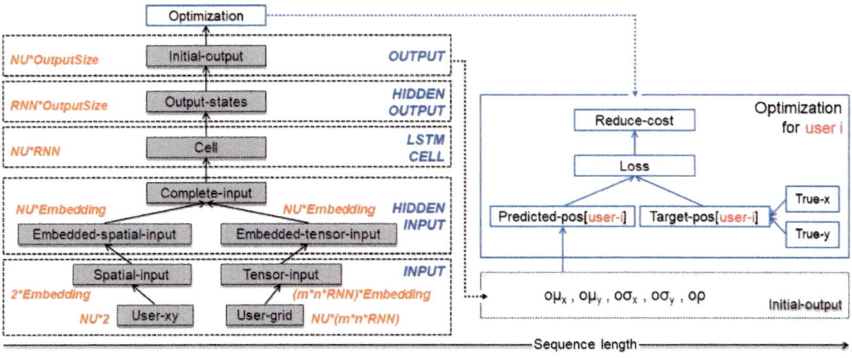

Fig. 1 Basic structure of the long short–term memory network. *NU*: number of users, $2 * Embedding$: embedding dimensions of weight W_S for spatial input, $(m * n * RNN) * Embedding$: embedding dimensions of weight W_T for tensor input, *RNN*: recurrent neural network size, $OutputSize = 5$

view (see Fig. 3c) and distinguishes user types. Equation (1) describes the process. $G_t^i(m, n, :)$ stands for the hidden state at time t for user i with a $m \times n$ cell bounding grid. This grid monitors all neighboring users whose positions are within ego-user i's grid and sight of view, and also stores the user type information for the ego- and neighboring users. Here, user j is from set N_i containing all user i's neighbors within $G_t^i(m, n, :)$. $\text{View}_t^i(\text{pos}_t^j)$ is a binary function that filters the neighboring users based on their positions in ego-user i's sight of view at time t—a value one is assigned if the neighboring user is in the ego-user's sight of view, otherwise zero is assigned. $(\text{type}^i, \text{type}^j)$ stores the pairwise user type information for ego-user i and neighboring user j. In total, there are nine different pairwise user types and they are coded in distinctive numerical values and stored in the $m \times n$ cell.

Since we can easily differentiate these two features—user type and sight of view—in Eq. (1), it empowers us to build controlled experiments to analyze the incorporation of user type and/or sight of view into different models. In order to guarantee valid comparisons, all the models defined in Sect. 5.1 have the same dimensions as described here but only with different pooling values in the bounding grid.

$$G_t^i(m, n, :) = \sum_{j \in N_i} (\text{type}^i, \text{type}^j)[\text{View}_t^i(\text{pos}_t^j)]. \tag{1}$$

From the input layer to the hidden input layer in Fig. 1, the spatial input and tensor input are embedded separately with Rectified Linear Unit (ReLU) as depicted by Eq. 2. W_S and W_T stand for the embedding weights for the spatial input and the tensor input respectively.

$$S_t^i = \text{ReLU}(W_S \cdot (x_t^i, y_t^i)); \quad T_t^i = \text{ReLU}(W_T \cdot G_t^i). \tag{2}$$

The embedded spatial input S_t^i and the embedded tensor input T_t^i are concatenated to form a complete input for the LSTM cell. Equation (3) denotes the forward propagation. h_{t-1}^i is the hidden state at time $t-1$ and W stands for the corresponding weights for the LSTM.

$$h_t^i = \text{LSTM}[h_{t-1}^i, (S_t^i + T_t^i), W]. \tag{3}$$

We apply the same method to train our models as (Alahi et al. 2016), which was initially used in (Graves 2013). Depicted by Fig. 1, the initial output of the neural network is a 5-dimensional vector ($o\mu_x$, $o\mu_y$, $o\sigma_x$, $o\sigma_y$, and $o\rho$) learned at time t, which is used to predict the position of user i for the next time-step $t+1$ using a bivariate Gaussian distribution (see Eq. (4)). μ^i is a 2-dimensional vector for the arithmetic means of the respective distributions in x and y coordinates. σ^i is a 2-dimensional vector for the corresponding standard deviations, and ρ^i is the correlation.

$$(\hat{x}^i, \hat{y}^i)_{t+1} \sim \mathcal{N}(\mu^i, \sigma^i, \rho^i)_t, \tag{4}$$

where

$$\mu^i = (\mu_x^i, \mu_y^i) = (o\mu_x, o\mu_y), \tag{5}$$

$$\sigma^i = (\sigma_x^i, \sigma_y^i) = (\exp(o\sigma_x), \exp(o\sigma_y)), \tag{6}$$

$$\rho^i = \tanh(o\rho). \tag{7}$$

The cost between the predicted position and the target position (true position) is computed by a negative log–likelihood loss function using Eqs. (4) and (8), and the complete loss for the user is the sum of all the costs in predicted time-steps.

$$\text{Loss} = -\sum \log \Pr(x_{t+1}^i, y_{t+1}^i | \mu_t^i, \sigma_t^i, \rho_t^i), \tag{8}$$

where

$$\mathcal{N}(\mu^i, \sigma^i, \rho^i) = \frac{1}{2\pi\sigma_x^i\sigma_y^i\sqrt{1-(\rho^i)^2}} \exp\left[\frac{-Z}{2(1-(\rho^i)^2)}\right], \tag{9}$$

$$Z = \frac{(x_{true}^i - \mu_x^i)^2}{(\sigma_x{}^i)^2} + \frac{(y_{true}^i - \mu_y^i)^2}{(\sigma_y{}^i)^2} - \frac{2\rho(x_{true}^i - \mu_x^i)(y_{true}^i - \mu_y^i)}{\sigma_x^i\sigma_y^i}. \tag{10}$$

To avoid overfitting, least square errors (L2) are used as the regularization to penalize all the learned weights. Hence, the total loss is the sum of the Loss computed by Eq. (8) and L2, and is optimized using Stochastic Gradient Decedent.

4 Dataset and Evaluation Metrics

4.1 Dataset

In this paper, LSTM–based models are evaluated on a real–world dataset provided by Pascucci et al. (2017). The whole area of a shared space is close to a busy train station in the German city of Hamburg and the shared space of a street is 63 m long (see Fig. 2a). There were two cameras positioned at C1 and C2 with an elevation of 7 m towards both directions of the street for incoming vehicles and cyclists. Vehicles are allowed to drive at a maximum speed of 20 km/h with a priority over other types of road users in the shared zone. Meanwhile, pedestrians and cyclists are allowed to cross the street at any point from both sides of the street. However, the captured data shows that rather than strictly followed the regulation, vehicles, cyclists, and pedestrians negotiated the space spontaneously and often gave priority to each other to share the space. More details can be found in Pascucci et al. (2017).

In a 30 min video, there were 1115 pedestrian, 22 cyclist, and 338 vehicle (331 cars and 7 motorcycles) trajectories. Figure 2b shows the corresponding velocity distributions. This video was divided into 1800 time-steps with each time-step lasting 0.5 s. After calibration, all the trajectories were tracked manually and projected onto a 2D plane with the help of video analysis and modeling tool Tracker.[1] After preprocessing, each trajectory contains information of user positions with time-step and user type. The first 10 min (31% of the dataset) are saved as a test set and the last 20 min (69% of the dataset) are used as a training set. 20% of the trajectories in the training set are selected as a validation set for tuning the models. Please note that the number of trajectories were not perfectly evenly distributed and none of the users returned to the shared space in the 30 min video footage.

4.2 Evaluation Metrics

To measure the performance between the predicted and true trajectories of each model, we use four metrics as follows:

1. *Euclidean distance*—The measurement used here is similar to the one used in (Alahi et al. 2016) and (Pellegrini et al. 2009). It is the mean square error (MSE) over all predicted positions and true positions.
2. *Hausdorff distance*—Unlike the Euclidean distance that gives a pointwise average displacement error between each predicted trajectory and true trajectory, the Hausdorff distance measures the largest distance from the set of predicted positions (X_{pred}) to the set of true positions (X_{true}, see Eq. (11)) (Munkres 2000). It can more explicitly show how far a predicted trajectory deviates from the true trajectory and also gives less penalty than the Euclidean distance when errors

[1] http://physlets.org/tracker.

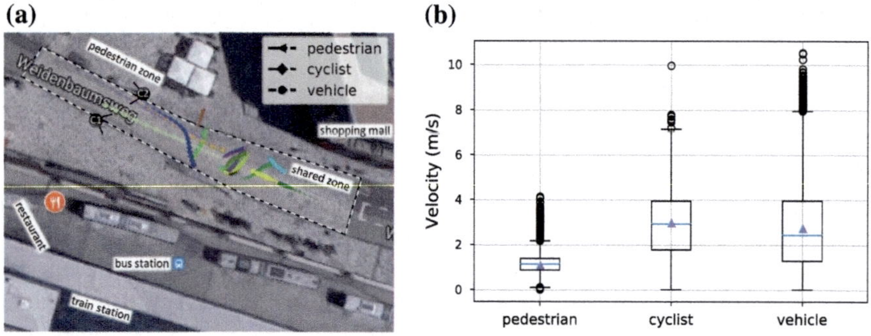

Fig. 2 **a** Layout of the shared space. Trajectories are denoted by color coded dot-lines with respective markers for different types of users. A color with larger size and opacity denotes a later time point. (Background image: Imagery ©2017 Google, Map data ©2017 GeoBasis–DE/BKG (©2009), Google); **b** Velocity distributions

are caused by time offsets. For example, in order to avoid collisions, the predicted trajectory for a user which depicts less accurate deceleration or acceleration compared with the true trajectory should be penalized less if the displacement error is small.

$$d_H(X_{\text{pred}}, X_{\text{true}}) = \max\{\sup_{x_{\text{pred}} \in X_{\text{pred}}} \inf_{x_{\text{true}} \in X_{\text{true}}} d(x_{\text{pred}}, x_{\text{true}})\}. \quad (11)$$

3. *Speed deviation*—Instead of measuring the MSE over all predicted positions and true positions, the speed deviation measures the MSE over all predicted speeds and true speeds in every time-step.
4. *Heading error*—This measurement computes the average degree for the angles between the predicted final heading directions and the true final heading directions over all the trajectories.

Altogether, these metrics allow comprehensive performance analyses for mixed traffic trajectory prediction in terms of positions, speeds, and heading directions.

5 Experiments

To analyze the contributions of incorporating user type and sight of view as described in Sect. 3, five LSTM-based models are tested on the aforementioned real-world dataset with the same configuration.

5.1 LSTM-Based Models

Table 1 lists the features which are used to feed each model respectively. Social-LSTM is the baseline model, which only considers the ego-user's position coordinates and corresponding pre-defined bounding grid. This model does not distinguish user types. In other words, pedestrians, cyclists, and vehicles are treated equally. It also does not consider the effect of the user's sight of view. The grid is the same for all four sides (right, left, front, and back, see Fig. 3a).

The average speeds of cyclists and vehicles in this shared space are about two times faster than the average pedestrian speed. Compared with cyclists and pedestrians, vehicles also occupy larger areas and generate bigger speed deviation (see Fig. 2b). Therefore, the bounding grid should be defined in a way that takes the type of ego-user into account. We call the corresponding model User-LSTM. To be more specific, vehicles and cyclists have twice the distance and 1.5–times the distance to the boundary of their bounding grid than pedestrians, respectively (see Fig. 3b). Given the reality that a user may have different levels of awareness regarding the type of neighbouring users (Rinke et al. 2017), UserType-LSTM not only defines a user-type aware bounding grid for the ego-user, but also accounts for the neighboring users' type. Therefore, the ego-user's interactions with different types of neighboring users are handled differently by this model.

However, the aforementioned models all have ego-user centralized grids. Unlike vehicles which are equipped with rear mirrors and sensors, pedestrians and cyclists normally do not have a good view of their back side. Humans have a maximum horizontal field of view of approximately 190° with two eyes, 120° of which make up the so-called binocular field of view (Henson 1993). As studied in personal space, people tend to preserve an elliptic protective zone around their body. Collision risks in front will be perceived higher than from the side (Gérin-Lajoie et al. 2005). Treating back and front sides of pedestrian or cyclist ego-users equally may lead to noisy information. Hence, on top of User-LSTM and UserType-LSTM, and for computational simplicity, we truncate the bounding grid according to the sight of view with 180° centralized towards the heading direction for pedestrian and cyclist ego-users (see Fig. 3c). The adjusted User-LSTM and UserType-LSTM are then called

Fig. 3 Bounding grids in different models: **a** *Social-LSTM*, **b** *User-LSTM/UserType-LSTM*, and **c** *View-LSTM/ViewType-LSTM*

Table 1 LSTM–based models

Model name	Input features
Social-LSTM (**baseline**)	Coordinates, bounding grid
User-LSTM	Coordinates, user-type aware bounding grid
UserType-LSTM	Coordinates, user-type aware bounding grid, user-type aware interaction
View-LSTM	Coordinates, user-view aware bounding grid
ViewType-LSTM	Coordinates, user-view aware bounding grid, user-type aware interaction

Table 2 Details of hyper–parameters

Training	Testing
Sequence length: 12 time-steps	Observed sequence: 6 time-steps
Mini–batch size: 16	Predicted sequence: 6 time-steps
Learning rate: 0.003	
Number of epochs: 100	

View-LSTM and ViewType-LSTM, respectively. It is worth mentioning that even the back side is treated passively from the ego-user's perspective for pedestrians and cyclists as this area is still in the sight of approaching users.

5.2 Setup

The experiments were performed on a PC with the Intel(R) Core(TM) i5-6600T CPU and 16 RAM using the framework of TensorFlow.[2] This can be optimized with a more powerful machine with GPU(s) in the future work.

To achieve an optimal configuration for all of the models, 20% of the trajectories in the training set are selected as a validation set to tune hyper–parameters (e.g. learning rate, mini-batch size, the lengths of observed and predicted trajectories), which play an important role in controlling the algorithm's behavior but cannot be directly learned through training (Goodfellow et al. 2016). Table 2 lists the values for the hyper–parameters that are applied in our experiments.

All the models observe six positions in historical trajectories as the input to predict the next six positions. In other words, the models observe 3 s trajectories and try to predict the trajectories of the next 3 s. This can be easily scaled up for longer term prediction by modifying the sequence parameters accordingly. In general, 2.4 s are sufficient for most drivers for a brake reaction (Taoka 1989). Hence, here we only report performances for the next 3 s prediction.

[2] https://www.tensorflow.org.

6 Results

6.1 Evaluation of the Models

Our assumption is that a model that mimics how a human sees and reacts to different transport modes in a shared space (ViewType-LSTM) can well predict human behavior. To validate this assumption, the performance of such a model is compared with other models (Social-LSTM, User-LSTM, UserType-LSTM, and View-LSTM) which do not or do not fully utilize human characteristics in this regard (see Sect. 5.1).

In many situations, road users make decisions based on narrow gaps between the approaching users. For example, a pedestrian may decide to continue crossing the street when the distance of an incoming vehicle is slightly above his or her expected safety distance. Hence, the evaluation metrics should be able to capture small but non-negligible differences of the models. For a close observation of how the models can be used for predicting trajectories of the next 3 s, here we take a look at average values of Euclidean distance, Hausdorff distance, speed deviation, and heading error between the true trajectories and the predicted trajectories (see Sect. 4.2).

From Table 3 we can see the average Euclidean distances from the predicted trajectories to the true trajectories for mixed traffic (all transport modes), pedestrians, cyclists, and vehicles, respectively. UserType-LSTM and User-LSTM generate larger Euclidean distances than the baseline model for all road users, and more profound errors for cyclists and vehicles. On the other hand, ViewType-LSTM gives the best performance, beating the baseline model and View-LSTM. The average Euclidean distance for all transport modes is reduced by 9%, from 0.93 to 0.85 m, for ViewType-LSTM compared with the baseline model. For vehicles, the Euclidean distance is reduced by 11%, from 1.15 to 1.02 m.

The differences of performances are more pronounced when measured by the Hausdorff distance. ViewType-LSTM reduces the error by 13%, from 1.30 to 1.13 m for all transport modes compared with the baseline. Similar improvements can be found for pedestrians, cyclists, and vehicles. However, UserType-LSTM and User-LSTM fall behind the baseline model remarkably.

The average speed deviation of the predicted trajectories to the true trajectories for ViewType-LSTM is 0.25 m/s for all transport modes, which is slightly smaller than the baseline model (0.26 m/s). But more profound improvements can be found for cyclists (ViewType-LSTM 0.34 m/s vs. baseline 0.37 m/s) and vehicles (ViewType-LSTM 0.37 m/s vs. baseline 0.41 m/s). Interestingly, the speed deviations for pedestrians are almost identical for View-LSTM, ViewType-LSTM, and the baseline. This can be explained by pedestrians traveling at a relatively slow and constant speed compared with cyclists and vehicles (see Fig. 2b). In the dataset we use, there are many more pedestrians than cyclists and vehicles (see Sect. 4.1). Therefore, the overall improvement measured by the speed deviation for all transport modes for ViewType-LSTM is not as profound as the ones measured by the Euclidean distance or the Hausdorff distance.

Table 3 Prediction errors for LSTM–based models. Euclidean and Hausdorff distances are measured by meter, speed deviation is measured by meter per second, and heading error is measured by degree. The best values are highlighted in boldface

Metrics	User type	*Social-LSTM*	*User-LSTM*	*UserType-LSTM*	*View-LSTM*	*ViewType-LSTM*
Avg. Euclidean distance (m)	Mixed	0.93	0.98	1.11	0.91	**0.85**
	Pedestrian	0.77	0.87	0.97	0.75	**0.71**
	Cyclist	1.08	1.18	1.23	1.08	**1.01**
	Vehicle	1.15	1.09	1.30	1.11	**1.02**
Avg. Hausdorff distance (m)	Mixed	1.30	1.44	1.65	1.32	**1.13**
	Pedestrian	1.24	1.41	1.66	1.24	**1.08**
	Cyclist	1.39	1.73	1.60	1.33	**1.25**
	Vehicle	1.48	1.56	1.74	1.52	**1.26**
Avg. speed deviation (m/s)	Mixed	0.26	0.27	0.31	0.26	**0.25**
	Pedestrian	**0.17**	0.20	0.22	**0.17**	**0.17**
	Cyclist	0.37	0.37	0.44	0.38	**0.34**
	Vehicle	0.41	0.40	0.47	0.41	**0.37**
Avg. heading error (°)	Mixed	32.72	31.99	38.91	31.68	**27.74**
	Pedestrian	36.79	38.81	45.78	36.44	**31.79**
	Cyclist	6.28	5.49	7.77	5.99	**5.09**
	Vehicle	26.39	**20.95**	28.33	24.90	22.28

The last lines in Table 3 show how far the predicted trajectories rotate from the true trajectories regarding final heading directions. The smallest average errors between the predicted and the true trajectories for all user types, pedestrians, and cyclists are again given by ViewType-LSTM. Interestingly, the best performance for vehicles is given by User-LSTM, which slightly outperforms the second best one—ViewType-LSTM. Overall, the heading errors are much smaller for cyclists than for the other user types across all models. This is caused by the small cyclist set and their similar behaviors (see Sect. 4.1).

To summarize, incorporating user types simply by extending bounding grids, i.e. by increasing the potential influence area for different user types cannot lead to a better performance. To the contrary, it can even degrade the model's performance by including noisy information, especially from the back side of road users. This is further proven by truncating the bounding grids regarding the sight of view for pedestrians and cyclists. Moreover, acknowledgement of neighboring users' transport modes along with sight of view can further boost the accuracy of predictions for the next 3 s trajectories.

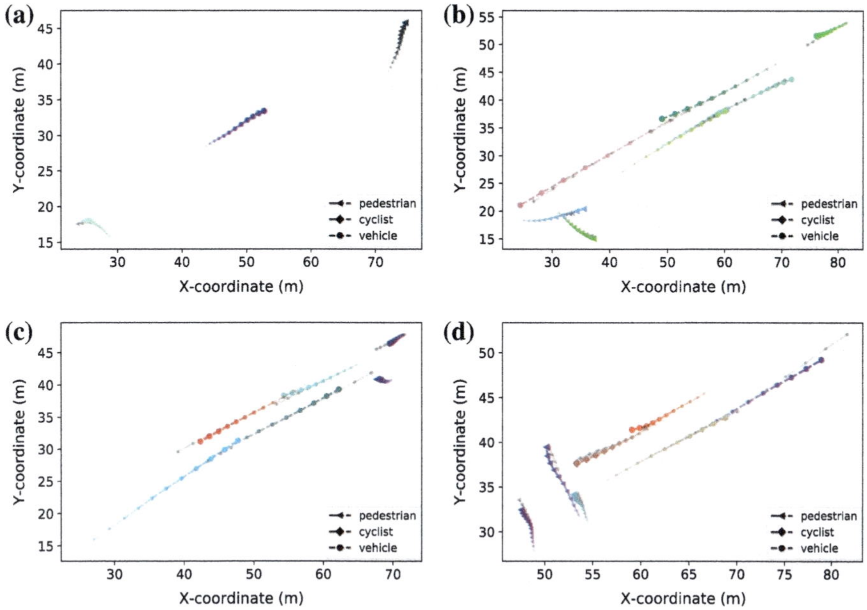

Fig. 4 Trajectory prediction in different situations: **a** Free flows of a vehicle and two pedestrians, **b** complicated situation with multiple users, **c** vehicles avoid an incoming pedestrian, **d** pedestrians avoid incoming vehicles and cyclist. True trajectories are denoted by black dot-lines with respective markers for different types of users. Predicted trajectories are color coded and a color with larger size and opacity denotes a later time point

6.2 Predicted Trajectories

In this section we show that ViewType-LSTM is able to mimic how a human sees and reacts to different transport modes in a shared space in the next 3 s and is also robust in complicated scenarios. Figure 4 shows different scenarios modeled by ViewType-LSTM.

From Fig. 4a we can see that the predicted trajectories for two pedestrians and one vehicle overlay their respective true trajectories. Since they are far from each other, their trajectories are barely impacted by interaction. On the lower left corner, ViewType-LSTM is able to correctly predict a left–turn for the pedestrian using a 3 s observed trajectory.

Figure 4b denotes a more complicated situation with multiple vehicles and pedestrians going in different directions. There is no collision in such busy mixed traffic. With only slight speed deviation and displacement for the upper right corner vehicle, predicted trajectories for the others are very close to the true trajectories.

Figure 4c, d depict different situations of how interactions happen between different road users and how ViewType-LSTM deals with potential collisions. The displacements from the predicted trajectories to the true trajectories in those two situations are barely noticeable, but most of them are caused by collision avoidance. In the upper right corner in Fig. 4c, there is a pedestrian waiting to cross the street. From the prediction, two approaching vehicles decelerate their speed to reduce the risks of hitting this pedestrian. On the lower left corner in Fig. 4d, three pedestrians are crossing the street. As a cyclist and a vehicle are approaching, ViewType-LSTM predicts a detour to the left side for the pedestrian who is very close to the incoming cyclist. It also predicts deceleration and slight left detours for the following two pedestrians to reduce the risks of potential collisions.

To more intuitively show how ViewType-LSTM can predict trajectories that have equal lengths as the observed trajectories, a scenario with mixed road users is depicted second by second in Fig. 5, in which a cyclist overtakes a vehicle from the right side to the left side after a pedestrian crossed the street in front of them. In this case, the prediction is also scaled up from a fixed length (six steps in 3 s) to a range of different lengths (1 s up to 6 s).

After only 1 s or 2 s observations, there are very few historical steps that can be referred to for the prediction. ViewType-LSTM, however, still predicts precise heading directions for each user, with average heading errors being 8.3° and 5.3°, respectively (see Fig. 5a, b).

After an appropriate length of observation (3 s), the performance for predicted trajectories is enhanced further. The predicted trajectories overlap their respective true trajectories (see Fig. 5c).

On the other hand, when the observed and predicted trajectory lengths are further increased to 4 and 5 s, the performances for the predictions of the cyclist and the pedestrian fall down. The reason is that, from the fifth second to the sixth second, both the cyclist and the pedestrian make a small right turn. Without observing the changes (the observed time point is only up to the fifth second), ViewType-LSTM keeps predicting straight trajectories for them (see Fig. 5d, e).

When the observation is extended to 6 s, the changes mentioned above are detected by ViewType-LSTM. In response to the changes, ViewType-LSTM calibrates the predicted trajectories to the right side for the cyclist and the pedestrian. In addition, the deceleration of the vehicle at later time points is also predicted by ViewType-LSTM, with the error for its speed deviation being 0.32 m/s (see Fig. 5f).

In summary, ViewType-LSTM generates reasonable and collision–free predictions in mixed traffic. Even with little information, it can estimate precise heading directions of road users. Moreover, ViewType-LSTM can be easily scaled up for longer term (e.g. up to 6 s) trajectory prediction. However, it is difficult to decide appropriate lengths for observed and predicted trajectories. A long observation for a short prediction might not be feasible in real–world trajectory prediction, but a short

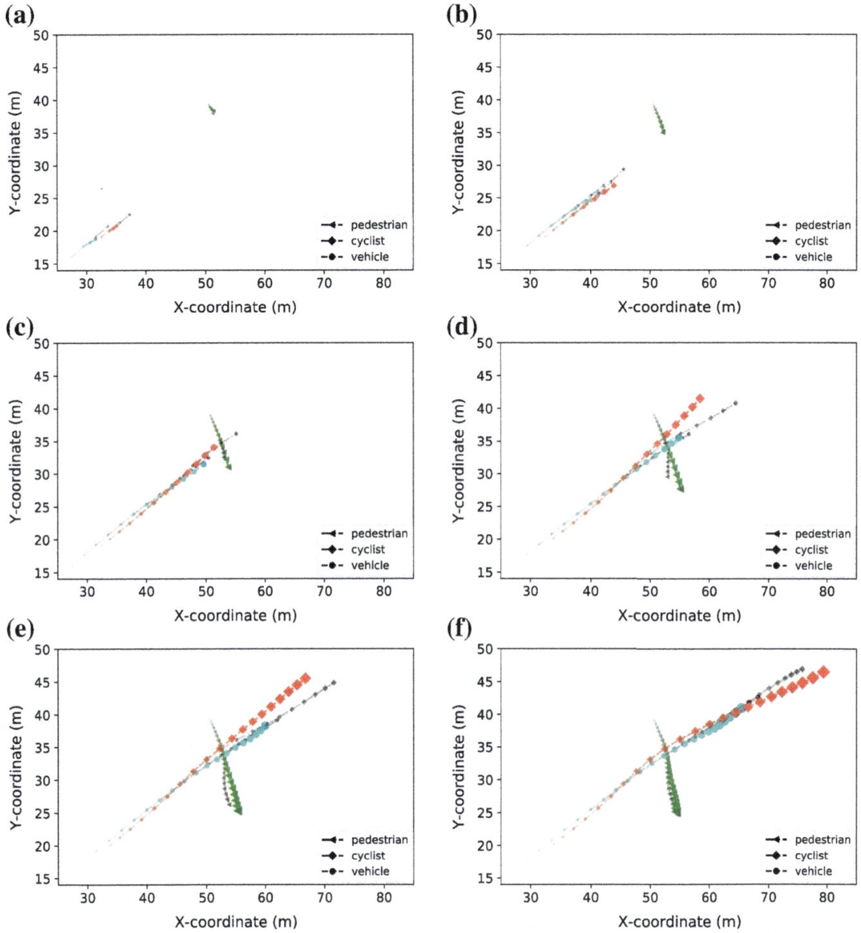

Fig. 5 Predictions of future trajectories that have equal lengths as the observed trajectories from 1 to 6 s: **a–f** Observing 1 s and predicting 1 s trajectories to observing 6 s and predicting 6 s trajectories, respectively. True trajectories are denoted by black dot-lines with respective markers for different types of users. Predicted trajectories are color coded and a color with larger size and opacity denotes a later time point

observation for a long prediction may fail to handle sudden changes made by road users at a later time. Hence, finding optimal observation and prediction lengths needs to be further investigated in future work.

7 Conclusion and Future Work

In this work we showed that LSTM–based models are capable of mixed traffic trajectory prediction in shared spaces. Spatio–temporal features—coordinates, sight of view, and interactions between different types of neighboring users—are encoded to mimic how a human sees and reacts to different transport modes. Instead of manual settings, LSTM–based models can be trained using real–world data for complicated traffic situations and can be easily scaled up for long term trajectory prediction.

In addition to the Spatio–temporal features mentioned above, user behaviors in shared spaces are also impacted by environment and context. An online survey shows that context– and design–specific factors significantly impact the comfort perceived by pedestrians and the willingness of car drivers to share road resources with others in shared spaces (Kaparias et al. 2012). Investigation of context–aware behavior modeling in shared spaces is a promising direction to further increase the accuracy of mixed traffic prediction.

Moreover, in order to extend our models on multiple and more balanced mixed trajectories in shared spaces with divergent space layouts, and make such data available for other studies, object detection and deep learning trajectory tracking techniques (i.e. Fully-Convolutional Siamese Networks, (Bertinetto et al. 2016)) will largely be employed for data acquisition and pre-processing procedures in our future work.

Acknowledgements The authors cordially thank the funding provided by DFG Training Group 1931 for SocialCars and the participants of the research project MODIS (Multi mODal Intersection Simulation) for providing the dataset of road user trajectories used in this work.

References

Alahi A, Goel K, Ramanathan V, Robicquet A, Fei-Fei L, Savarese S (2016) Social LSTM: human trajectory prediction in crowded spaces. In: Proceedings of the IEEE conference on computer vision and pattern recognition, pp 961–971

Andrienko G, Andrienko N, Bak P, Keim D, Kisilevich S, Wrobel S (2011) A conceptual framework and taxonomy of techniques for analyzing movement. J Vis Lang Comput 22(3):213–232

Bertinetto L, Valmadre J, Henriques JF, Vedaldi A, Torr PH (2016) Fully-convolutional Siamese networks for object tracking. In: European conference on computer vision. Springer, pp 850–865

Clarke E (2006) Shared space: the alternative approach to calming traffic. Traffic Eng. Control 47(8):290–292

Gérin-Lajoie M, Richards CL, McFadyen BJ (2005) The negotiation of stationary and moving obstructions during walking: anticipatory locomotor adaptations and preservation of personal space. Motor control 9(3):242–269

Goodfellow I, Bengio Y, Courville A (2016) Deep learning. MIT Press. http://www.deeplearningbook.org

Graves A (2013) Generating sequences with recurrent neural networks. arXiv:13080850

Graves A, Jaitly N (2014) Towards end-to-end speech recognition with recurrent neural networks. In: Proceedings of the 31st international conference on machine learning (ICML-14), pp 1764–1772

Helbing D, Molnar P (1995) Social force model for pedestrian dynamics. Phys Rev E 51(5):4282

Henson DB (1993) Visual fields. Oxford Medical Publications, Butterworth-Heinemann Ltd (1772)
Kaparias I, Bell MG, Miri A, Chan C, Mount B (2012) Analysing the perceptions of pedestrians and drivers to shared space. Trans Res part F Traffic Psychol Behav 15(3):297–310
Karndacharuk A, Wilson DJ, Dunn R (2014) A review of the evolution of shared (street) space concepts in urban environments. Trans Rev 34(2):190–220
Lerner A, Chrysanthou Y, Lischinski D (2007) Crowds by example. In: Computer graphics forum, vol 26, no 3. Wiley Online Library, pp 655–664
Long JA, Nelson TA (2013) A review of quantitative methods for movement data. Int J Geogr Inf Sci 27(2):292–318
Morris B, Trivedi M (2009) Learning trajectory patterns by clustering: experimental studies and comparative evaluation. In: 2009 IEEE conference on computer vision and pattern recognition CVPR 2009. IEEE, pp 312–319
Munkres JR (2000) Topology. Prentice Hall
Pascucci F, Rinke N, Schiermeyer C, Friedrich B, Berkhahn V (2015) Modeling of shared space with multi-modal traffic using a multi-layer social force approach. Trans Res Procedia 10:316–326
Pascucci F, Rinke N, Schiermeyer C, Berkhahn V, Friedrich B (2017) A discrete choice model for solving conflict situations between pedestrians and vehicles in shared space. arXiv:170909412
Pelekis N, Kopanakis I, Kotsifakos EE, Frentzos E, Theodoridis Y (2011) Clustering uncertain trajectories. Knowl Inf Syst 28(1):117–147
Pellegrini S, Ess A, Schindler K, Van Gool L (2009) You'll never walk alone: modeling social behavior for multi-target tracking. In: 2009 IEEE 12th international conference on computer vision. IEEE, pp 261–268
Reid S (2009) DfT shared space project stage 1: appraisal of shared space. MVA Consultancy
Rinke N, Schiermeyer C, Pascucci F, Berkhahn V, Friedrich B (2017) A multi-layer social force approach to model interactions in shared spaces using collision prediction. Trans Res Procedia 25:1249–1267
Schiermeyer C, Pascucci F, Rinke N, Berkhahn V, Friedrich B (2016) A genetic algorithm approach for the calibration of a social force based model for shared spaces. In: Proceedings of the 8th international conference on pedestrian and evacuation dynamics (PED)
Schönauer R, Stubenschrott M, Huang W, Rudloff C, Fellendorf M (2012) Modeling concepts for mixed traffic: steps toward a microscopic simulation tool for shared space zones. Trans Res Rec: J Trans Res Board 2316:114–121
Taoka GT (1989) Brake reaction times of unalerted drivers. ITE J 59(3):19–21
Trautman P, Ma J, Murray RM, Krause A (2013) Robot navigation in dense human crowds: the case for cooperation. In: 2013 IEEE international conference on robotics and automation (ICRA). IEEE, pp 2153–2160
Wang X, Jiang R, Li L, Lin Y, Zheng X, Wang FY (2017) Capturing car-following behaviors by deep learning. IEEE Trans Intell Trans Syst
Yamaguchi K, Berg AC, Ortiz LE, Berg TL (2011) Who are you with and where are you going? In: 2011 IEEE conference on computer vision and pattern recognition (CVPR). IEEE, pp 1345–1352
Yi S, Li H, Wang X (2016) Pedestrian behavior understanding and prediction with deep neural networks. In: European conference on computer vision. Springer, pp 263–279

A Framework for the Management of Deformable Moving Objects

José Duarte, Paulo Dias and José Moreira

Abstract There is an emergence of a growing number of applications and services based on spatiotemporal data in the most diverse areas of knowledge and human activity. The representation of the continuous evolution of moving regions, i.e., entities (or objects) whose position, shape and extent change continuously over time, is particularly challenging and the methods proposed in the literature to obtain such representation still present some issues. In this paper we present a framework for moving objects, in particular, moving regions, that uses the concept of mesh, i.e., a triangulated polygon, compatible triangulation and rigid interpolation methods to represent the continuous evolution of moving regions over time. We also present a spatiotemporal database extension for PostgreSQL that uses this framework and that allows to store moving objects data in a PostgreSQL database and to analyze and manipulate them using SQL. This extension can be smoothly integrated with PostGIS. Experiments show that our framework works with real data and provides a base for further work and investigation in this area.

Keywords Moving objects · Spatiotemporal data management
Morphing

1 Introduction

Currently, there are many tools and technologies able to monitor and record the evolution of real-world phenomena over time, such as: (1) satellite or aerial images tracking the movement of icebergs at the Antarctic, the propagation of forest fires or coastal erosion, and (2) video microscopy and fluorescence microscopy recording the evolution of the size and shape of cells over time. However, the modelling, management and processing of spatiotemporal data are complex tasks. Despite recent advances, the support provided by existing solutions to perform these tasks is

J. Duarte (✉) · P. Dias · J. Moreira
DETI/IEETA, Universidade de Aveiro, Aveiro, Portugal
e-mail: hfduarte@ua.pt

A. Mansourian et al. (eds.), *Geospatial Technologies for All*,
Lecture Notes in Geoinformation and Cartography,
https://doi.org/10.1007/978-3-319-78208-9_17

insufficient and does not cover the whole spectrum of potential applications. It is often required to implement time-consuming and complex programs tailored to solve a specific problem from a particular domain, which cannot be easily applied to other problems and these custom solutions often have limited functionality because development is difficult and costly. Thus, the study of general solutions to represent, manage and process large spatiotemporal datasets has the potential to enable an easier implementation of those programs and boost the development of applications in several domains, e.g., environmental and climate sciences, agriculture and medical biology.

The databases research community is working on spatiotemporal data models and query languages for the management of spatiotemporal data, including the representation of moving objects, i.e., objects whose position or shape change continuously over time. Their main focus is on efficient data management and on topological issues concerning the representation of spatiotemporal data in databases. However, the methods proposed do not always obtain a realistic representation of the actual evolution of the spatiotemporal phenomena (Moreira et al. 2016). On the other hand, there are numerous works on morphing techniques, e.g., Alexa et al. (2000) and Liu et al. (2015), to represent the transformation of planar shapes, e.g., polygons and polylines, over time. Their focus is on achieving a continuous representation of the spatial transformations of a shape that is smooth and realistic at all times. Thus, the use of morphing techniques has the potential to enable obtaining representations of the evolution of spatiotemporal phenomena that are more realistic than the corresponding representations obtained using the techniques proposed in the spatiotemporal databases literature.

This paper presents the design, architecture and implementation of a spatiotemporal data management framework. The data model and operations that it implements follow the guidelines proposed in Güting et al. (2000), Forlizzi et al. (2000) and Pelekis et al. (2006). The novelty that it introduces is the use of the concept of mesh to represent the evolution of moving regions over time. A mesh is constructed using compatible triangulation techniques (Gotsman and Surazhsky 2004) and its evolution is represented using the rigidity preserving interpolation method proposed in Alexa et al. (2000) and Baxter et al. (2008).

The framework has two components: (1) a library that is independent of any client using it and that provides a framework to represent, analyze and manipulate moving objects, in particular moving regions, and (2) a spatiotemporal database extension for PostgreSQL that allows moving objects to be stored in a PostgreSQL database and analyzed and manipulated in a convenient way using the Structured Query Language (SQL).

This paper is organized as follows. Section 2 presents an overview on spatiotemporal data models and query languages. Section 3 presents the compatible triangulation method, the interpolation method and the concepts on continuous and discrete models proposed in the spatiotemporal databases literature that are used in this work. Section 4 presents the implementation details of a library to represent, analyze and manipulate moving regions that uses triangulation-based morphing techniques. Section 5 presents a spatiotemporal database extension for PostgreSQL that uses the library presented in Sect. 4. Section 6 presents the experimental results obtained when using the proposed framework and Sect. 7 presents the conclusions and guidelines for future work.

2 Related Work

The most well-known data model and query language for representing and querying moving objects, i.e., spatiotemporal data about entities (or objects) whose position or shape and extent change continuously over time, uses Abstract Data Types (ADTs) (Güting et al. 2000; Cotelo Lema et al. 2003). ADTs can be smoothly built into extensible Database Management Systems (DBMSs), such as the Secondo prototype (Güting et al. 2010), and object-relational DBMSs (Pelekis et al. 2006; Matos et al. 2012; Zhao et al. 2011). Moving objects, e.g., moving points, moving lines and moving regions, are represented using the *sliced representation* proposed in Forlizzi et al. (2000), i.e., they are represented as an ordered collection of units where each unit represents the evolution (i.e., the changes in position or shape and extent) of the moving object between two consecutive observations. Therefore, methods are required to model the spatial behavior of moving objects between observations. These methods should provide a realistic approximation of the actual evolution of the moving object at all times. Additionally, storage and computation time should be low to enable the processing of big datasets.

The first algorithm proposed in the literature to create the so-called moving regions from observations in spatiotemporal databases is presented in Tøssebro and Güting (2001). This algorithm is used and extended in recent works (Mckenney and Webb 2010; Mckennney and Frye 2015; Heinz and Güting 2016). However, the methods using this algorithm produce important deformations of the geometries estimated during interpolation and the approximation errors are too big to be neglected in scientific work, namely: (1) if a rotation exists, the geometries tend to inflate at the middle point of the interpolation, and (2) the methods used to deal with concavities either do not perform well with noisy data (Tøssebro and Güting 2001; Heinz and Güting 2016; Moreira et al. 2016) or make the shapes approximately convex during interpolation, causing deformation (Mckenney and Webb 2010; Mckennney and Frye 2015).

In this work we propose the use of morphing techniques (Alexa et al. 2000; Baxter et al. 2008) to represent the evolution of moving regions and eliminate these issues.

3 Background

This section presents the main methods and concepts used in this work. First we present the data models that the proposed framework has as a reference. Then we present the compatible triangulation and interpolation methods that it uses.

3.1 Spatiotemporal Data Models

The framework proposed in this work uses the data models presented in Güting et al. (2000) and Forlizzi et al. (2000) as a reference. In Güting et al. (2000) the authors present an abstract data model to represent and query moving objects and in Forlizzi et al. (2000) the authors present a discrete data model to implement the model proposed in Güting et al. (2000). This discrete model proposes the following type constructors: the base types *int*, *real*, *string* and *bool*, the spatial types *point*, *points*, *line* and *region*, the *instant* type, the *range* type, the unit types *ureal*, *upoint*, *upoints*, *uline* and *uregion* and the *mapping* type. Type *instant* represents a point in time, *range* represents sets of pairwise disjoint intervals (in the set base types ∪ *instant*), the unit types represent the continuous evolution of an entity (or object) during a time interval, e.g., the position and shape changes of an iceberg, and *mapping* assembles sets of units according to the constraints that have been defined. It also proposes the concept of *sliced representation* used to represent the moving types. The *sliced representation* (Fig. 1) decomposes the development (or evolution) of a value into fragments called slices (or units) and the evolution within a unit is described by a function. So, a unit is a pair $(I, v(t))$ where I is the time interval where the unit is defined and $v(t)$ is a function that gives the value of the unit during I.

3.2 Morphing

The representation of the evolution of a moving region during a unit is given by a function that should preserve the physical characteristics of the object (or event) being represented and generate only valid geometries. That is, we are interested in an as realistic as possible representation of the continuous evolution of moving regions over time. Several works in the literature propose solutions to achieve this (Tøssebro and Güting 2001; Heinz and Güting 2016; Mckenney and Webb 2010; Mckennney and Frye 2015). These works use the concept of moving segments to interpolate the evolution of a moving region during a unit. They provide good results but they have some issues as discussed in Sect. 2. In Amaral (2015) the

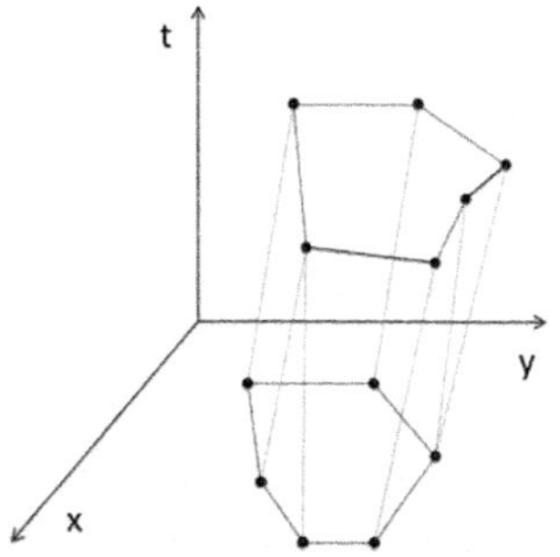

Fig. 1 Sliced representation

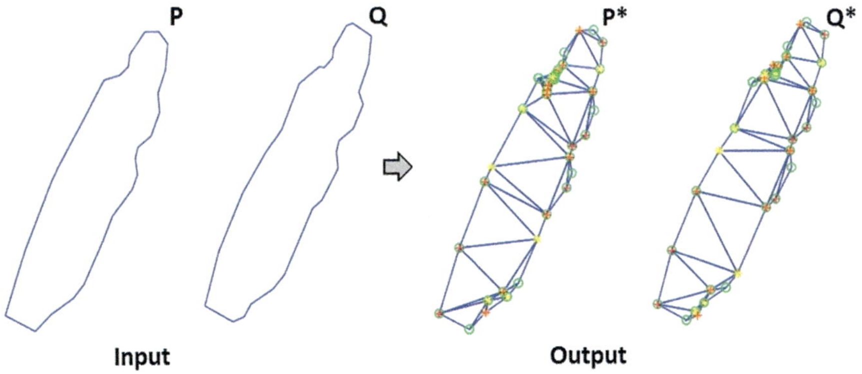

Fig. 2 Compatible triangulation method input (*left*) and output (*right*)

author uses the compatible triangulation method proposed in Gotsman and Surazhsky (2004) and the rigid interpolation method proposed in Alexa et al. (2000) and Baxter et al. (2008) to represent the evolution of moving regions over time, i.e., to solve the Region Interpolation Problem. The use of these methods implies that a region is represented by a mesh (Fig. 2) and they are used to obtain a more realistic representation of the continuous evolution of moving regions over time.

When using these methods, there are two main steps to compute the evolution of a moving region during a unit. In step 1 we find a compatible triangulation between the polygons (observations) representing the moving region at the begin and end instants of the unit's interval. The method proposed in Gotsman and Surazhsky (2004) receives two polygons, a source, P, and a target, Q, polygons and triangulates them generating two meshes, P* and Q*. P and Q must have a one-to-one correspondence between their vertices and P* and Q* have a one-to-one correspondence between their triangles (Fig. 2). This method is known to have complexity $O(N^3 \log N)$ (Liu et al. 2015), where N is the number of boundary vertices of the source polygon.

In step 2 a rigid interpolation method is used to compute the transformation of the moving region during a unit. In the method proposed in Alexa et al. (2000) and Baxter et al. (2008) the affine transformation of a triangle is decomposed into a product of three matrices given by the Single Value Decomposition (SVD). Since there is an affine matrix for each triangle, shared vertices may have more than one position during a transformation. To ensure that their position is the same during transformations, Alexa et al. (2000) proposes a solution based on the least squares method, while Baxter et al. (2008) proposes a solution based on normal equations. Indeed, the two methods are equivalent, but the latter presents an elegant formulation that allows estimating the position of the vertices using a linear system of the form $V(t) = -H^{-1}G(t)$, where $V(t)$ is a vector denoting the position of the vertices of a mesh at time t, and H and G are matrices calculated using SVD and normal equations. Briefly, the rigid interpolation method receives two meshes, a source, P*, and a target, Q*, with a one-to-one correspondence between their triangles and computes the interpolation components used to transform (morph) mesh P* into Q*. Each triangle of P* will be rotated, scaled, translated and transformed to its corresponding triangle in Q*, using a continuous function of time.

4 A Framework for Moving Objects

We implemented a framework for moving objects, in particular, for moving regions, that is independent of any client or application using it, called SPTMesh. SPTMesh is a C++ library that enables the analysis and manipulation of moving objects. It implements the compatible triangulation and rigid interpolation methods presented in Sect. 3.2. Introduces a new spatial type called *mesh*, uses the architecture of the GEOS[1] C++ library as a reference and Armadillo (Sanderson and Curtin 2016) to perform linear algebra operations. Table 1 shows its external dependencies.

The set of data types implemented by SPTMesh has as its base the types proposed in Forlizzi et al. (2000) with the following changes. (1) SPTMesh introduces the following new types: the spatial type *mesh* and the types *umesh*, *function* and *mmesh*. Type *mesh* is a triangulated polygon used to represent a region, *umesh* represents the evolution of a *mesh* during a time interval, *function* represents a function in the mathematical sense and *mmesh* represents the evolution of a *mesh* over time. (2) The spatial types implemented in SPTMesh (except for the *mesh* type) are compliant with the Open Geospatial Consortium (OGC) standards for spatial objects. (3) It defines the types *interval* and *period* instead of the type *range* and defines the MOVING type instead of the *mapping* type constructor. (4) SPTMesh does not consider lines, collections and regions with holes and it uses the GEOS C++ library for spatial operations.

4.1 Data Structures

Table 2 presents the SPTMesh data structures used to implement the data types that it considers. Figure 3 shows the SPTMesh type system.

Types *mesh* and *umesh* are the most important data types in SPTMesh. A *mesh* has a boundary, a set of interior points, i.e., the Steiner[2] points added by the compatible triangulation method, and a set of non-overlapping triangles. The Mesh class uses two vectors to hold the mesh's boundary and Steiner points, respectively, and a vector to hold the indices of the points of the mesh's triangles.

[1]http://geos.osgeo.org/.

[2]Points added to the original geometry to help in finding a compatible triangulation.

Table 1 SPTMesh external dependencies

Dependency	Usage
Armadillo	Linear algebra operations
BLAS,[a] LAPACK[b]	Armadillo dependencies to provide various matrix decompositions, e.g., SVD
GEOS	Manipulation and analysis of 2D geometries in the Cartesian plane and spatial operations

[a]http://www.openblas.net/
[b]http://www.netlib.org/lapack/

Table 2 SPTMesh data structures

Data type(s)	Data structure implementation
BASE	Types *int*, *real* and *bool* are implemented using the C++ *int*, *double* and *bool* data types
instant	Type *instant* is implemented using the C++ *long long* data type (we are not using dates with locales and time zones, yet)
SPATIAL	Type *mesh* is implemented in class Mesh. The other SPATIAL types: *point*, *linestring*, *polygon*, *multipoint*, *multilinestring*, *multipolygon* and *geometrycollection* are provided by the GEOS C++ library
interval, *period*	Type *interval* represents intervals of values, e.g., a time interval in the form $[instant_i, instant_j[$ with $instant_i < instant_j$, and *period* represents sets of intervals. They are implemented in the template classes Interval and Period, respectively. Interval supports the 4 types of intervals but we only use closed-open intervals
function	Type *function* represents functions in the mathematical sense. It considers the constant function, $F = c$, the linear function, $F = bx + c$ and the quadratic function, $F = ax^2 + bx + c$. It is implemented in the UnitFunction class
UNIT	UNIT types represent the evolution of moving objects (points and regions) and the changes of bool (true or false) and real values during a time interval (between two consecutive observations). They are implemented in the classes: UnitBool, UnitReal, UnitPoint and UnitMesh
MOVING	MOVING types represent moving objects and values that change continuously (*mreal*) and in discrete steps (*mbool*) over time. They are implemented in the classes: MovingBool, MovingReal, MovingPoint and MovingMesh

Type *umesh* is defined as the set of tuples $\{(I, t_b, t_e, c, P^*, Q^*, pStar, rScale, rGamma) \mid I \in$ Interval, $t_b, t_e \in long$, $c \in$ UnitPoint, $P^*, Q^* \in$ Mesh, $pStar, rScale \in mat$, $rGamma \in vector$ <*double*>}, such that:

- I is the time interval in which the unit is defined.
- t_b and t_e represent the begin and end instants of the original time interval, respectively. They are an implementation detail that we do not discuss in this paper.
- c is a UnitPoint that represents the evolution of the position of the mesh's centroid during I.

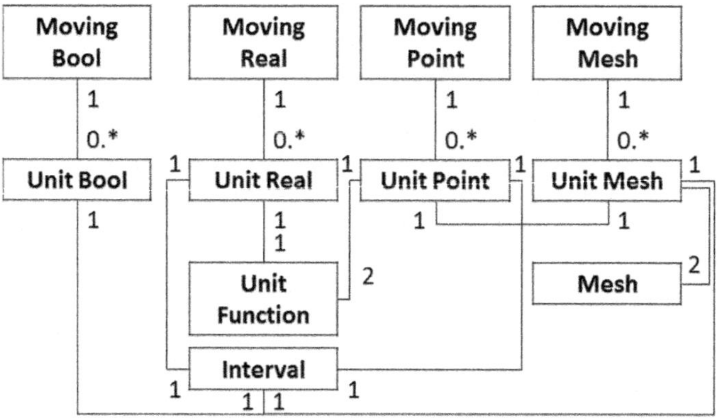

Fig. 3 The SPTMesh type system

- P^* and Q^* represent the original source and target meshes at the begin and end instants of the original interval, respectively. P^* and Q^* must have a one-to-one correspondence between their triangles.
- *mat* is an Armadillo data type that represents a matrix.
- *vector* is the C++ standard library vector data type.
- *pStar*, *rScale* and *rGamma* are the interpolation components, computed by the interpolation method, used to interpolate the mesh during *I*. *pStar* holds the mixed and pure quadratic coefficients of the quadratic form which minimizes the quadratic error between the actual transformation and the individual ideal transformation of a triangle (see Sect. 3.2), *rScale* is the mesh's triangles global scale-shear components matrix and *rGamma* is the list of the mesh's triangles global rotation angles. That is, *pStar* represents the matrix *H* and *rScale* and *rGamma* are used to compute the matrix *G*(*t*) presented in Sect. 3.2.

4.2 Operations on Moving Types

SPTMesh implements only a small subset of the spatiotemporal operations proposed in Güting et al. (2000) (Table 3). We chose to implement a set of operations

Table 3 Operations on MOVING types defined in SPTMesh

Class of operation	Operation
Predicates	Equals, intersects
Set operations	Intersection
Numeric	Area
Projection to domain and range	deftime
Interaction with domain and range	atinstant, atperiod, present
Constructors	Unit, moving

according to the following criteria: (1) implement atomic operations, i.e., operations that can be used as building blocks to implement other operations, (2) use all the implemented moving types and (3) implement operations from the different classes of operations proposed in the literature.

4.3 *Continuity for the Unit Types*

Two units can meet at an instant. To avoid situations like instantaneous appearance and disappearance or instantaneous growing and shrinking we need to establish continuity for the UNIT types. We establish continuity for UnitPoint and UnitMesh as follows:

- Two UnitPoint objects, u and v, are continuous if the distance between their positions at the instant, t, where they meet is less than or equal to a ξ, i.e., distance $(u, v, t) \leq \xi$. SPTMesh defines $\xi = 0.00001$. This value was not strictly established and needs to be validated.
- Two UnitMesh objects are continuous if distance$(m_i, m_j, t) \leq \xi_p$ (1) $\wedge$ $\frac{\text{area}\,(m_i, m_j, \cap, t)}{\text{area}\,(m_i, m_j, \cup, t)} \leq \delta_s$ (2), $m_i, m_j \in$ *umesh*, $t \in$ *instant*. That is, if the distance between their centroids is less or equal to a constant value (1) and the ratio between the intersection and the union of their areas is less or equal to some other constant value (2), at the time instant, t, where they meet. SPTMesh defines $\xi_p = 0.5$ and $\delta_s = 0.95$. These values were empirically set during the tests phase using real data. However, given the limited size of our dataset the constants and formulas used to establish continuity need further evaluation and may need to be adapted for larger datasets.

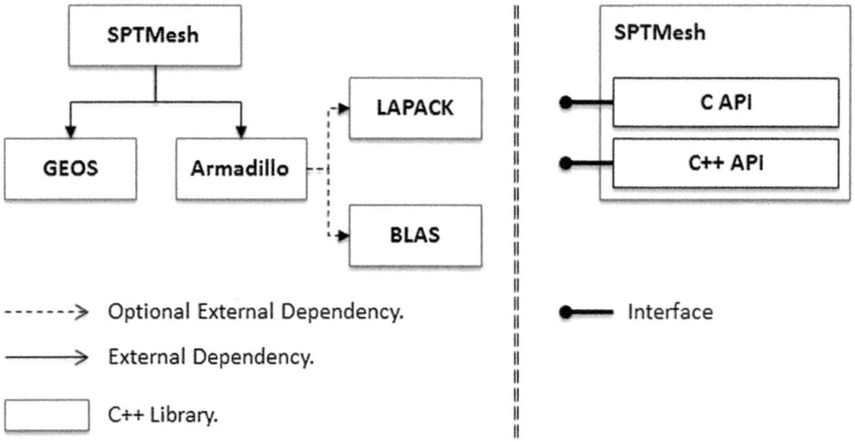

Fig. 4 SPTMesh architecture (*left*) and APIs (*right*)

We do not establish continuity for UnitBool because it represents a value that changes in discrete steps over time. Continuity for UnitReal may be established in the future.

4.4 Architecture

The SPTMesh architecture has as a reference the GEOS C++ library architecture (Fig. 4). SPTMesh provides a C Application Programming Interface (API) on top of a C++ implementation.

4.5 Spatiotemporal Well-Known Text Form

SPTMesh provides a Spatiotemporal Well-Known Text (STWKT) form for the UNIT and MOVING types (Tables 5 and 6). The STWKT provides a convenient way of expressing unit and moving objects and it is particularly useful to create moving objects and to update them with new evolutions. The STWKT form was inspired by the OGC standard Well-Known Text (WKT) form for spatial objects.

This section uses the notation presented in Table 4.

5 A Spatiotemporal Database Extension for PostgreSQL

We implemented a spatiotemporal database extension for PostgreSQL called MeshGIS. MeshGIS is a C library that uses SPTMesh to analyze and manipulate moving objects. It allows the moving objects provided by SPTMesh to be stored in a PostgreSQL database and manipulated using SQL.

Table 4 Notation

Symbol	Description
t_b, t_e $\in$ *instant*	Begin and end instants of an interval
$v \in$ *bool*	A value in {*true*, *false*}
v_b, v_e $\in$ *double*	Begin and end values, respectively
type, *typeX*, *typeY* $\in$ *int*	Type of interpolation function and type of interpolation function for the x and y components, respectively. An interpolation function can be of the following types: constant, linear and quadratic. An interpolation function is represented by the *function* data type
x_b, y_b, x_e, y_e $\in$ *double*	Coordinates of a moving point at the begin and end instants of the unit's interval
x_i y_i $\in$ *double*	The x and y coordinates of a vertex

Table 5 STWKT form for the UNIT types

Data type	STWKT form
ubool	UNITBOOL(t_b t_e v)
ureal	UNITREAL(t_b t_e v_b v_e *type*)
upoint	UNITPOINT(t_b t_e x_b y_b x_e y_e *typeX typeY*)
umesh	UNITMESH(t_b t_e, (x_1 y_1, …, x_n y_n), (x'_1 y'_1, …, x'_n y'_n))

Table 6 STWKT form for the MOVING types

Data type	STWKT form
mbool	MOVINGBOOL(UNITBOOL_1, …, UNITBOOL_n)
mreal	MOVINGREAL(UNITREAL_1, …, UNITREAL_n)
mpoint	MOVINGPOINT(UNITPOINT_1, …, UNITPOINT_n)
mmesh	MOVINGMESH(UNITMESH_1, …, UNITMESH_n)

5.1 Data Structures

The PostgreSQL backend is implemented in C. It may, however, be possible to load functions written in other languages into PostgreSQL and we can write PostgreSQL extensions using C++, if certain guidelines[3] are followed. PostgreSQL also supports procedural languages,[4] e.g., PL/Python, but they have several significant limitations.[5] Following the examples of PostGIS[6] and Hermes (Pelekis et al. 2006) and in order to avoid unnecessary complexity we decided to implement MeshGIS as a C library.

The MeshGIS data structures used to represent SPTMesh types are defined as C typedef structs and they are presented in Table 7.

MeshGIS also defines data structures to represent the interpolation components (Table 8). This is to be changed in the future so that MeshGIS is decoupled from interpolation method details.

MeshGIS defines operations that use all the operations implemented in SPTMesh for the moving types, namely constructors, operations to add and delete units from moving objects, operations to test and obtain the intersection of two moving regions at an instant and operations to get a moving region at an instant or period.

We provide SQL operations and types that bind to the operations and types provided by MeshGIS. The SQL types provided are: MovingBool, MovingReal, MovingPoint and MovingMesh.

[3]https://www.postgresql.org/docs/9.4/static/xfunc-c.html.

[4]https://www.postgresql.org/docs/9.4/static/xplang.html.

[5]https://www.postgresql.org/docs/9.4/static/extend.html.

[6]http://postgis.net/.

Table 7 MeshGIS data structures used to represent SPTMesh types

Data structure	Description
ArrayOfX	A generic array to hold units or other elements, e.g., UnitReal, UnitBool, UnitPoint and Matrix 2×2
UnitFunction	A SPTMesh UnitFunction
UnitInterval	A SPTMesh Interval
UnitBool	A SPTMesh UnitBool
UnitReal	A SPTMesh UnitReal
UnitPoint	A SPTMesh UnitPoint
UnitMesh	A SPTMesh UnitMesh
SerializedPostgreSQLObject	This data structure is used to send and retrieve moving objects to/from PostgreSQL for storage, analysis and manipulation, respectively. It has the same structure as the GSERIALIZED data structure. GSERIALIZED is a serialized form used primarily by PostGIS. It is a PostgreSQL data type for variable size user-defined data types
SerializedMovingObject	An abstract data type that represents any type of moving object
SerializedMovingX	Represents the SPTMesh MovingBool, MovingReal and MovingPoint types
SerializedMovingMesh	A SPTMesh MovingMesh

Table 8 MeshGIS data structures used to represent the interpolation components

Data structure	Description
Matrix 2×2	Holds the symmetric matrix in the decomposition used to find the rotation and the scale-shear components of a triangle
Matrix 2×3	Holds the coordinates' transformation matrix of a triangle
Triangle	Holds the vertices' ids of a triangle

5.2 *Architecture*

The MeshGIS architecture has as a reference the PostGIS architecture (Fig. 5). MeshGIS uses SPTMesh through its C API.

MeshGIS has 2 main use cases: (1) store a moving object in a PostgreSQL database and (2) analyze or manipulate a moving object stored in a PostgreSQL database.

In (1) a client can send a STWKT form of a moving object to MeshGIS using SQL. MeshGIS will use the SPTMesh C API to create a valid moving object from its STWKT form. That object will be converted to the corresponding MeshGIS object, serialized to a SerializedPostgreSQLObject and sent to PostgreSQL for storage.

Fig. 5 MeshGIS architecture

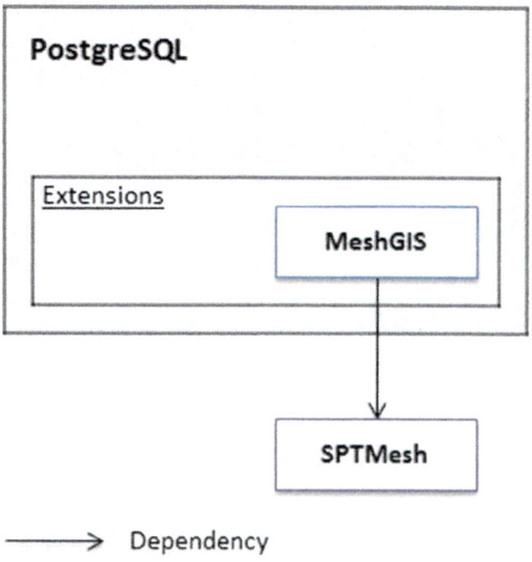

In (2) MeshGIS receives the moving object from PostgreSQL in a serialized form, i.e., in a SerializedPostgreSQLObject. This object will be converted to the corresponding MeshGIS object. Then MeshGIS will use SPTMesh to create the moving object and to analyze and manipulate it as needed.

6 Experimental Results

This section presents application examples to demonstrate the use of SPTMesh and MeshGIS. Two datasets were used for validation purposes:

- Synthetic data, i.e., data created manually to test specific conditions that do not occur when using real data.
- A set of real data extracted from a sequence of satellite images tracking the movement of two icebergs in the Antarctic (Anon 2004) using the methods implemented in Mesquita (2013).

6.1 Tests Using Synthetic Data

We used synthetic data to test specific conditions that we consider relevant. Here, we consider a case with rotations $\geq 2\pi$ (Fig. 6) and the 180° rotation of an object (Fig. 7).

These examples show that by using compatible triangulation and rigid interpolation methods we are able to deal with complex cases involving the rotation of objects and concavities, two issues that can be found when using the methods

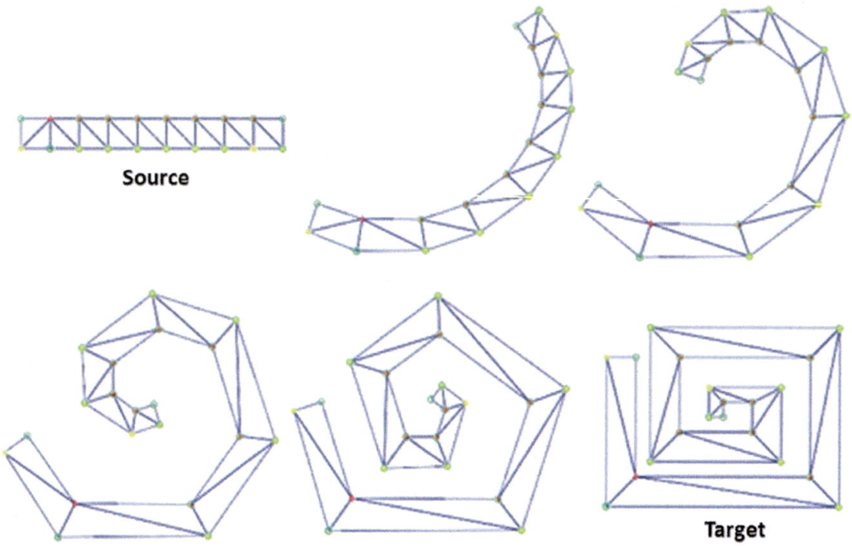

Fig. 6 Coil interpolation test

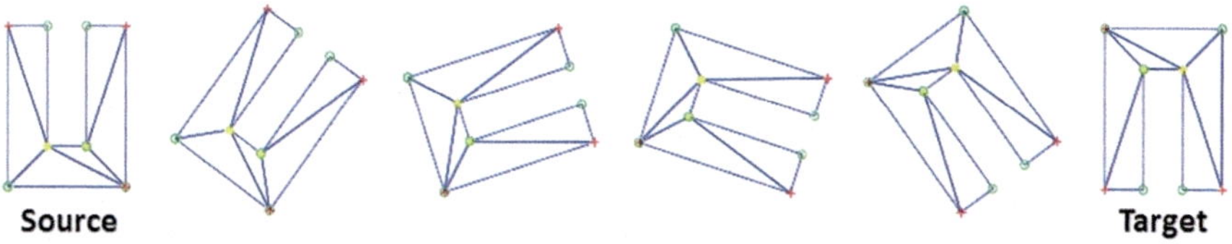

Fig. 7 180° rotation test

proposed in Tøssebro and Güting (2001), Mckenney and Webb (2010), Mckennney and Frye (2015) and Heinz and Güting (2016).

6.2 Tests Using Real Data

To demonstrate MeshGIS we start by creating a table in PostgreSQL to store the spatiotemporal data about the evolution of two icebergs. For this we use the MovingMesh data type. Note that the values of the time intervals used for testing purposes do not represent valid dates.

```
CREATE TABLE db.icebergs(
   id    integer,
   name  varchar(50),
   mobj  movingmesh
)
```

We can insert data into the icebergs table using the functions ST_MovingMesh_FromSTWKT and ST_MovingMesh_CreateEmpty.

```
INSERT INTO db.icebergs(id, name, mobj) VALUES(1,
'ice 1', ST_MovingMesh_FromSTWKT('MOVINGMESH((1000
2000, (1052 987,...,1034 941), (1055 999,...,1001
875)))'));
```

```
INSERT INTO db.icebergs(id, name, mobj) VALUES(2,
'ice 2', ST_MovingMesh_CreateEmpty());
```

The first command creates a moving object called 'ice 1' with a unit describing its evolution during the time interval [1000, 2000]. The two sequences of coordinates given in the MovingMesh STWKT form represent the position and the shape of the iceberg at instants 1000 and 2000, respectively. Internally, MeshGIS uses SPTMesh to construct the moving object from its STWKT form, i.e., SPTMesh will apply the compatible triangulation method and compute the interpolation components for this unit. The second command creates a moving object with an empty spatiotemporal component.

The command below displays the contents of the table after the execution of the previous commands (Table 9).

```
SELECT * FROM db.icebergs;
```

Note that the representation of the moving objects uses the STWKT form presented in Table 6.

We can add data to a record in the icebergs table independently of the method used to create it. For example, we can add a unit, i.e., a new evolution of the iceberg during a time interval, to the record with id = 1.

```
UPDATE db.icebergs SET mobj =
ST_Add_UnitMesh((SELECT mobj FROM db.icebergs
WHERE id=1), 'UNITMESH(2000 3000, (1001 875,…,979
848), (1030 942,…,996 896))',false) WHERE id=1;
```

Table 9 Result of the select command in the icebergs table

Id	Name	mobj
1	ice 1	MOVINGMESH((1000 2000, (1052 987, 1090 1037, …, 1034 941), (1055 999, …, 1001 875)))
2	ice 2	MOVINGMESH EMPTY

After inserting the icebergs' data in the icebergs table we can ask questions about the icebergs' properties and the relationships that they establish with each other.

We can obtain information about the icebergs at a specific period (Fig. 8) or instant (Fig. 9), assuming each record has 9 units corresponding to a period between 1000 and 10000.

```
SELECT  ST_Get_AtPeriod(mobj,'PERIOD(1100  4500)')
FROM db.icebergs WHERE id=2;

SELECT    ST_Get_AtInstant(mobj,    1500)    FROM
db.icebergs WHERE id=1;
```

The result of the query depicted in Fig. 8 is also a moving region and represents the shape of the object at all times between 1100 and 4500. Figure 8 shows only 5 snapshots for demonstration purposes. The result of the query depicted in Fig. 9 is a projection of an *umesh* at time instant 1500, which SPTMesh converts automatically into a SPATIAL type that can be used, for example, as the input of a PostGIS operation.

Fig. 8 Iceberg 2 at instants 1100, 2000, 3000, 4000 and 4500

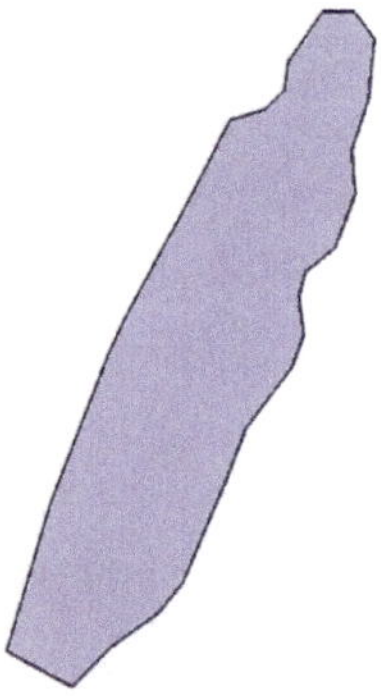

Fig. 9 Iceberg 1 at instant 1500

We can obtain the intersection of the two icebergs at a specified instant (Fig. 10). As in the real dataset the two icebergs do not intersect, we created two icebergs based on the real dataset that intersect each other at time instant 1000.

```
SELECT ST_Intersection((SELECT mobj FROM
db.icebergs WHERE id=3), (SELECT mobj FROM
db.icebergs WHERE id=4), 1000);
```

4 We can use PostGIS to get the area of the intersection between the two icebergs at instant 1000 (Table 10). That is, we can use PostGIS functions to process spatial results obtained by MeshGIS.

```
SELECT
ST_Area(ST_GeomFromText(ST_Intersection((SELECT
mobj FROM db.icebergs WHERE id=3), (SELECT mobj
FROM db.icebergs WHERE id=4),1000)));
```

Finally, we can use the operations provided by MeshGIS to see the evolution of a moving object during a time interval, e.g., during a unit (Fig. 11). We can choose the interpolation step, i.e., the frequency used to interpolate the evolution of the moving object.

Fig. 10 Intersection of two icebergs at instant 1000

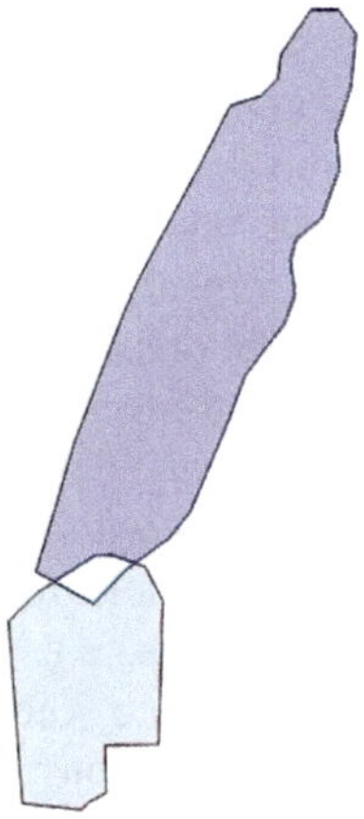

Table 10 Using PostGIS to get the area of the intersection of two moving regions at instant 1000

Intersection area (abstract units)	1815.20

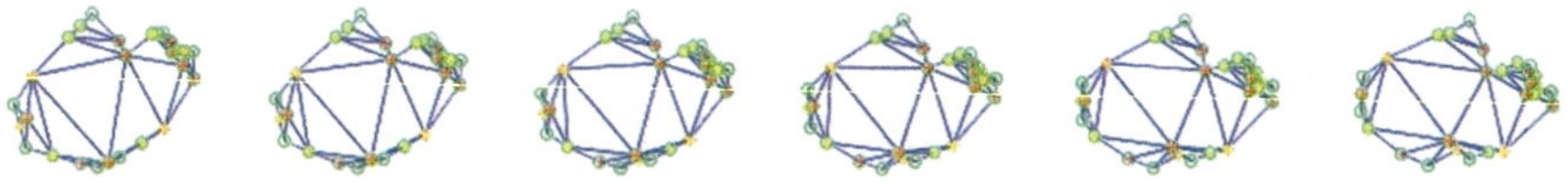

Fig. 11 Iceberg 2 interpolation during a time interval

7 Conclusions and Future Work

We implemented a framework for moving objects, in particular moving regions, called SPTMesh in a library that is independent from any application using it. A region is represented by a mesh, i.e., a triangulated polygon, and the framework uses compatible triangulation and rigid interpolation methods to provide a representation of the continuous evolution, i.e., the continuous changes of the position, shape and size, of moving regions over time. Our framework for moving objects does not consider lines, collections and regions with holes and implements only a subset of the spatiotemporal operations proposed in the literature. We implemented a spatiotemporal database extension for PostgreSQL called MeshGIS as a C library that uses SPTMesh to analyze and manipulate moving objects and we provide Structured Query Language (SQL) types and operations that bind to the MeshGIS types and operations. MeshGIS makes it possible to store moving objects in a PostgreSQL database and to analyze and manipulate them in a convenient way using SQL. It is also possible to use PostGIS to further process MeshGIS spatial results obtained from operations on moving types. We also provide a convenient representation of the moving types called Spatiotemporal Well-Known Text (STWKT). Our experiments show that SPTMesh can: (1) work with real data, (2) provide a representation of the evolution of moving regions and (3) establishes a base for future work and investigation in this area.

We intend to continue the development of SPTMesh and MeshGIS in the future. We want to test SPTMesh with larger and diverse datasets and extend it so that it can represent regions with holes and collections. Working with polygons with holes poses additional challenges, e.g., a hole can be formed, collapse or evolve to n holes. To the best of our knowledge, there is no well-defined algorithm to compatibly triangulate polygons with holes. The complexity of the compatible triangulation method that we are using is known to be $O(N^3 \log N)$, where N is the number of boundary vertices of the source polygon, and we intend to perform a formal analysis on the complexity of our system. We also want to implement a larger set of spatiotemporal operations, e.g., query the time at which two moving regions intersect and the region traversed by a moving region.

Acknowledgements We thank the anonymous reviewers for their comments and suggestions. This work is partially funded by National Funds through the FCT—Foundation for Science and Technology, in the context of the project UID/CEC/00127/2013.

References

Alexa M, Cohen-or D, Levin D (2000) As-rigid-as-possible shape interpolation. In: SIGGRAPH '00 Proceedings of the 27th annual conference on computer graphics and interactive techniques, pp 157–164

Amaral A (2015) Representation of spatio-temporal data using compatible triangulation and morphing techniques. Aveiro University

Anon (2004) RossSea subsets. http://rapidfire.sci.gsfc.nasa.gov/imagery/subsets/?project=antarctica&subset=RossSea&date=11/15/20. Accessed 20 Sept 2016

Baxter W, Barla P, Anjyo K (2008) Rigid shape interpolation using normal equations. In: NPAR '08 Proceedings of the 6th international symposium on Non-photorealistic animation and rendering, pp 59–64. http://dl.acm.org/citation.cfm?id=1377993

Cotelo Lema J et al (2003) Algorithms for moving objects databases. Comput J 46(6):680–712

Forlizzi L et al (2000) A data model and data structures for moving objects databases. In: Proceedings of the 2000 ACM SIGMOD international conference on management of data, pp 319–330

Gotsman C, Surazhsky V (2004) High quality compatible triangulations. Eng Comput 20(2):147–156

Güting RH et al (2000) A foundation for representing and querying moving objects. ACM Trans Database Syst 25(1):1–42. http://doi.acm.org/10.1145/352958.352963

Güting RH, Behr T, Düntgen C (2010) SECONDO: a platform for moving objects database research and for publishing and integrating research implementations. Bull IEEE Comput Soc Tech Comm Data Eng 33(2):56–63

Heinz F, Güting RH (2016) Robust high-quality interpolation of regions to moving regions. GeoInformatica 20(3):385–413. http://link.springer.com/10.1007/s10707-015-0240-z

Liu Z et al (2015) High quality compatible triangulations for 2D shape morphing. In: VRST '15 Proceedings of the 21st ACM symposium on virtual reality software and technology, Beijing, China, pp 85–94

Matos L, Moreira J, Carvalho A (2012) Representation and management of spatiotemporal data in object-relational databases. In: Proceedings of the 27th annual ACM symposium on applied computing, SAC '12. ACM, New York, NY, USA, pp 13–20. http://doi.acm.org/10.1145/2245276.2245280

Mckenney M, Webb J (2010) Extracting moving regions from spatial data. In: Proceedings of the 18th SIGSPATIAL international conference on advances in geographic information systems, San Jose, California, pp 438–441. http://doi.acm.org/10.1145/1869790.1869856

Mckennney M, Frye R (2015) Generating moving regions from snapshots of complex regions. ACM Trans Spat Algorithms Syst 1(1):1–30. http://doi.acm.org/10.1145/2774220

Mesquita P (2013) Morphing techniques for representation of geographical moving objects. Universidade de Aveiro

Moreira J, Dias P, Amaral P (2016) Representation of continuously changing data over time and space. In: 2016 IEEE 12th international conference on e-science, Baltimore, MD, USA. IEEE, pp 111–119

Pelekis N et al (2006) Hermes—a framework for location-based data management. In: EDBT'06 Proceedings of the 10th international conference on advances in database technology, Munich, Germany, pp 1130–1134. https://doi.org/10.1007/11687238_75

Sanderson C, Curtin R (2016) Armadillo: a template-based C++ library for linear algebra. J Open Source Softw 1:26

Tøssebro E, Güting R (2001) Creating representations for continuously moving regions from observations. In: Proceedings of the 7th international symposium on advances in spatial and temporal databases. Springer, Berlin, Heidelberg, pp 321–344. https://doi.org/10.1007/3-540-47724-1_17

Zhao L et al (2011) STOC: extending oracle to support spatiotemporal data management. Springer, Berlin, Heidelberg, pp 393–397. http://link.springer.com/10.1007/978-3-642-20291-9_43. Accessed 30 Sept 2017

Part IV
Quality and Uncertainty of Geographic Information

A Positional Quality Control Test Based on Proportions

Francisco Javier Ariza-López, José Rodríguez-Avi and Virtudes Alba-Fernández

Abstract This paper presents a new method for positional accuracy quality control of spatial data. This method is valid for 1D, 2D, 3D and nD dimensional data, where data can follow any kind of distribution function. Normality of errors, or any other assumption are not required. The method is an exact statistical hypothesis testing based on multinomial distribution. The proportions of the multinomial distribution are defined by means of several metric tolerances. The proposed statistical test is exact, so the p-value can be derived by exploring a space of solutions and summing up the probabilities of each isolated case of this space. The performance of the test has been analyzed by means of a simulation procedure. The validity and the power of the contrast seem to be good enough. An application example is presented for the 3D case of working with two tolerances. In all cases, H_0 is the same, but in the first one, its hypothesis is true, in the second, the true distribution has larger errors than assumed by H_0 (it is *worse*) and, in the third case, the true distribution implies smaller errors than that stated by H_0 (it is *better* in the sense of the error magnitude). In all three cases, the behavior of the proposed method is acceptable.

1 Introduction

The phrase "80% of data is geographic" is one of those commonly-cited facts (Dempsey 2013) in order to highlight the importance of spatial data in our society. One specific and differential fact of spatial data in relation to other kinds of data is the fact of being located, in a certain position. So spatial data are data referring to

F. J. Ariza-López (✉) · J. Rodríguez-Avi · V. Alba-Fernández
Universidad de Jaén, Paraje de Las Lagunillas S/N, E-23071 Jaén, Spain
e-mail: fjariza@ujaen.es

J. Rodríguez-Avi
e-mail: jravi@ujaen.es

V. Alba-Fernández
e-mail: mvalba@ujaen.es

A. Mansourian et al. (eds.), *Geospatial Technologies for All*,
Lecture Notes in Geoinformation and Cartography,
https://doi.org/10.1007/978-3-319-78208-9_18

features that have a position in space, and for this reason, positional quality is one of the most desirable characteristics of spatial datasets (SDS).

The positional accuracy of SDS has traditionally been evaluated using control points. Following this idea there are many statistical Positional Accuracy Assessment Methodologies (PAAM). A general analysis of these PAAM is presented in Ariza-López and Atkinson-Gordo (2008), this analysis indicates that majority of PAAMS are based on the assumption of normal distributed errors. An interesting analysis was developed by Zandbergen (2008), who presents a general view of the NMAP (USBB 1947) and the NSSDA (FGDC 1998), and a specific critique of the NSSDA and some significant recommendations for the use of the NSSDA. Some of the recommendations are: (i) the need for consideration of alternative approaches to characterizing positional accuracy, (ii) the need for reconsideration of normality of errors, (iii) the need to increase sample size, and (iv) the need for techniques to characterize other commonly-observed distributions (e.g. Rayleigh, log-normal). Analyzing the errors of Digital Elevation Models (DEM), Liu et al. (2012) pointed out that DEM errors are not normally distributed and not identically distributed.

In general, there are several causes for the non-normality of errors, for instance: the presence of too many extreme values, the overlap of two or more processes, an insufficient data discrimination (e.g. round-off errors, poor resolution), the elimination of data from the sample, distribution of values close to zero or the natural limit, and data following a different distribution. See Zandbergen (2008) for a more complete revision of the non-normality issue, and for a complete example of the characterization of errors of four types of spatial data (GPS locations, street geocoding, TIGER roads, and LIDAR elevation data). In this way, for the case of vertical errors in DEMs it is proposed to perform and express the results of quality control checks by means of percentiles (Maune 2007) of the observed distribution. The most recent PAAM, which has been proposed by the American Society of Photogrammetry and Remote Sensing (ASPRS) (ASPRS 2015), follows this method by informing separately vegetated vertical accuracy (supposed to be non-Gaussian distributed) and non-vegetated vertical accuracy (supposed to be normally distributed). Nevertheless, the work of Zandbergen (2011) puts into question the normality of errors for bare earth observations. Moreover, this situation, where two different methods and measures are in use (ASPRS 2015), is somewhat confusing for non-expert users. The non-normal distribution of errors has a direct consequence for the results of PAAMs based on the hypothesis of normality of errors. But non-normality of errors has implications beyond spatial data accuracy control and assessment, since most error propagation techniques are based on an assumption of normality (Zandbergen 2011). Non-normality can imply the invalidation of such results.

Parametric distribution functions are very convenient when dealing with scarce data and limited computing capabilities, as in the past. However, we live surrounded by big-data sets, and by very large computing capabilities nowadays, thus we can use the observed distribution functions (Free-Distribution Functions or

Parameters-Free-Distribution Functions) without major problems. The PAAMs based on Free-Distribution Functions are scarce in positional accuracy assessment. Until now, the use of percentiles is merely descriptive, in a statistical sense, and no method for quality control is proposed. Recently, Ariza-López and Rodríguez-Avi (2015a) proposed a method that can be applied to error data belonging to free-distribution functions, but a step further would require a statistical method allowing control in the same way (mean value, variation, control of outliers) as performed when data come from a Normal Distribution Function.

The objective of this paper is to propose a general positional accuracy control method for dealing with error data following any kind of distribution function (e.g. non-normal errors, distribution-free errors, etc.). It is a method based on the observed distribution function of the data and, in this way, it can be applied to 1D, 2D or 3D data without the limitations of normality and homogeneity of variances stated in traditional PAAM. To achieve this aim we propose a method based on proportions of a multinomial distribution function in order to establish a strict control over data coming from any distribution function. The control is multiple and can test jointly proportions corresponding to tolerances related with, for instance, median values (50% percentile), extreme values (e.g. 95% percentile), or the amount of outliers existing in the data set. The proposed control is based on an exact test, the same as Fisher's exact test (Fisher 1922; Freeman and Halton 1951; Müller 2001).

The paper is organized as follows: after this introduction, a literature revision centered on count-based control methods is performed; next, in Sect. 3 the proposed method is established in a general manner. To illustrate the procedure in a particular case, section four presents the case of two tolerances (three proportions). Section five presents a 3D positional control under three different compliances of the hypothesis. Finally, the main conclusions are stated.

2 Related Work

Our work proposes the statistical bases for a new PAAM, which is based on the idea of defective counting, which is not a very common approach between the PAAMs. Given a SDS with N elements of interest (e.g. points, lines, polygons, etc.), defective counting can be related to positional accuracy by means of a decision rule. If we establish a metric tolerance, for instance $T = 5$ m, we can compare all positional error values E_i (in 1D, 2D, 3D or n-D) of a control sample of size n with this tolerance. In this way we define a positional defective as those cases where $E_i > T$. The total counting of (positional) defectives defines a proportion in relation to N, and this is the key element for the positional accuracy control because it can be performed by a test on a binomial distribution (Eq. 1). In consequence, attention must be paid to the few cases of PAAMs based on the same idea.

$$P[F > mc | F \rightarrow B(n, \pi)] = \sum_{k=mc+1}^{n} \binom{n}{k} \pi^k (1-\pi)^{n-k} \quad (1)$$

where

- F Number of sampling cases of the fail success.
- mc Maximum number of fails allowed, and that is previously specified.
- n Sampling size.
- π Population probability that the error be greater than the tolerance ($\pi = P[E_i > T]$).

The NMAS (USBB 1947) is the first comprehensive standard developed in modern history for the United States of America (Abdullah 2008), and remains a widely employed standard (Zandbergen 2008; Abdullah 2008); but has been severely criticized (e.g. see Acharya and Bell 1992). Essentially, the steps for the realization of the NMAS in the planimetric (2D) case are: (i) selection of a metric tolerance *tol*, (ii) selection of a random sample of size n (there is no specific sample size requirement); (iii) calculation of errors (planimetric); (iv) determination of the pass/fail. The last step is achieved by a simple rule, counting the number of sample control elements $\#E_i$ where $E_i > T$, and if this number exceeds 10% of the sample, the product is rejected. As demonstrated by Ariza-López and Rodríguez-Avi (2014) this PAAM can be seen as the realization of a statistical contrast test based on a binomial distribution assumption.

The very well-known International Standard ISO 2859: Sampling procedures for inspection by attributes is a series of standards designed for its application in the industry that has been in use since the times of the World War II. This series is a set of acceptance sampling methods based on defective counting, ISO 2859 part 1 (ISO 1985) and 2 (ISO 1999) and have been proposed by ISO 19157 (ISO 2013) for quality control in spatial data (e.g. completeness, consistency, etc.). The underlying statistical model is the binomial or the hypergeometric, depending on the population size, infinite or finite. Recently, the Spanish Standard UNE 148002 (AENOR 2016) has defined a positional accuracy quality control procedure based on this International Standard ISO 2859.

Finally, we must mention the NIST (National Institute of Standards and Technology) method, developed by Cheok et al. (2008) in order to control 2D and 3D building plans. It is also an acceptance sampling method but, in this case, they consider two conditions in order to classify an error as positional defective. As in the previous cases, the statistical model applied is the binomial.

Figure 1 shows the situation presented above where the metric tolerance T establishes a dichotomous division. All of the control elements with errors greater than the T value ($E_i > T$) are positional defectives, and it does not matter which is the underlying distribution, that in the case of the Figure is clearly non-normal. On the other hand, all control elements with errors less than the T are considered correct from the positional quality point of view.

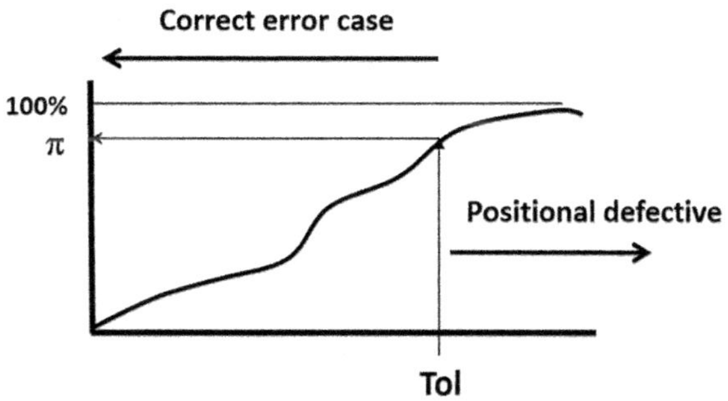

Fig. 1 Observed distribution function controlled by means of one metric tolerance

3 Proposed Method

In order to analyse a pass/fail situation, the first useful approximation is given by a statistical hypothesis test in terms of the Binomial distribution, as evidenced by the NMAS, ISO 2859 and NIST methods. In this case, given a metric tolerance T (e.g. 5 m) (it does not matter if it is a linear or quadratic value), once a sample of size n is taken, the number of defectives x (elements with error greater than the metric tolerance T), can be seen as a realization of a Binomial distribution with parameters n and π. In this case, if n_1 is the number of positional defective elements found in the control sample and π^0 is the the maximum proportion of positional defectives allowed, the hypothesis test is:

- $\mathbb{H}_0$: The proportion of defective control elements (n_1/n) is less than or equal to π^0.
- $\mathbb{H}_1$: The proportion of defective control elements (n_1/n) is greater than π^0.

So a right hand unilateral test is proposed and rejection of $\mathbb{H}_0$ occurs when the p-value $p = P[X \geq n_1 | X \sim B(n, \pi^0)] < \alpha$. The rejection of $\mathbb{H}_0$ implies that actually the number large errors is greater than the percentage specified under $\mathbb{H}_0$.

This approximation is simple and has been applied by the NMAS, ISO 2859 and NIST methods, but this approximation implies that we are not able to distinguish the "degree" of an error of a control element because is a binary situation (pass/fail, good/bad) derived from the binomial model. However, in several circumstances it may be interesting to propose an ordination in the degree of fail. For example, if we consider two metric tolerances T_1 and T_2 with $T_1 \leq T_2$, they define three intervals I_1, I_2, I_3 or levels and we can establish the positional quality requirements of a product by the desired proportion of error cases on each level. *Example gratia*, in relation to a control over a normal distribution of errors, we can control the mean value, establish a confidence interval of 95% and limit the presence of outliers. Following this example, it can be considered 50% as a minimum percentage of errors on the first level, 45% as a maximum percentage on the second level and 5% as a maximum percentage on the third level. These values establish a vector of proportions for the control ($\pi_1 = 0.5$, $\pi_2 = 0.45$, $\pi_3 = 0.05$). This produces an ordered classification on the positional errors of control elements in a similar form to Likert's scale, from the best to the worst class. Thus, in general there are k categories, each of them with a probability π_j, with $j = 1, \ldots, k$, under the null hypothesis where $\pi_1 + \cdots + \pi_k = 1$. In this case, and supposing that sampling control elements are taken independently, the distribution of the variable "number of error of elements belonging to each category in a sample of size n", is a multinomial distribution with parameters $(n, \pi_1, \ldots, \pi_k)$.

In this scenario, where there are n independent trials, each of them leading to success for exactly one of k categories and with each category having a given fixed success probability, the multinomial is the appropriate model. So if we realize n independent experiments where we classify the results for exactly one of k categories, with probabilities $\boldsymbol{\pi} = (\pi_1, \ldots, \pi_k)$, and $\pi_1 + \cdots + \pi_k = 1$, they follow a multinomial law, and the mass probability function of a multinomial $\mathcal{M}(n, \pi_1, \ldots, \pi_k)$ is given by (Eq. 2):

$$P(X_1 = n_1, \ldots, X_k = n_k) = \frac{n!}{n_1! \ldots n_k!} \pi_1^{n_1} \ldots \pi_k^{n_k} \tag{2}$$

where n_i is the number of errors of control elements that belongs to the category i, which has a probability π_i. In this case (Fig. 2), with two tolerances, the distribution is split into three groups (instead of two, as in Fig. 1). Here it is possible to establish a gradation for the size of positional errors of control elements. Extremes are good (left, $E_i \leq T_1$) and unacceptable (right, $E_i \geq T_2$), but between them there is an interval where they are neither good nor bad (middle, $T_1 \leq E_i \leq T_2$).

In consequence, a hypothesis test based on a multinomial distribution is adequate to prove the null hypothesis mentioned above. In general, and once the number of categories, k, has been fixed in order to determine a statistical hypothesis test, the following considerations apply:

1. The sampling statistic is $\boldsymbol{\nu}^* = (n_1, n_2, \ldots, n_k)$, so that $n_1 + \cdots + n_k = n$ is the sampling size. Thus it is assumed that the sampling statistic is distributed

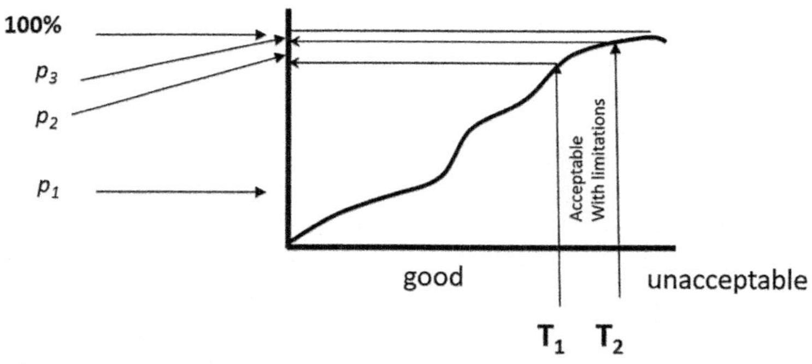

Fig. 2 Control of an observed distribution function by means of two metric tolerances

according to a multinomial distribution with parameters n and $\pi_1, \ldots, \pi_{k-1}, \pi_k = 1 - \pi_1 - \cdots - \pi_{k-1}$.

2. The null hypothesis is that at least $P_1\%$ (expressed in natural language as, for example $P_I = 80\%$), of observed values have a measured error less than T_1 and no more than $P_j\%$ of observed values have a measured error between T_{j-1} and $T_j, j = 2, \ldots k$ (assuming that $P_1\% + \cdots + P_k\% = 100$). So we can explain it thus:

   - $\mathbb{H}_0$: Z The sampling statistics, $\boldsymbol{\nu}^*$ follows a multinomial distribution with parameters n, $\boldsymbol{\pi}^0 = \left(\pi_1^0, \ldots, \pi_k^0\right)$.

3. To determine the alternative hypothesis we have to take into account that $\mathbb{H}_0$ will be rejected if the proportion of elements with measured error less than T_1 is less than P_1, or, if this proportion is equal to P_1, then the proportion of elements between T_1 y T_2 has to be less than P_2, and so on, because this implies a worsening in tails. So a given case $m = (m_1, \ldots, m_k)$ is worse than the observed case $\boldsymbol{\nu}^*$, if the number of elements in the best category decreases while the number in the remaining categories increase.
4. In a statistical test, the p-value is calculated as the probability of obtaining something as harmful, or more than the test statistic. To calculate the p-value, proceed as follow: given the test statistics $\boldsymbol{\nu}^* = (n_1, \ldots, n_k)$, we calculate the probability in the multinomial fixed by the null hypothesis to the obtained value and those cases $m = (m_1, \ldots, m_k)$ that we consider *worse* because there are less cases in the first category and more cases in the remaining categories, that is to say, the number of cases with large errors increases. For us, all those cases who meet any of the following conditions are worse:

   - $m_1 < n_1$ or
   - $m_1 = n_1$ and $m_2 < n_2$ or
   - $m_1 = n_1; m_2 = n_2$ and $m_3 < n_3$ or
   - ....
   - $m_1 = n_1; \ldots m_{k-2} = n_{k-2}$ and $m_{k-1} \leq n_{k-1}$

   Note that the above conditions established, from the last to the first (inverse order), an orderly preference of what we consider to be getting worse. Finally, the p-value is calculated as the sum of the individual probabilities of all these cases.

5. The null hypothesis is rejected if the p-value obtained is less than α.
   It is important to notice here that the above-mentioned procedure is an exact (significance) test where the p-value is determined by scanning the space of solutions that have greater errors than the given one, and adding the probability of each of these possible solutions (Fisher 1935). No demonstration is needed in this case, In Mehta and Patel (1983) or in Storer and Choongrak (1990), among others, examples of exact tests can be found.

4 The Two Tolerances Case

In order to illustrate the general procedure, and to introduce the examples in the next section, we present here the case of two metric tolerances T_1 and T_2 and three proportions (π_1, π_2, π_3). In this case we can classify the positional error E_i in a control element into three categories:

i. small errors if $E_i \leq T_1$,
ii. moderate errors if $T_1 < E_i \leq T_2$, and
iii. excessive errors if $T_2 < E_i$.

In this case, the multinomial model is adequate. Therefore, in order to pass the control it is desired that the proportion of elements where $E_i \leq T_1$ has to be greater than π_1, and the proportion of elements where $T_1 < E_i \leq T_2$ has to be less than π_2 or the proportion of elements $E_i > T_2$ has to be less than π_3. To prove this a sample of size n is taken from a population of size N. n_1 the number of elements where $E_i \leq T_1$; n_2 is the number of elements with $T_1 < E_i \leq T_2$, and n_3 the number of elements with $T_2 < E_i$. And following the considerations stated before:

A. The sampling statistics is: $\boldsymbol{\nu}^* = (n_1, n_2, n_3)$, so that $n_1 + n_2 + n_3 = \text{n}$.
B. The parameters of the multinomial distributions are: N, $\pi_1, \pi_2, \pi_3 = 1 - \pi_1 - \pi_2$.
C. The null hypothesis is:

- $\mathbb{H}_0$: The sampling statistics, $\boldsymbol{\nu}^*$, has a multinomial distribution with parameters $(n, \boldsymbol{\pi}^0) = (\pi_1^0, \pi_2^0, \pi_3^0)$ where $\pi_k^0 = P_k/100$ and $\pi_1^0 + \pi_2^0 + \pi_3^0 = 1$.

D. The alternative hypothesis is that the true distribution of errors presents more large errors than the specified under $\mathbb{H}_0$:

- $\mathbb{H}_1$: At least one of these conditions: $\pi_1 \geq \pi_1^0$ or $\pi_2 \leq \pi_2^0$, or $\pi_3 \leq \pi_3^0$, is false. Here the alternative hypothesis specifies what we consider a *worse* situation, and this situation takes place when the proportion of elements with tolerance less than T_1 is less than P_1, or when the other two proportions account for more than P_2 or P_3, because this implies a worsening in tails.

E. This is an exact test, so the p-value is calculated as follows: given the test statistics $\boldsymbol{\nu}^* = (n_1, n_2, n_3)$ we calculate the probability in the multinomial fixed by the null hypothesis to the obtained value and those counting of elements $m = (m_1, m_2, m_3)$ that verify the conditions:

   a. $m_1 < n_1$
   b. $m_1 = m_1$ and $m_2 \leq n_2$

   Adding up the p-values of all the cases that verify these conditions and rejecting the null hypothesis if the p-value obtained (the sum) is less than α.

5 Application Examples

5.1 A Theoretical Example

Now we are going to show this procedure working with three controlled and different data cases: In the first case (C#1) $\mathbb{H}_0$ is true, and we are going to see what happens if a sample confirms this hypothesis or not. In the second case (C#2) the true model is less good, in the sense of generating a higher number of positional defectives than the desired model, and the third case (C#3) shows the case where the sample refers to a situation which is better, in the sense of a larger number of small errors than that stated by $\mathbb{H}_0$.

Supposing that a spatial data product establishes that errors in X, Y and Z are distributed according to three Normal and independent distributions with $\mu = 0$ m and $\sigma = 1.5$ m. This is the null hypothesis for all the three cases (C#1, C#2 and C#3). Under this hypothesis, the quadratic error in each element (e.g. point, line or whatever kind) is (Eq. 3):

$$QE_i = Ex_i^2 + Ey_i^2 + Ez_i^2 \quad (3)$$

which is distributed according to a Gamma distribution with parameters of shape $K = 3/2$ and scale $\theta = 4.5$. In this case, the probability that an element has a $QE \leq 9.243$ m^2 is 0.75, and the probability that an element has a $QE \leq 14.065$ m^2 is 0.90. In consequence, if we take a sample of control elements of size n and calculate the errors and count the number of elements whose $QE \leq 9.243$ m^2; the number of elements with $9.243\,\text{m}^2 \leq QE \leq 14.065\,\text{m}^2$ and the number of elements whose $QE > 14.065$ m^2, these quantities will follow a multinomial distribution with parameters (n, 0.75, 0.15, 0.10).

Let $n = 20$, which is a very usual minimum recommendation for sampling sizes in PAAMs. Under the labels C#1, C#2 and C#3, Table 1 shows 20 rows with the measured errors for each component (Ex_i, Ey_i, Ez_i) for whatever kind of control element (e.g. point, line, etc.). Also, for each case, Table 1 shows the type of QE in each row under the column which is labelled with T. The symbol "□" means $QE \leq 9.243$ m^2, the symbol "▪" means $9.243\,\text{m}^2 \leq QE \leq 14.065\,\text{m}^2$, and the symbol "█" means $QE > 14.065$ m^2.

Case #1. Now let us suppose a 3D sample coming from three N(0, 1.5) (hypothesized case) with errors presented in Table 1 (columns under C#1). In this case, the sampling statistic is $\boldsymbol{\nu}^* = (15, 4, 1)$ because there are 15 cases identified with the symbol "□", 4 cases identified with the symbol "▪" and 1 case identified with the symbol "█". To obtain the p-value we calculate the probability in a multinomial (20, 0.75, 0.15, 0.1) for case $\boldsymbol{\nu}^*$ and for all possible cases m where $m_1 < 15$ and when $m_1 = 15, m_2 \leq 4$, so that $m_1 + m_2 + m_3 = 20$. For this $\boldsymbol{\nu}^*$ case, Table 2 presents several cases compliant with these conditions (see columns labeled with m_1, m_2 and m_3),

Table 1 Data of the three cases

C#1 Case of the hypothesis N(μ = 0, σ = 1,5)				C#2 Worse than the hypothesis N(μ = 0, σ = 2)				C#3 better than the hypothesis N(μ = 0, σ = 1)			
e_X[m]	e_Y[m]	e_Z[m]	T	e_X[m]	e_Y[m]	e_Z[m]	T	e_X[m]	e_Y[m]	e_Z[m]	T
−0.371	−1.672	2.755	▪	4.263	2.439	3.298	■	0.745	−0.001	−0.892	□
−3.359	−0.815	1.454	▪	−2.547	−0.959	−0.483	□	−1.174	−0.299	0.527	□
0.251	0.340	0.467	□	0.876	3.985	−0.851	■	0.938	0.031	0.993	□
−0.308	−0.324	0.718	□	−0.010	0.352	2.098	□	0.219	−1.092	0.651	□
1.172	−0.320	−0.411	□	−1.920	0.744	4.166	■	−1.533	−2.152	−1.834	▪
−0.206	−3.074	−1.651	▪	2.313	−1.372	−3.335	■	0.481	−0.010	0.497	□
3.873	0.304	2.280	■	−2.584	2.832	0.049	■	−1.551	−0.163	0.902	□
0.394	−0.442	0.989	□	0.830	−0.932	−0.510	□	−0.383	0.239	−1.118	□
2.322	−1.667	0.623	□	1.895	−0.902	−2.230	▪	−1.267	2.032	−0.887	□
−1.380	−2.260	0.342	□	2.242	−1.206	1.741	▪	1.555	2.436	−0.998	▪
1.384	−1.444	−1.730	□	−1.341	1.723	−0.745	□	−0.371	−0.219	1.323	□
−1.131	−0.549	−0.930	□	−1.457	−1.699	−4.995	■	−0.217	0.438	0.003	□
0.423	0.627	−1.257	□	−0.541	4.164	1.924	■	1.606	−1.278	−0.310	□
1.494	−1.359	−2.168	□	2.818	2.699	−0.834	■	−1.338	−0.733	0.132	□
−1.740	0.017	−1.281	□	0.772	−0.099	−2.907	□	−0.365	1.711	0.526	□
−1.397	−0.196	0.214	□	3.217	−1.191	1.744	■	−1.115	−1.208	−0.971	□
1.670	0.262	−2.015	□	0.343	−0.024	4.905	■	0.004	−0.203	0.307	□
−0.399	−1.194	1.553	□	−4.844	0.044	0.493	■	−1.031	0.998	−0.232	□
−0.309	−1.106	1.562	□	−0.050	−0.657	0.206	□	0.740	−0.638	−0.397	□
−1.329	0.014	2.745	▪	−1.096	−1.909	−1.731	□	0.861	−0.080	0.879	□

Table 2 Worse values for the calculation of the exact significance of the test

Worse value	m_1	m_2	m_3	Probability	Accumulated probability
1	15	4	1	0.05244	0.05244
2	15	3	2	0.06992	0.12236
3	15	2	3	0.04661	0.16898
4	15	1	4	0.01553	0.18452
5	15	0	5	0.00207	0.18659
6	14	6	0	0.00786	0.19446
7	14	5	1	0.03146	0.22593
8	14	4	2	0.05244	0.27837
9	14	3	3	0.04661	0.32499
10	14	2	4	0.02330	0.34830
11	14	1	5	0.00621	0.35451
12	14	0	6	0.00069	0.35520
13	13	7	0	0.00314	0.35835
14	13	6	1	0.01468	0.37303
….	….	….	….	….	….
….	….	….	….	….	….
….	….	….	….	….	….
195	0	20	0	3.32E-17	0.56942
196	0	19	1	4.43E-16	0.56942
….	….	….	….	….	….
214	0	1	19	3E-19	0.56942
215	0	0	20	1E-20	0.56942

with the idea of facilitating the understanding of the process. The probability of each case can be calculated using the function *dmultinom* of R (R Core Team 2012). This value is presented in the column labeled "Probability" of Table 2. Finally, the p-value is the sum of the probabilities of all cases. This probability, the p-value, is 0.5694, so if the considered significance level is $\alpha = 0.05$ (5%), we cannot reject the null hypothesis. The result of the test is the acceptance of the product from the positional quality control point of view.

Case #2. Now let us suppose we have a 3D sample of control elements whose errors follow Normal Distributions (N(0, 2)) (the true distribution presents more elements with larger errors than the hypothesised case). Error data are presented in Table 1. We proceed in a similar way to the previous one; the sampling statistic is $\nu^* = (7, 2, 11)$ and the p-value is 0.00003 ($<\alpha = 5\%$). So, we will reject the null hypothesis. The result of the test is the rejection of the product from the positional quality control point of view.

Case #3. Now the actual population of sampled errors is N(0, 1) for each component, so it is better than the desired distribution under the null hypothesis (N (0, 1.5)). Let us suppose we have a 3D sample of control elements whose errors follow Normal Distributions (N(0, 1)). Error data are presented in Table 1 and we proceed in a similar way to the previous two cases. The sampling statistic is $\nu^* = (18, 2, 0)$ and the corresponding p-value is 0.9756 ($\geq 5\%$). So we cannot reject the null hypothesis. The result of the test is the acceptance of the product from the positional quality control point of view.

5.2 A Real Example

A real case data coming from the Cartographic and Geological Institute of Catalonia ICGC (Spain) is presented here. The ICGC has an infrastructure of control points materialized in the territory. The ICGC uses this infrastructure to control the positional quality of the supplies it acquires from private companies. In this case, the data with which we are going to work are the errors of a given supply. The descriptive statistics of the population of errors is presented in Table 3 (more details in Ariza-López FJ and Rodríguez-Avi 2015b). Figure 3 shows the lack of normality of the three components X, Y and Z.

In this population we can calculate the RSQ error, E_i, for each point, where (Eq. 4):

Table 3 Descriptive parameters

Parameter	E_X	E_Y	E_Z
Size	641	641	641
Minimum	−3.1178	4.06016	−3.72464
1st Quartile	−0.2519	−0.33018	−0.57083
Median	0.0948	0.04772	−0.04422
Mean	0.1135	0.10720	−0.11497
3rd Quartile	0.5289	0.60150	0.32724
Maximum	3.5674	4.14792	3.40110

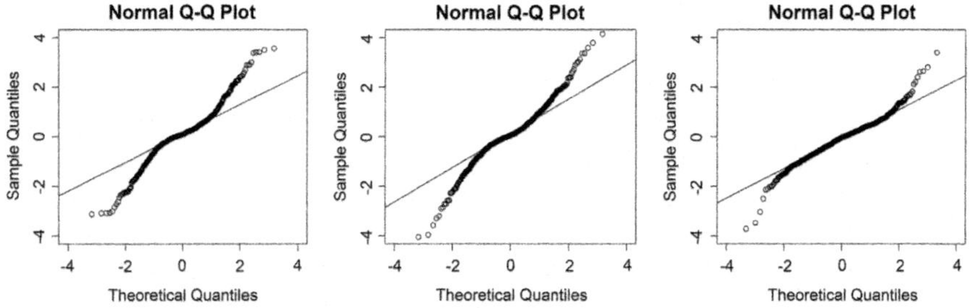

Fig. 3 QQ plots for normality

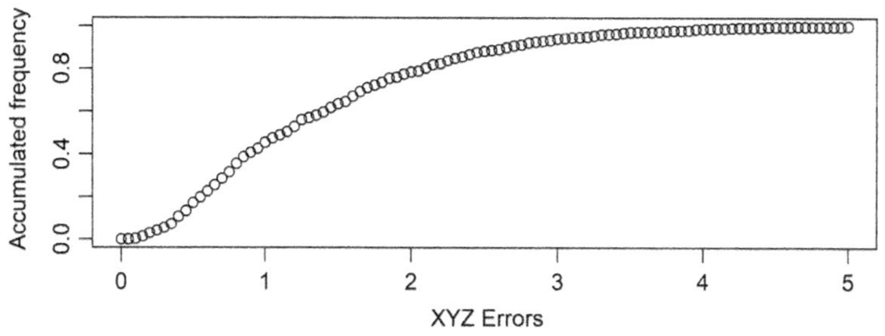

Fig. 4 Distribution function of observed errors

$$E_i = \sqrt{Ex_i^2 + Ey_i^2 + Ez_i^2} \tag{4}$$

The distribution of population errors, as defined above, appears in Fig. 4.

For this population it can be seen that the following specifications are assumable and can be reached:

- At least the 80% of errors E_i have to present a value less than 2.10 m (first metric tolerance).
- Only the 5% of errors E_i can present a value greater than 3.20 m (second metric tolerance).

To analyse the consistence of our proposal we have developed a simulation: from the given population we have taken 1000 random samples of size 25, and in each one we have counted the number of errors values E_i that fall into the three classes stablished by the two tolerances (the estimator $\boldsymbol{\nu}^*$). In all cases, the null hypothesis implies that $\boldsymbol{\pi}^0 = (0.8, 0.15, 0.05)$, and the estimator is supposed to be distributed according to a multinomial of parameters $(25, 0.8, 0.15, 0.05)$, so that the p-value can be calculated applying the procedure above described. Once the 1000 simulations have run, we have counted the proportions of times that the null hypothesis is rejected. This proportion is $0.054 \approx \alpha = 0.05$, that is the expected value in this case of validity of the null hypothesis.

We have made a second simulation but in this case we have specified stricter specifications for the positional accuracy:

- At least the 80% of errors E_i have to present a value less than 1.90 m (first metric tolcrance).
- Only the 5% of errors E_i can present a value greater than 3.00 m (second metric tolerance).

Now, the proportion of times where the null hypothesis is rejected is 12.2%, that is higher than 0.05.

6 Conclusions

As has been argued in the introduction, the majority of PAAMs take as an underlying hypothesis the Gaussian distribution of positional errors, but several studies indicate that this hypothesis is not true. Another main underlying hypothesis is the homoscedasticity of error components, which is also not strictly true. So when analyzing positional errors, we are contradicting one major hypothesis of the PAAMs being applied.

In this paper we have proposed a general positional accuracy control method for dealing with error data following any kind of distribution function. It is a method based on the observed distribution function of the error data and, in this way, can be applied to 1D, 2D, 3D or nD error data without the limitations indicated previously.

The method is an exact statistical hypothesis testing based on multinomial distribution. The proportions of the multinomial distribution are defined by means of several metric tolerances that are used to define error intervals. By means of this test we can control, jointly, the exact number of errors in each of the intervals defined by the tolerances. The proposed statistical test is exact, so the p-value can be derived by exploring a space of solutions (*worse values*) and summing up the probabilities of each isolated case of this space. The method has been proposed in a general form, so it can be applied to 2, 3 or whatever number tolerances one would need.

Three laboratory examples and an example with real data have been presented for the case of working with two tolerances. All the examples demonstrated that the application is direct and that the results are consistent with data. The given examples are for the case of two tolerances, but the same idea can be extended to more tolerances (intervals).

We think that this proposal opens up new possibilities for the positional accuracy control of spatial data. The application of these ideas allows for a strict control of positional accuracy errors by means of proportions when normality and other hypothesis are not assured in the errors of the control sample. This is a way to control the distribution of actual errors versus our desires regarding the distribution of errors expressed by means of proportions.

Acknowledgements Research in this paper has been partially funded by grant CTM2015-68276-R of the Spanish Ministry on Science and Innovation (European Regional Development Funds).

References

Abdullah QA (2008) Mapping matters: the layman's perspective on technical theory and practical applications of mapping and GIS. Photogramm Eng Remote Sens 74(5):683–685

Acharya B, Bell W (1992) Designing an optimal and scientific GIS project. ISPRS archives—volume XXIX Part B3. XVIIth ISPRS Congress. Technical Commission III: Mathematical analysis of data. August 2–14, 1992, Washington, D.C., USA. pp 627–632
AENOR (2016) UNE 148002:2016 Metodología de evaluación de la exactitud posicional de la información geográfica. Asociación Española de Normalización, Madrid
Ariza-López FJ, Atkinson-Gordo AD (2008) Analysis of some positional accuracy assessment methodologies. Surv Eng 134(2):45–54
Ariza-López FJ, Rodríguez-Avi J (2014) A statistical model inspired by the national map accuracy standard. Photogramm Eng Remote Sens 80(3):271–281
Ariza-López FJ, Rodríguez-Avi J (2015a) A method of positional quality control testing for 2D and 3D line strings. Trans GIS 19(3):480–492
Ariza-López FJ, Rodríguez-Avi J (2015b) Informe sobre la aplicación de la propuesta de norma UNE 148002 sobre un conjunto de datos real. Universidad de Jaén, Informe para el Comité Técnico 148 de AENOR
ASPRS (2015) ASPRS positional accuracy standards for digital geospatial data. Photogramm Eng Remote Sens 81(4):53–63
Cheok G, Filliben J, Lytle AM (2008) NISTIR 7638. Guidelines for accepting 2D building plans. National Institute of Standards and Technology, Gaithersburg, Maryland
Dempsey C (2013) Where is the Phrase "80% of Data is Geographic" From? https://www.gislounge.com/80-percent-data-is-geographic/. Accessed 4 Nov 2016
FGDC (1998) FGDC-STD-007: geospatial positioning accuracy standards, Part 3. National standard for spatial data accuracy. Federal Geographic Data Committee, Reston, VA
Fisher RA (1922) On the interpretation of $\chi 2$ from contingency tables, and the calculation of P. J Roy Stat Soc 85(1):87–94
Fisher RA (1935) The design of experiments. Oliver & Boyd, Edinburgh, Scot
Freeman GH, Halton JH (1951) Note on an exact test treatment of contingency, goodness of fit and other problems of significance. Biometrika 38(1/2):141–149
ISO (1985) ISO 2859-2:1985. Sampling procedures for inspection by attributes—Part 2: sampling plans indexed by limiting quality (LQ) for isolated lot inspection. Genebre, Switzerland
ISO (1999) ISO 2859-1:1999. Sampling procedures for inspection by attributes—Part 1: sampling schemes indexed by acceptance quality limit (AQL) for lot-by-lot inspection. Genebre, Switzerland
ISO (2013) ISO 19157:2013. Geographic information—data quality. Genebre, Switzerland
Liu X, Hu P, Hu H, Sherba J (2012) Approximation theory applied to DEM vertical accuracy assessment. Trans GIS 16(3):397–410
Maune DF (ed) (2007) Digital elevation model technologies and applications: the Dem user's manual. American Society for Photogrammetry and Remote Sensing, Bethesda, MD
Mehta CR, Patel NR (1983) A network algorithm for performing Fisher's exact test in r $\times$ c contingency tables. J Am Stat Assoc 78(382):427–434
Müller MJ (2001) Exact tests for small sample 3 $\times$ 3 contingency tables with embedded fourfold tables: rationale and application. German J Psychiatry 4(1):57–62
R Core Team (2012) R: a language and environment for statistical computing. R Foundation for statistical computing, Vienna, Austria. ISBN 3-900051-07-0; http://www.R-project.org/
Storer BE, Choongrak K (1990) Exact properties of some exact test statistics for comparing two binomial proportions. J Am Stat Assoc 85(409):146–155
USBB (1947) United States National Map Accuracy Standards. US Bureau of the Budget, Washington DC
Zandbergen PA (2008) Positional accuracy of spatial data: non-normal distributions and a critique of the national standard for spatial data accuracy. Trans GIS 12(1):103–130
Zandbergen PA (2011) Characterizing the error distribution of lidar elevation data for North Carolina. Int J Remote Sens 32(2):409–430

Animation as a Visual Indicator of Positional Uncertainty in Geographic Information

Carsten Keßler and Enid Lotstein

Abstract Effectively communicating the uncertainty that is inherent in any kind of geographic information remains a challenge. This paper investigates the efficacy of animation as a visual variable to represent positional uncertainty in a web mapping context. More specifically, two different kinds of animation (a 'bouncing' and a 'rubberband' effect) have been compared to two static visual variables (symbol size and transparency), as well as different combinations of those variables in an online experiment with 163 participants. The participants' task was to identify the most and least uncertain point objects in a series of web maps. The results indicate that the use of animation to represent uncertainty imposes a learning step on the participants, which is reflected in longer response times. However, once the participants got used to the animations, they were both more consistent and slightly faster in solving the tasks, especially when the animation was combined with a second visual variable. According to the test results, animation is also particularly well suited to represent positional uncertainty, as more participants interpreted the animated visualizations correctly, compared to the static visualizations using symbol size and transparency. Somewhat contradictory to those results, the participants showed a clear preference for those static visualizations.

1 Introduction

Uncertainty is inherent in any kind of geographic information (Couclelis 2003). It can stem from issues such as measurement errors, the limited precision of the sensors in use, processing errors, or the involvement of lay contributors in the collection of Volunteered Geographic Information, to name but a few examples. At the

C. Keßler (✉)
Department of Planning, Aalborg University Copenhagen, Copenhagen, Denmark
e-mail: kessler@plan.aau.dk

E. Lotstein
Bronx Community College, City University of New York, New York, USA
e-mail: enid.lotstein@bcc.cuny.edu

A. Mansourian et al. (eds.), *Geospatial Technologies for All*,
Lecture Notes in Geoinformation and Cartography,
https://doi.org/10.1007/978-3-319-78208-9_19

conceptual level, uncertainty can also go back to vague or ambiguous object definitions (Fisher 1999); non-specific concepts such as *downtown* are a prime example of the latter (Montello et al. 2003). As Duckham et al. (2001, p. 89) put it, 'no observation of geographic phenomena will ever be perfect'. We adopt the broad definition by Longley et al. (2005, p. 128) here, who describe uncertainty as 'a measure of the user's understanding of the difference between the contents of a dataset, and the real phenomena that the dataset are believed to represent'. Several theoretical (Worboys 1998; Gahegan and Ehlers 2000; Roth 2009b) and practical (Williams et al. 2008) approaches have been proposed to address this issue. The effective communication of uncertainty of a dataset to its users through visualization remains a challenge, however, both in desktop GIS and in web-based and cloud-based solutions for geographic information.

In this paper, we address the issue of visualizing positional uncertainty for web mapping. Animated maps offer additional visual variables that go beyond the static variables introduced by Bertin (1973), for which DiBiase et al. (1992) proposed the three dynamic visual variables of duration, rate of change, and order. Animation can be useful when other visual variables have already been used to represent other aspects of the data and may also be able to convey a different notion of uncertainty. We have tested the efficacy of different static (symbol size, transparency) and dynamic (a *bouncing* and a *rubberband* effect) visual variables, as well as combinations of them in an online experiment with 163 participants. We have measured how fast and accurate participants were able to rate the uncertainty in point objects on a web map. Moreover, we have tested how they interpreted the different visual variables, and which kinds of visualizations they preferred. As such, the goal of this work is to address the 'lack of comprehensive empirical work that attempts to cognitively assess uncertainty visualization and decision-making through a human factors standpoint' (Smith et al. 2013, p. 1).

While the presented approaches can also be employed for the visualization of uncertainty in geographic information in other contexts, they are most easily replicated in web mapping. Data visualization frameworks such as D3.js,[1] which has also been used in our experiment, now allow for a relatively straight-forward implementation of variation in different visual variables, both static and dynamic. Moreover, the capabilities for animation are still limited in most desktop GIS systems, and not an option on printed maps for obvious reasons. We have therefore conducted and evaluated this study in the context of web mapping and designed the experiment to be conducted in a 'natural' environment for the user, i.e., on their own computer, using their preferred web browser.

The remainder of this paper is organized as follows: The next section provides an overview of relevant related work, followed by an introduction to the different visualization types tested in this research in Sect. 3. Section 4 discusses the design of the online participants test, followed by an evaluation (Sect. 5) and discussion (Sect. 6) of the results as well as concluding remarks in Sect. 7.

[1] See https://d3js.org.

2 Related Work

The effective communication of uncertainty in geographic information has concerned researchers in GIScience for almost as long as the modeling of this uncertainty. Davis and Keller (1997) have tested both static and dynamic interactive visualizations of uncertainty in slope stability. They conclude that a careful calibration of such visualizations against user specifications (e.g., of allowable uncertainty) is required, especially in risk assessment. Roth (2009a) conducted an online experiment to reveal potential differences between experts and novices assessing the flood risk at a given location. The participants were shown three floodplain delineations, each with a different degree of certainty, visualized with differently colored outlines. The results indicate that domain expertise is more important for decisions under uncertainty than map expertise. Domain experts with little map use experience were still able to accurately assess the risk, whereas map use experts with little or no domain experience underestimated the flood risk with low perceived assessment difficulty—a 'potential disaster' (Roth 2009a, p. 42). Riveiro (2016) evaluates a similar setting with novice and expert air traffic operators, where uncertainty about the exact position of aircrafts was visualized by circle size. Contrary to Roth (2009a), this study does not find significant differences in performance between the two groups.

Fisher (1993) was among the first to use animation to visualize uncertainty. He used 'flickering' colors on a soil map to represent the certainty of the correct classification of a pixel: the longer the pixel was shown in a color, the higher the certainty that it has been classified correctly. Ehlschlaeger et al. (1997) developed animations of elevation surfaces that reflect the uncertainty in elevation measurements. The animations would alternate between different possible realizations of the surface, whereas steps between the different realization scenarios were generated by interpolation. While these two papers proposed animation techniques without having them tested by users, Evans (1997) performed a usability study of a flickering technique to visualize uncertainty. The participants found the flickering on land cover maps a useful, albeit somewhat annoying means of communicating uncertainty in geographic information.

Concerning the visualization of positional uncertainty in point data, McKenzie et al. (2016) showed in a recent experiment that the kind of visualization employed does have an effect on the perceived positional uncertainty. Between four different static visualizations shown to participants—a gaussian blur and a solid circle, both with and without a centroid point—the participants responded fastest and most accurate to the solid circle without centroid. Moreover, the test results show that they employed different heuristics depending on the visualization used.

The literature discussed in this section is limited to the work most closely related to the research presented in this paper. Based on advances in both hardware and software libraries to generate animated visualizations, recent research has investigated the use of animation for visualizations in air-traffic analysis (Buschmann et al. 2016), emergency management (Wang et al. 2017), storm surge flooding (Reyes and Chen 2017) and soil acidification (Russo et al. 2014), to name but a few examples.

Readers looking for more complete reviews of the work that has been done around the communication of uncertainty in geographic information are referred to the broad review papers by MacEachren et al. (2005) and Kinkeldey et al. (2014, 2017). Readers looking for practical recommendations as to which kind of visualization is suitable for which user group are referred to Senaratne et al. (2012), who conducted a usability study with different visualizations of uncertainty and users from different backgrounds.

3 Visualization Types

In the experiment conducted for this research, the participants were shown eleven different combinations of four different visual variables to represent uncertainty. We used the static visual variables of symbol size and transparency—described as *fog* by MacEachren (1992)—as shown in Fig. 1.

For the animated visualizations, we used a 'rubberband' and a 'bouncing' visualization. To the best of our knowledge, this is the first systematic attempt to evaluate the effectiveness of animation to convey information about the uncertainty in point data. These two techniques were used because they both provide a clear indication of the area in which the observation was actually made, both with ('rubberband') and without ('bouncing') a direct indication of the measured location.

In the rubberband visualization, the object is moving away from its measured position at a random angle, and then bounces back into that location. This bouncing movement is repeated indefinitely at a constant rate, with a new random angle at every iteration. The distance for the movement is derived from the uncertainty, i.e., the larger the uncertainty, the larger the distance. The speed at which the object moves varies with the distance: the further it gets away from its measured location, the slower it moves, before it bounces back quickly. Visually, this gives the

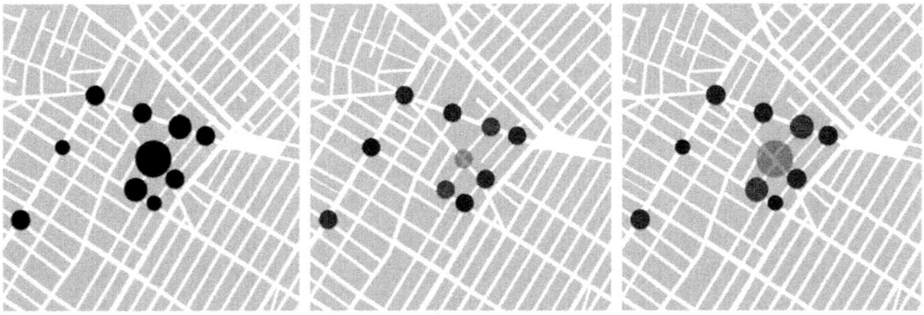

Fig. 1 Static visual variables, from left to right: Symbol size, transparency, and a combination of both

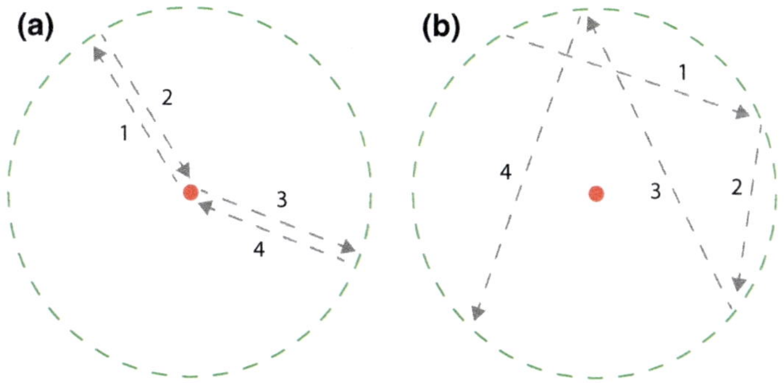

Fig. 2 Illustrations of the rubberband (**a**) and bouncing (**b**) animated visualization types. The maximum movement distance (dashed green line) from the measured location (red dot) is a linear function of uncertainty

impression of pulling on a rubberband and then letting go—hence the name. It has been implemented using the *elastic* easing function in D3.js.[2]

For the bouncing animation, the distance travelled is calculated the same way. The object, however, does not move back to its measured position, it rather moves indefinitely between random positions on the (invisible) circle around the measured position at constant speed. This creates a more 'chaotic' visual effect, as the objects are moving around randomly within that circle; the idea here is to convey the impression that the actual location of the object could be *anywhere* within this area. Figure 2 illustrates the two types of of animated visualization.

The animation techniques used here represent an abstract phenomenon. It is therefore worth noting that they do not fall into any of the types of time (linear, cyclic, or branching) discussed by Harrower and Fabrikant (2008), as the animation is used as a visual effect, and not as a representation of a process in reality. The uncertainty values were mapped to the different static and dynamic visual variables as follows:

- Symbol size: Linear mapping of the range of uncertainty values to circle radii between 3 and 10 pixels, i.e., $[u_{min}, u_{max}] \rightarrow [3, 10]$
- Transparency: Inverse linear mapping of the range of uncertainty values to opacity values between 100% and 30%, i.e., $[u_{min}, u_{max}] \rightarrow [1, 0.3]$
- Bouncing and rubberband animation: The size for the circle that marks the maximum distance travelled by an object is calculated the same way as the symbol size above, i.e., $[u_{min}, u_{max}] \rightarrow [3, 10]$.

The minimum radius (3 pixels) and opacity (30%) were set to make sure that the objects do not become too small and/or transparent and can still be easily recognized by the participants. The maximum radius was chosen pragmatically to minimize overlap between neighboring points. All visualizations were kept in gray scale colors, including the base map. This was done in order to prevent the participants from imposing any kind of interpretation onto the color choice. Moreover, this excluded

[2]See https://github.com/d3/d3-ease#easeElastic.

Table 1 Overview of the eleven different visualizations shown during the test

Page	Size	Transparency	Rubberband	Bouncing
1	•			
2		•		
3	•	•		
4			•	
5	•		•	
6		•	•	
7	•	•	•	
8				•
9	•			•
10		•		•
11	•	•		•

any potential effects of color blindness on the results. A minimal style for the base map without any labels was used to prevent any influence of the area shown on the results. Table 1 provides an overview of the test sequence, which consists of 11 pages, each showing a different combination of visual variables.

4 Test Design

The visualization types discussed in the previous section were shown to a group of participants in an online test.[3] This section describes the goals of the test, the test sequence and the data used to generate the visualizations discussed in the previous section.

4.1 Goals

The goal of this test was to evaluate the efficacy of the visual variables of size, transparency, and two different kinds of animation, as well as combinations thereof in an online test. More specifically, the following questions were driving the design of this test:

1. What kind of visualization allows the participants to identify the most and least uncertain objects on a map quickly and correctly?
2. What are the participants' interpretations of the different visual variables?

---

[3]The experiment is available online at http://carsten.io/uncertainty/. The corresponding source code and data produced is available at https://github.com/crstn/UncertD3/.

3. What are the participants' preferences concerning the different kinds of visualizations?

In order to answer those questions, we have measured how fast and accurate participants were able to rate the uncertainty in point objects on a web map based on the test sequence discussed in the following subsection.

4.2 Test Sequence

The test conducted for this study consisted of three parts. The first part was an introduction that explained the goals of the test to the participants. It also contained a preview of the data that were shown to them in different ways during the test (see Fig. 3). This was included to allow the participants to familiarize themselves with the map and the area shown before they start the actual test.

The second part of the test consisted of eleven pages that all looked like the one shown in Fig. 4. However, every page showed the data using a different visualization type. Every possible combination of symbol size, transparency, rubberband animation, and bouncing animation was tested in the order shown in Table 1. The order was thus not randomized in order to be able to observe a potential learning effect. The uncertainty value associated with each object was randomized for every page to prevent the participants from identifying the most uncertain object once, and then selecting it again on the following pages. The tasks on the eleven pages alternated between selecting the *most* and *least* uncertain object in order to test the visualization types both for maximal and minimal uncertainty. The participants selected the object that they believed to be most or least uncertain by placing a red circle around it and clicking. The circle was chosen as a selection tool because simply clicking the corresponding object is very difficult when the points are moving. The selection circle was larger than the biggest possible diameter for the animation of a point. Therefore, participants could select the point by placing the selection circle around it even when it was moving.

The third and final part of the test consisted of a questionnaire. The corresponding web page showed a small example of every one of the eleven visualizations for reference. Participants were then asked to select:

- the *best* visualization to identify the *most* uncertain object;
- the *worst* visualization to identify the *most* uncertain object;
- the *best* visualization to identify the *least* uncertain object; and
- the *worst* visualization to identify the *least* uncertain object.

Moreover, the participants were asked about their interpretation of the different symbol variations. For symbol size, transparency, and movement, they were asked whether they thought that the respective visual variable reflects positional uncertainty, uncertainty of a different attribute, or something else. Moreover, they were asked about their interpretation of the relationship between visual variable and uncertainty, e.g., whether larger symbols reflect more or less uncertainty. They were also

Testing Visualizations of Uncertainty

In the next minutes, you will be shown different visualizations of uncertainty. That data you will see is a series of points that could be generated from a GPS, for example, overlaid on top of a simple street map. You will be asked to identify the most or least uncertain of those points based on their visualization, which will indicate uncertainty in different ways.

Please not that this test **does not work on touchscreen devices** such as smart phones or tablets. Please make sure that you are using a laptop or desktop computer if you would like to participate.

Try moving your mouse pointer on the map. It will turn into a circle. You will be asked to place this circle on the point that you believe is the most or least uncertain on the map, and then click.

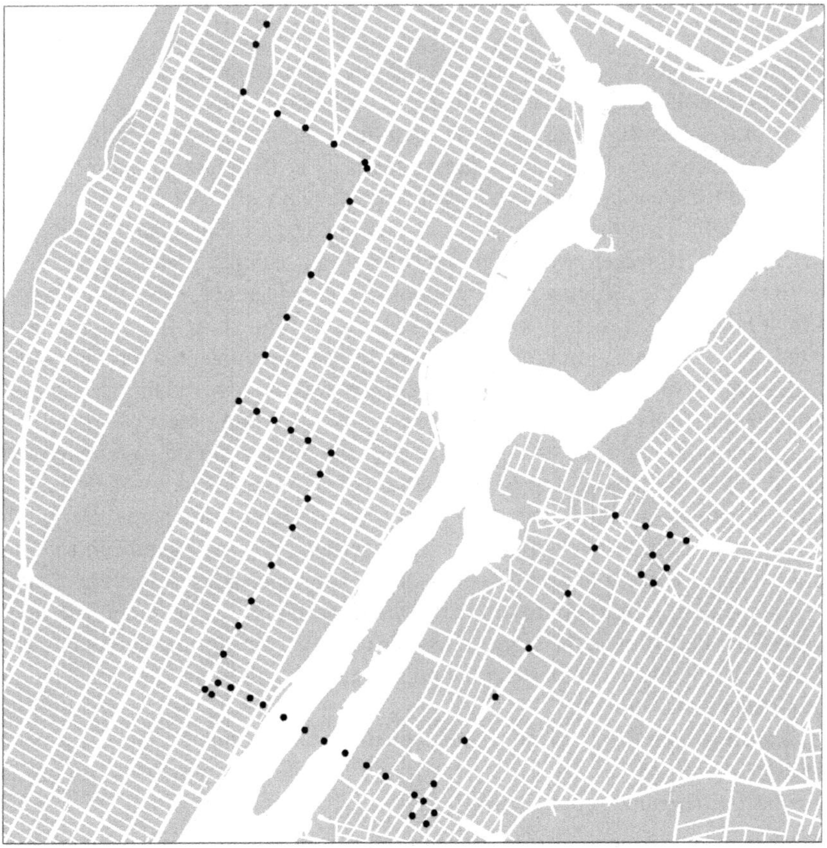

Try to solve this task quickly, but do not rush it – we want you to consider the different points shown and then make a decision. There are no right or wrong answers – please select the point that you believe is most or least uncertain, based on the respective visualization. Participation should not take more than 10 minutes.

Contact: If you have any questions about this experiment, please contact Carsten Kessler.

START

Fig. 3 Introduction page of the online test

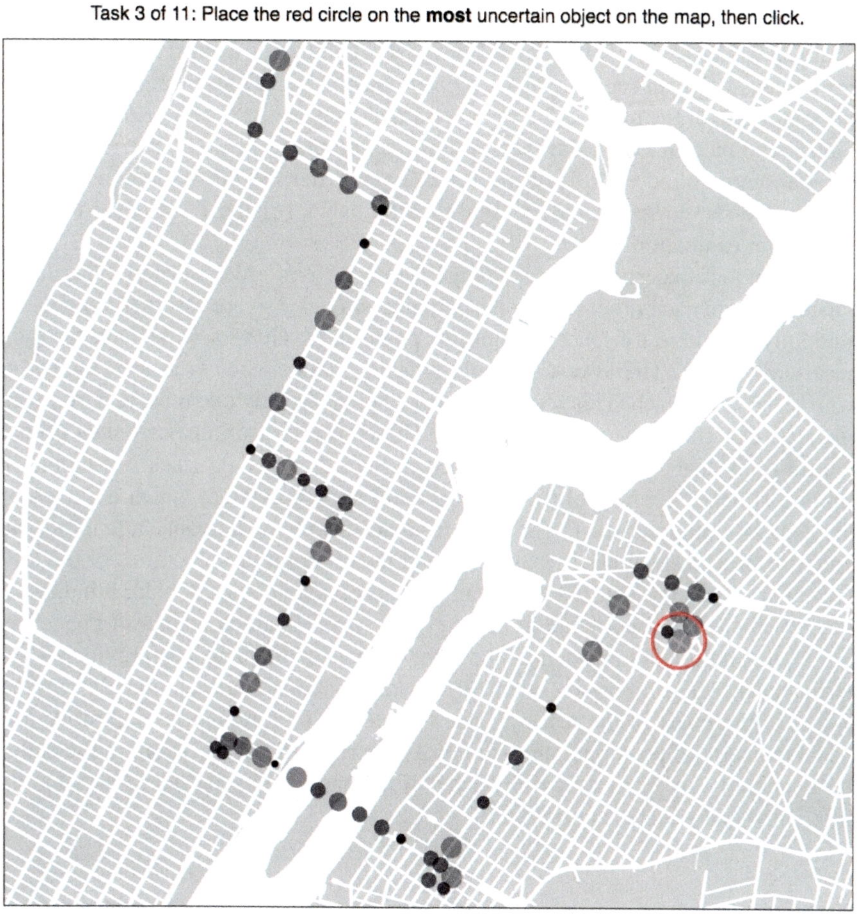

Fig. 4 Example test page with selection circle placed over the most uncertain object

asked how they interpreted the combination of different visual variables, i.e., whether they all reflected uncertainty, or just one of them. Finally, the participants were asked to provide some personal information: age, gender, and whether they work with GIS or maps at their job or study.

4.3 Data

The data used in the test is originally a GPS track of a car driving through Manhattan and Queens, New York City, which was collected by the enviroCar project (Broering et al. 2015). The uncertainty variable used for the visualization in this test is the accuracy of the GPS signal, i.e., positional uncertainty. The randomization of the accuracy values is performed every time a participant visits a page; that way, no two participants will ever see the exact same version of a page. This excludes any effects of the location of the most and least uncertain objects on the results. The uncertainty values are randomized via JavaScript in the participant's browser after loading the original, unmodified GPS track from the server. After the randomization it is sent back to the server and stored under the participant's session ID and current page number for evaluation.

The random uncertainty values range from 1 to 7 in order to keep them distinguishable from each other for the participants. Values 2–6 are assigned completely randomly, i.e., there are generally multiple points in the GPS track that have the same uncertainty value. The lowest (1) and highest (7) values, however, are assigned to exactly one point each. This makes sure that there is exactly one most uncertain and one least uncertain object on every page. These uncertainty values are then used to control the visual variables, as described in Sect. 3. The GPS track is overlaid on top of a simple base map of the city blocks in New York City, which is intentionally kept very minimalistic to prevent it from driving the participants' attention away from the GPS track. Both the base map and the different visualizations of the GPS track are kept in grayscale to exclude any effects of color choice or color blindness in the participants. It is worth noting that besides the information on the start page (see Fig. 3), no additional information about the objects on the map was provided to the participants. A legend or any other further information were intentionally omitted, as one of the goals of the test was also to determine if animation is intuitive as a visual indicator of positional uncertainty. The corresponding questions at the end of the test could only be meaningfully asked if the participants did not know what the different visual variables represent.

5 Evaluation

This section evaluates the test results with respect to the three research questions from Sect. 4.1.

5.1 Participants

The test was advertised on social media and different mailing lists at the end of October 2016. The tweet announcing the test was retweeted by more than 100 Twitter users, which motivated 163 participants to complete the test within the following ten days. The participants remained anonymous and did not receive any compensation. Based on their responses on the final questionnaire, their mean age was 36 years ($\sigma = 10.9$). The majority of the participants were male (103 or 63%), 57 were female (34%), 1 selected other gender, and 2 did not disclose their gender. 133 (82%) stated that they work with maps, GIS, etc. at work and/or college, whereas 30 (18%) did not. While Twitter did prove very useful for participant recruitment, this overrepresentation of map literate participants is most likely a result of the authors' *filter bubble* on social media.

5.2 Visualization Efficacy

In order to answer research question 1—*What kind of visualization allows the participants to identify the most and least uncertain objects on a map quickly and correctly?*—we have analyzed the response times and locations where participants placed the red circle on the map to select the most/least uncertain object. Response times were measured in milliseconds, representing the time that passed from showing the map to the participant until they clicked to submit the location of their choice. The location was measured as the center of the red circle in geographic coordinates.

The density map in Fig. 5 has been generated from the locations of all 1,793 locations selected by the 163 participants over 11 pages each. It shows that some of them clicked on locations away from the GPS track, with some of them concentrating in the bottom left of the figure. We have still included these outliers in the analysis because it is not possible to tell whether those participants did not understand the instructions, or whether they clicked these locations for some other reason, e.g., accidentally or because they were not able to identify the most or least uncertain object.

Fig. 5 A density map of the participants' clicks

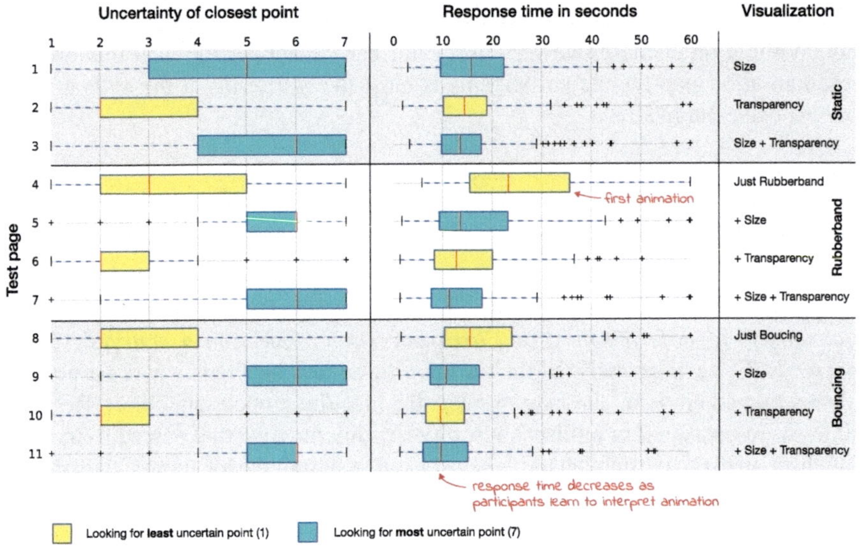

Fig. 6 Box plots of uncertainty of the closest point to the participant's click (left) and the response times (right) for each of the 11 test pages. The red line indicates the median and the whiskers are placed at 1.5 times the interquartile range

The boxplots shown in Fig. 6 give some indication as to which visualization allowed the participants to identify the most and least uncertain objects most confidently and most quickly. Both plots show all 11 test pages in the order given in Table 1, i.e., in the same order they were shown to the participants. The task for the participants was to identify the *most* uncertain object on all *uneven* test pages (light blue boxes in the figure), and the *least* uncertain one on the *even* test pages (yellow boxes in the figure). Therefore, the 'correct' answer for tests 1, 3, 5, … would be the object with uncertainty value 7, and for tests 2, 4, 6, … the object with uncertainty value 1.

The boxplot on the left of Fig. 6 shows the distribution of the uncertainty values of the point from our GPS track that was closest to the location selected by the user by placing the red circle. It shows that the boxes all have the right tendency—i.e., they are closer to 7 for the tests looking for the most uncertain object, and closer to 1 for the tests looking for the least uncertain object. The fact that the median is not on the 'correct' result for any of the 11 tests can most likely be attributed to the fact that participants could not distinguish between the visualizations of two adjacent uncertainty values (1 and 2, for example), and therefore selected an object with value 2 instead of 1 (or 6 instead of 7). The distribution of the results for the selected uncertainty values (shown in the left half of Fig. 6), however, shows that the participants were overall more consistent in their judgments on the pages that used animation and at least one other visual variable (pages 5, 6, 10, and 11), as these have the smallest interquartile range. In comparison to the other visualizations, this also means that fewer participants selected uncertainty values that were far off from the correct result. Somewhat counterintuitively, the interquartile range is largest for pages 1–3, which were limited to static visualizations. These results indicate that animation may help participants judge the uncertainty of an object correctly if it is used in combination with symbol size and/or transparency.

Concerning the response times[4] shown in the right part of Fig. 6, one apparent outlier is page 4, where the median response time (23.1 s) is more than 10 s higher than the average median response time on all other pages (12.5 s). This can be attributed to the fact that page 4 was the first page in the test where the participants encountered an animated visualization, which apparently took them some time to understand. After that, however, the response times are very similar to the response times for the static visualizations on pages 1–3, and even declining towards page 11. This speaks for a learning effect, allowing participants to respond slightly faster after solving several different tasks.

We also tested for correlation between response time and (a) distance from the placed circle to the correct object (with value 1 or 7, respectively), and (b) difference in uncertainty value between the correct answer and the point closest to the circle (along the lines of the left box plot in Fig. 6). Notably, there is no correlation whatsoever between these variables, i.e., the judgments made by participants who only looked at a test page for 5–10 s were as good as those who took much more time. This still applies if the outliers shown in the density map (Fig. 5) are excluded from the analysis.

5.3 *Participants' Interpretation*

Research question 2—*What are the participants' interpretations of the different visual variables?*—can be answered by evaluating the questionnaire at the end of the test, which explicitly asked for their individual interpretations. Figure 7 shows the counts for responses to the question which kind of uncertainty size, transparency, and animation represented: positional uncertainty, uncertainty in another attribute, or something else. As the bar chart shows, 79% of the participants interpreted movement as an indicator for positional uncertainty, whereas both symbol size and transparency were significantly more often interpreted to reflect uncertainty in another attribute, or something else. That indicates that animation seems to be an intuitive way to visualize positional uncertainty, and, most notably, more intuitive than the traditional static visual variables of symbol size and transparency.

Figure 8 shows that the relationship between the visual variable and the underlying data was interpreted correctly in most cases: more uncertainty was represented with larger symbols, more transparent symbols, and more movement (i.e., a linear

[4]Some of the participants had single pages open for several minutes; since it is more likely that they answered the phone or went to get a coffee instead of actually looking at the test page for such a long time, all values above 60 s were set to 60 s.

Fig. 7 Interpretation of different visual variables

Fig. 8 Interpretation of the relationship between visualization and uncertainty

relationship). However, movement also seems to be the most intuitive visual variable in this case, as 91% of the participants interpreted the relationship correctly. Transparency and especially symbol size were more often interpreted to follow a reverse relationship (i.e., more uncertainty being visualized with smaller symbols).

The interpretation of two or more visual variables in combination was less consistent. 36% participants thought that all visual variables represented uncertainty; 17% thought that just one of the visual variables represented uncertainty, while the others represented something else, and 42% found them very hard to interpret (5% did not answer the question). Consequently, while Fig. 6 indicates that representing uncertainty with multiple visual variables at once seems to help identify the most and least uncertain objects quickly and consistently, this approach seems to make it more difficult for users to interpret the data.

5.4 *Participants' Preferences*

Research question 3—*What are the participants' preferences concerning the different kinds of visualizations?*—can also be answered by analyzing the responses to the questionnaire. Four questions were asked about which visualizations the participants found best/worst to find the most/least uncertain object. The responses are summarized in the pie charts shown in Fig. 9 and paint a very clear picture: about two thirds of the participants think that the visualizations that only use static visual variables are best to identify the most or least uncertain object on the map. When asked about the worst visualizations for those tasks, the tendency is even stronger: 75–80% think that one of the visualizations using animation is the worst for the job. These results therefore show a clear preference of static visualizations over animated visualizations, even though the results presented in the previous two sections show certain advantages of using animation to visualize uncertainty.

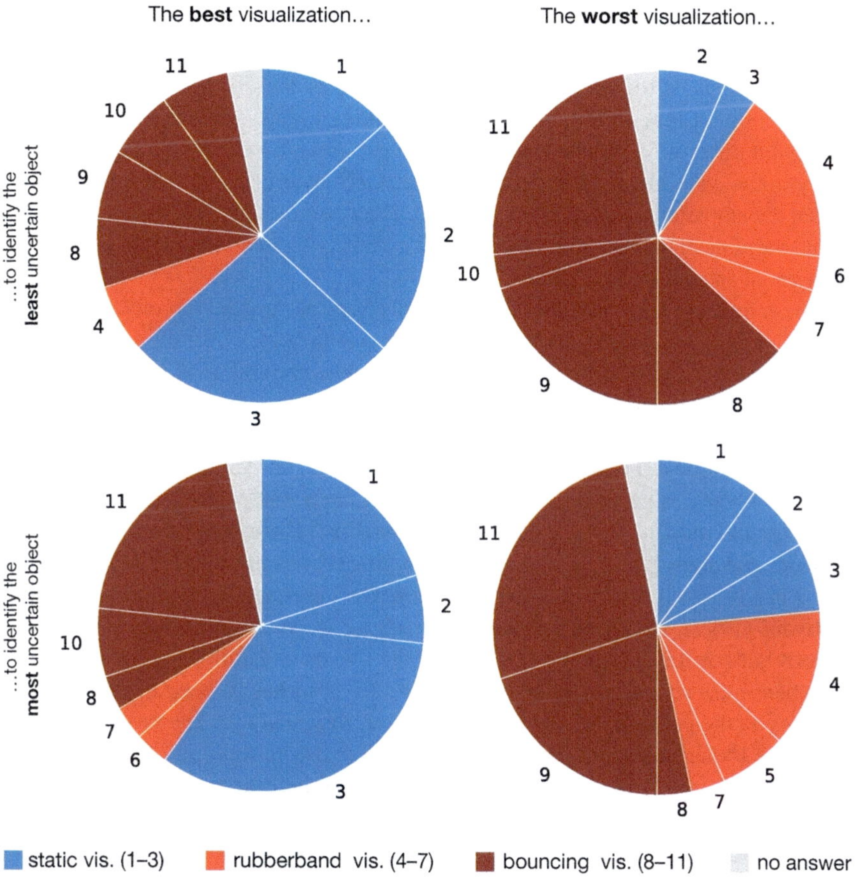

Fig. 9 Participants' selection of the best/worst visualizations to find the most/least uncertain object

6 Interpretation of Results and Discussion

Participants showed the most consistent behavior with fewer wrong answers for visualizations that used animation in combination with at least one other visual variable. The evaluation of response times, however, shows that animation seems to require an initial learning step, reflected in a significantly longer response time. After encountering the first visualization with animation response times were even slightly lower than for the static visualizations. This gives an indication that a combination of animation with at least one other visual variable may indeed be an effective method to convey information about the uncertainty of geographic information to its users. One potential explanation for these results is that the animated visualizations allow users to inspect parts of the data that are hard to see in a static visualization because of overlapping symbols. In the animated visualizations, these are revealed temporarily as the symbols are shuffled around.

The analysis of the participants' interpretation of the different visual variables, however, shows that animation works most intuitively as a means to communicate *positional* uncertainty. Even though the participants were never told that the uncertainty in the test did indeed reflect positional GPS accuracy, the vast majority interpreted the animated visualizations this way. They also rated the relationship between visual variable and uncertainty value more consistently and more often correct than for symbol size and transparency.

These findings seem to contradict the analysis of participants' preferences, which show that the vast majority preferred the static visualizations over the animated ones, particularly visualization 3, which combines symbol size and transparency. This may be attributed to experimental nature of the animations and that they take some time to get used to. Moreover, the map used in the test did not contain a legend or any other explanations of the meaning of the different visual variables. While this was part of the test design to be able to learn about the participants' intuition, having a legend would most likely have changed this outcome. Designing meaningful legends for animated maps, however, remains challenging (Kraak et al. 1997). One indication that it is possible to use animation with more positive user feedback is the research by Evans (1997), who used a flickering technique to communicate uncertainty. The majority of participants in his study found the flickering helpful.

When interpreting the results of this test, one also has to keep in mind that the majority of the participants can be described as 'map literate'. Therefore, the results may have looked different for a group of participants that do not work with maps on a regular basis. It is worth noting, however, that there were no significant differences in any of the results (response time, correctness, participant preferences and interpretation) between expert and non-expert participants. We did not find any significant differences between the female and the male participant groups either, in contrast to previous studies looking at gender differences in the perception of spatial visualizations (Battista 1990; Maeda and Yoon 2013).

7 Conclusions

We have tested and evaluated the combination of different visual variables to represent uncertainty in geographic information in an online test. To the best of our knowledge, this is the first systematic analysis of the use of animation to represent positional uncertainty in point data. The results give several indications for the effective visualization of positional uncertainty on web maps, especially concerning the use of animation, which allowed participants to solve the tasks slightly faster and more consistently when combined with other visual variables. The participants' preferences are in stark contrast to these findings, as the majority preferred the static visualizations. Future work needs to evaluate whether this rejection of animation disappears as users get accustomed to it. Another avenue for future work is the distinction of different kinds of uncertainty, such as precision, completeness, or currency. Visualizing them in a way so that users can intuitively distinguish them is a research problem in its own right (MacEachren et al. 2005; Andrienko et al. 2010), which may only be solvable for professionals working with geographic information on a regular basis. Finally, the results presented here are limited to point data so far. While the different visualizations can certainly be adapted for lines and polygons, the findings of this paper need to be confirmed for those other geometry types.

References

Andrienko G, Andrienko N, Demsar U, Dransch D, Dykes J, Fabrikant SI, Jern M, Kraak MJ, Schumann H, Tominski C (2010) Space, time and visual analytics. Int J Geogr Inf Sci 24(10):1577–1600

Battista MT (1990) Spatial visualization and gender differences in high school geometry source. J Res Math Educ 21(1):47–60

Bertin J (1973) S'emiologie graphique: Les diagrammes-Les réseaux-Les cartes. Gauthier-Villars Mouton & Cie, Paris

Broering A, Remke A, Stasch C, Autermann C, Rieke M, Moellers J (2015) enviroCar: a citizen science plattform for analyzing and mapping crowdsourced car sensor data. Trans GIS 19:362–376

Buschmann S, Trapp M, Döllner J (2016) Animated visualization of spatial-temporal trajectory data for air-traffic analysis. Visual Comput 32(3):371–381. https://doi.org/10.1007/s00371-015-1185-9. http://link.springer.com/10.1007/s00371-015-1185-9

Couclelis H (2003) The certainty of uncertainty: GIS and the limits of geographic knowledge. Trans GIS 7(2):165–175

Davis TJ, Keller C (1997) Modelling and visualizing multiple spatial uncertainties. Comput Geosci 23(4):397–408

DiBiase D, MacEachren AM, Krygier JB, Reeves C (1992) Animation and the role of map design in scientific visualization. Cartogr Geogr Inf Syst 19(4):201–214

Duckham M, Mason K, Stell J, Worboys M (2001) A formal approach to imperfection in geographic information. Comput Environ Urban Syst 25(1):89–103

Ehlschlaeger CR, Shortridge AM, Goodchild MF (1997) Visualizing spatial data uncertainty using animation. Comput Geosci 23(4):387–395

Evans BJ (1997) Dynamic display of spatial data-reliability: does it benefit the map user? Comput Geosci 23(4):409–422

Fisher PF (1993) Visualizing uncertainty in soil maps by animation. Cartogr: Int J Geogr Inf Geovisual 30(2–3):20–27

Fisher PF (1999) Models of uncertainty in spatial data. Geogr Inf Syst 1:191–205

Gahegan M, Ehlers M (2000) A framework for the modelling of uncertainty between remote sensing and geographic information systems. ISPRS J Photogram Remote Sens 55(3):176–188

Harrower M, Fabrikant S (2008) The role of map animation for geographic visualization. In: Dodge M, McDerby M, Turner M (eds) Geographic visualization. Wiley, pp 49–65

Kinkeldey C, MacEachren AM, Schiewe J (2014) How to assess visual communication of uncertainty? A systematic review of geospatial uncertainty visualisation user studies. Cartogr J 51(4):372–386

Kinkeldey C, MacEachren AM, Riveiro M, Schiewe J (2017) Evaluating the effect of visually represented geodata uncertainty on decision-making: systematic review, lessons learned, and recommendations. Cartogr Geogr Inf Sci 44(1):1–21

Kraak MJ, Edsall R, MacEachren AM (1997) Cartographic animation and legends for temporal maps: exploration and or interaction. In: Proceedings of the 18th international cartographic conference, international cartographic association, vol 1, pp 253–261

Longley P, Goodchild MF, Maguire DJ, Rhind DW (2005) Geographic information systems and science, 2nd edn. Wiley

MacEachren AM (1992) Visualizing uncertain information. Cartogr Perspect 13:10–19

MacEachren AM, Robinson A, Hopper S, Gardner S, Murray R, Gahegan M, Hetzler E (2005) Visualizing geospatial information uncertainty: what we know and what we need to know. Cartogr Geogr Inf Sci 32(3):139–160

Maeda Y, Yoon SY (2013) A meta-analysis on gender differences in mental rotation ability measured by the Purdue spatial visualization tests: visualization of rotations (PSVT:R). Educ Psychol Rev 25(1):69–94. https://doi.org/10.1007/s10648-012-9215-x. arXiv:0507464v2

McKenzie G, Hegarty M, Barrett T, Goodchild M (2016) Assessing the effectiveness of different visualizations for judgments of positional uncertainty. Int J Geogr Inf Sci 30(2):221–239

Montello DR, Goodchild MF, Gottsegen J, Fohl P (2003) Where's downtown?: behavioral methods for determining referents of vague spatial queries. Spat Cogn Comput 3(2–3):185–204

Reyes MEP, Chen SC (2017) A 3D virtual environment for storm surge flooding animation. In: 2017 IEEE third international conference on multimedia big data (BigMM), pp 244–245. https://doi.org/10.1109/BigMM.2017.54

Riveiro M (2016) Visually supported reasoning under uncertain conditions: effects of domain expertise on air traffic risk assessment. Spat Cogn Comput 16(2):133–153

Roth RE (2009a) The impact of user expertise on geographic risk assessment under uncertain conditions. Cartogr Geogr Inf Sci 36(1):29–43

Roth RE (2009b) A qualitative approach to understanding the role of geographic information uncertainty during decision making. Cartogr Geogr Inf Sci 36(4):315–330

Russo P, Pettit C, Çöltekin A, Imhof M, Cox M, Bayliss C (2014) Understanding soil acidification process using animation and text: an empirical user evaluation with eye tracking. Springer, Berlin, pp 431–448. https://doi.org/10.1007/978-3-642-32618-9_31

Senaratne H, Gerharz L, Pebesma E, Schwering A (2012) Usability of spatio-temporal uncertainty visualisation methods. In: Gensel J, Josselin D, Vandenbroucke D (eds) Bridging the geographic information sciences: international AGILE'2012 conference, Avignon (France), 24–27 April 2012. Springer, Berlin, pp 3–23

Smith J, Retchless D, Kinkeldey C, Klippel A (2013) Beyond the surface: current issues and future directions in uncertainty visualization research. In: Buchroithner MF, Prechtel N, Burghardt D, Pippig K, Schröter B (eds) Proceedigs of the 26th international cartographic conference, international cartographic association, pp 1–10

Wang D, Guo D, Zhang H (2017) Spatial temporal data visualization in emergency management: a view from data-driven decision. In: Proceedings of the 3rd ACM SIGSPATIAL workshop on emergency management using ACM, New York, NY, USA, EM-GIS'17. pp 8:1–8:7. https://doi.org/10.1145/3152465.3152473. http://doi.acm.org/10.1145/3152465.3152473

Williams M, Cornford D, Bastin L, Ingram B (2008) UncertML: an XML schema for exchanging uncertainty. In: Proceedings of GISRUK, Manchester, UK 44

Worboys M (1998) Computation with imprecise geospatial data. Comput Environ Urban Syst 22(2):85–106

Printed by Printforce, the Netherlands